The Agriculture of the Sudan

Centre for Agricultural Strategy Series

Editors: C. R. W. Spedding and L. J. Peel

Published in conjunction with the Centre for Agricultural Strategy, University of Reading

1. G. M. Craig (editor) *The Agriculture of the Sudan*
2. Xu Guohua and L. J. Peel (editors) *The Agriculture of China*

THE AGRICULTURE OF THE

Sudan

Edited by
G. M. CRAIG
Centre for Agricultural Strategy
University of Reading

OXFORD UNIVERSITY PRESS
1991

Oxford University Press, Walton Street, Oxford OX2 6DP
Oxford New York Toronto
Delhi Bombay Calcutta Madras Karachi
Petaling Jaya Singapore Hong Kong Tokyo
Nairobi Dar es Salaam Cape Town
Melbourne Auckland
and associated companies in
Berlin Ibadan

Oxford is a trade mark of Oxford University Press

Published in the United States
by Oxford University Press, New York

British Library Cataloguing in Publication Data
The agriculture of the Sudan.
1. Sudan. Agricultural industries
I. Craig, Gillian M.
338.109624
ISBN 0 19 859210 8

Library of Congress Cataloging in Publication Data
The Agriculture of the Sudan/edited by Gillian M. Craig.
p. cm.—(CAS (Series); no. 1)
1. Agriculture—Sudan. I. Craig, G. M. II. Series.
S473.S7A49 1990
338.1'09624—dc20 90–21423 CIP

ISBN 0 19 859210 8

Set by
Footnote Graphics, Warminster, Wiltshire

Printed by St. Edmundsbury Press,
Bury St. Edmunds, Suffolk

Foreword

This is one of a series of volumes on the agriculture of different countries. It is a generalist's book, written to provide both background information and a wider understanding of the agriculture of the Sudan in all its facets. It does not follow any particular disciplinary approach, but considers the farmer and his family, the systems of production he employs, the physical and environmental constraints on his activities, the economic system within which he must operate, and the historical and cultural backgrounds which help to determine all these things.

The book is intended to be useful to all those who wish to go to the Sudan for professional or vocational reasons connected with agricultural development, or trade in agricultural and related products; we hope that it may also assist those who, as tourists, want to learn more about local circumstances, and enlighten those who simply wish, without visiting the Sudan, to discover more about Sudanese agriculture and rural life. This is why we are particularly concerned to provide the sort of general picture and breadth of understanding to which individuals can relate their own specific activities and interests. For agriculture as experienced by the farmer has many aspects, and those who would help him or trade with him must first understand something of the world as he sees it.

C.R.W.S.
L.J.P.

Preface

Forty-five years have passed since J. D. Tothill's *Agriculture in the Sudan*, published in 1948, was written 'to give a picture of progress or otherwise made in the agricultural life of the country from the time of Lord Kitchener's reoccupation in 1898 to June 1945'. This volume is a classic and, despite its age, remains a much-quoted and authoritative work of reference on the Sudan's agriculture up to the mid-twentieth century.

Since Tothill's book a great deal has been written on the Sudan and its agriculture, but the information is scattered in books, government reports, and in the occasional publications of numerous other organizations with an interest in the Sudan. These can be difficult to obtain, and it is not always easy to acquire an overall picture as well as detailed factual information from them.

The Agriculture of the Sudan brings together in one volume the contributions of twenty-one authors from several different countries, each with a specialist knowledge of the Sudan. The volume may be regarded as a successor to Tothill's *Agriculture in the Sudan* in that it continues the picture of Sudanese agriculture into the early 1990s.

The book has, however, reached completion at a time when the Sudan's economy is in turmoil because of an escalating debt crisis and because the civil war in southern Sudan continues to result in death, devastation, homelessness, and misery for millions of Sudanese people. In recent years, the Sudan has also suffered from drought leading to widespread human starvation and, more recently, disastrous flooding in the Khartoum area.

In 1948, Tothill claimed that his book told 'a remarkable story of sound economic progress' and he saw 'no reason why a highly civilized state cannot be evolved in the Sudan on a purely agricultural foundation.' Today, regrettably, it is a widely-held view that sustained development will not be attained in the Sudan as a whole until the civil war is ended. As the war has progressed, large areas of the countryside have been ravaged by armies and it is hoped that when peace is restored, the information given in this volume will be of value in creating new agricultural foundations.

The individuality of the contributions to this book reflects the wide-ranging experiences, interests, and nationality of their authors, all of whom have first-hand experience of the Sudan. The editor has aimed for reasonable consistency and uniformity of style wherever possible, but it has been assumed that the terminology used by authors in their contributions is, for the most part, that which would be encountered by any newcomer to the Sudan. The Arabic and vernacular words used most frequently have been standardized and are included in the glossary, together with a list of abbreviations.

I wish to record my thanks to all the authors who have contributed to the book, responded to my queries, updated material, and have waited patiently while I have dealt with the editing procedures for all the chapters. Acknowledgements are also due to Dr Lynnette J. Peel for her continuous support and frequent advice, and to colleagues in the University of Reading for comments and advice on individual manuscripts.

Reading
March 1990 G.M.C.

Contents

Contributors

A. H. Abdoun, Amin Enterprises Ltd., Khartoum

M. Z. Ali, Department of Food Science, University of the Gezira

G. M. Craig, Research Fellow, Centre for Agricultural Strategy, University of Reading

H. R. J. Davies, Department of Geography, University College of Swansea

A. Dickie, Overseas Development Administration, Lusaka

C. G. Gurdon, MENAS Services Ltd., London

G. Haaland, Professor of Social Anthropology, University of Bergen

D. Hansohm, Sudan Economy Research Group, University of Bremen

L. Kapteijns, Assistant Professor, Department of History, Wellesley, College, USA

C. W. Mitchell, Honorary Fellow, Department of Geography, University of Reading

P. Oesterdiekhoff, Sudan Economy Research Group, University of Bremen

W. I. Robinson, University College of North Wales, Bangor

I. G. Simpson, Honorary Lecturer, School of Business and Economic Studies, University of Leeds

M. C. Simpson, Honorary Lecturer, School of Business and Economic Studies, University of Leeds

G. M. Sørbø, Director, Centre for Development Studies, University of Bergen

J. Spaulding, Associate Professor, Department of History, Kean College, USA

A. Trilsbach, Sudan Studies Society of the UK, c/o 10 Thornlea, Godinton Park, Ashford, Kent TN23 3JX

R. P. D. Walsh, Department of Geography, University College of Swansea

G. E. Wickens, previously of the Royal Botanic Gardens, Kew

R. T. Wilson, International Livestock Centre for Africa, Addis Ababa, Ethiopia

P. R. Woodward, Department of Politics, University of Reading

K. Wohlmuth, Professor of Comparative Economic Systems, University of Bremen

1

Physiography, geology, and soils

C. W. MITCHELL

The country and its limits

The Sudan forms a part of the extensive, flat, peneplained surface which dominates the whole of north-east Africa. It falls from about 800 m above sea level in the west and south to about 300 m on the Egyptian border and is defined by a ring of uplands. The physiography of the country is shown schematically in Fig. 1.1. In the north-east, the Red Sea Hills, composed of basement granites and metamorphosed sediments, rise to over 2000 m. They form a sharply defined escarpment less than 50 km from the coast, which is the drainage divide between the Red Sea and the Nile. The locations of places mentioned in the text are shown in Fig. 1.2.

In the south-east, the plains meet the high Ethiopian Plateau, composed of Basement Complex rocks covered by Mesozoic sediments and Tertiary lavas, whose margins are cut into deep canyons by a series of rivers: the Gash, Atbara, Settit, Blue Nile, Angar, Didessa, Akobo, and Kurun.

The southern and western frontiers as far as Jebel Marra are defined by Basement Complex uplands, rising to above 3000 m in the Imatong Mountains on the Ugandan border, and then running along the Nile–Congo divide, which delimits the Sudan's boundary with Zaire and the Central African Republic. The high ground of western Darfur is formed by the volcanic masses of Jebel Marra and the Meidob Hills, the former rising to 3042 m above the Deriba Crater, the highest point in western Sudan. Streams drain east to the While Nile, south to the Bahr El Arab and west to the Lake Chad drainage system. North of this the border passes east of the Ennedi and Erdi Plateaux and the Murdi Depression to reach the north-western corner of the country at Jebel Uweinat. The frontier with Egypt lies across relatively low, horizontally bedded sandstone hills, mantled in the western parts by dunes and bisected by the Nile at Lake Nasser.

The central Sudan gives the impression of a vast plain becoming increasingly vegetated southwards. It is dotted with inselbergs, invariably formed on rock types more resistant to weathering than their surroundings. Granite is a frequent cause and forms steep slopes. The Nuba Mountains, for

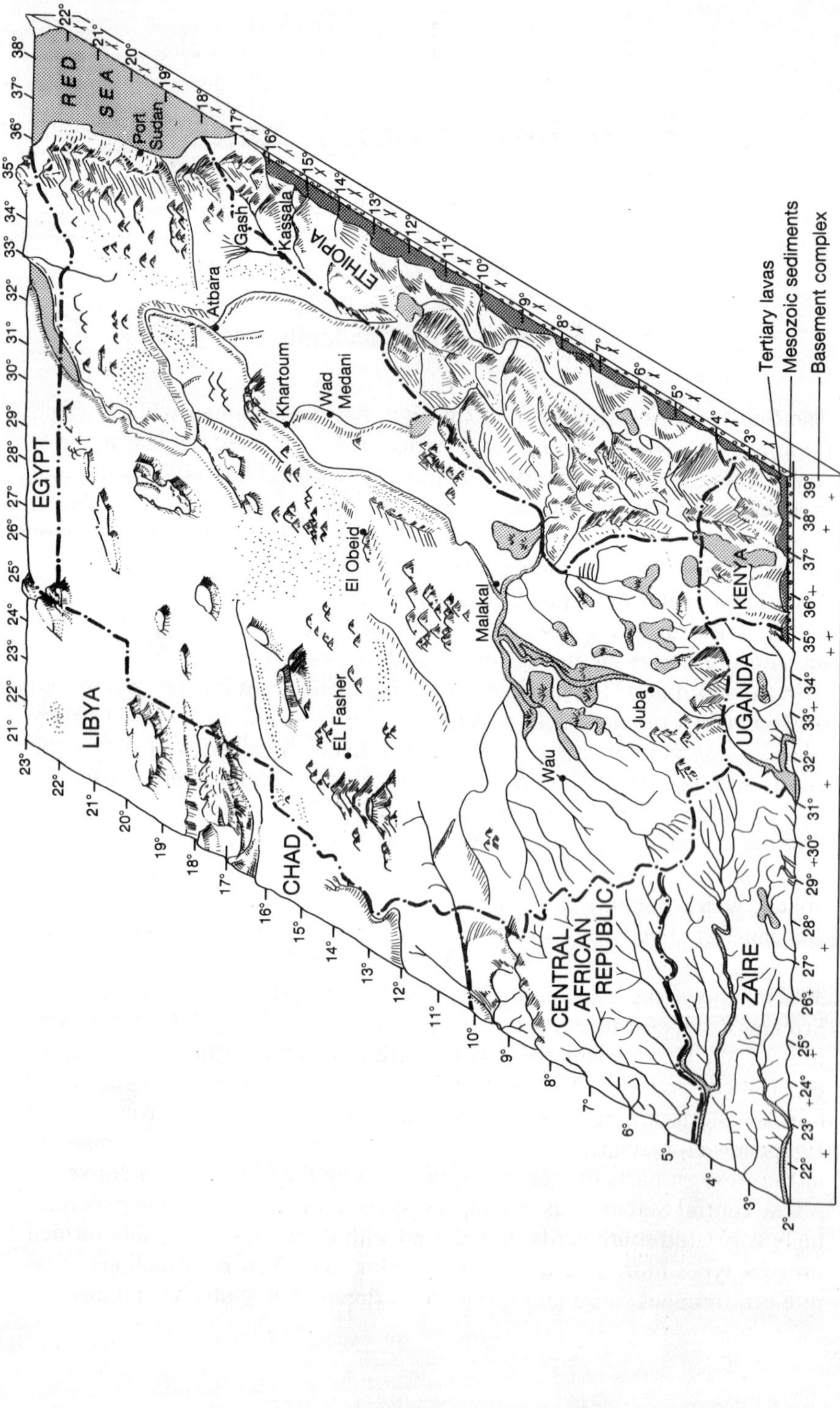

FIG. 1.1. Sudan: schematic physiography.

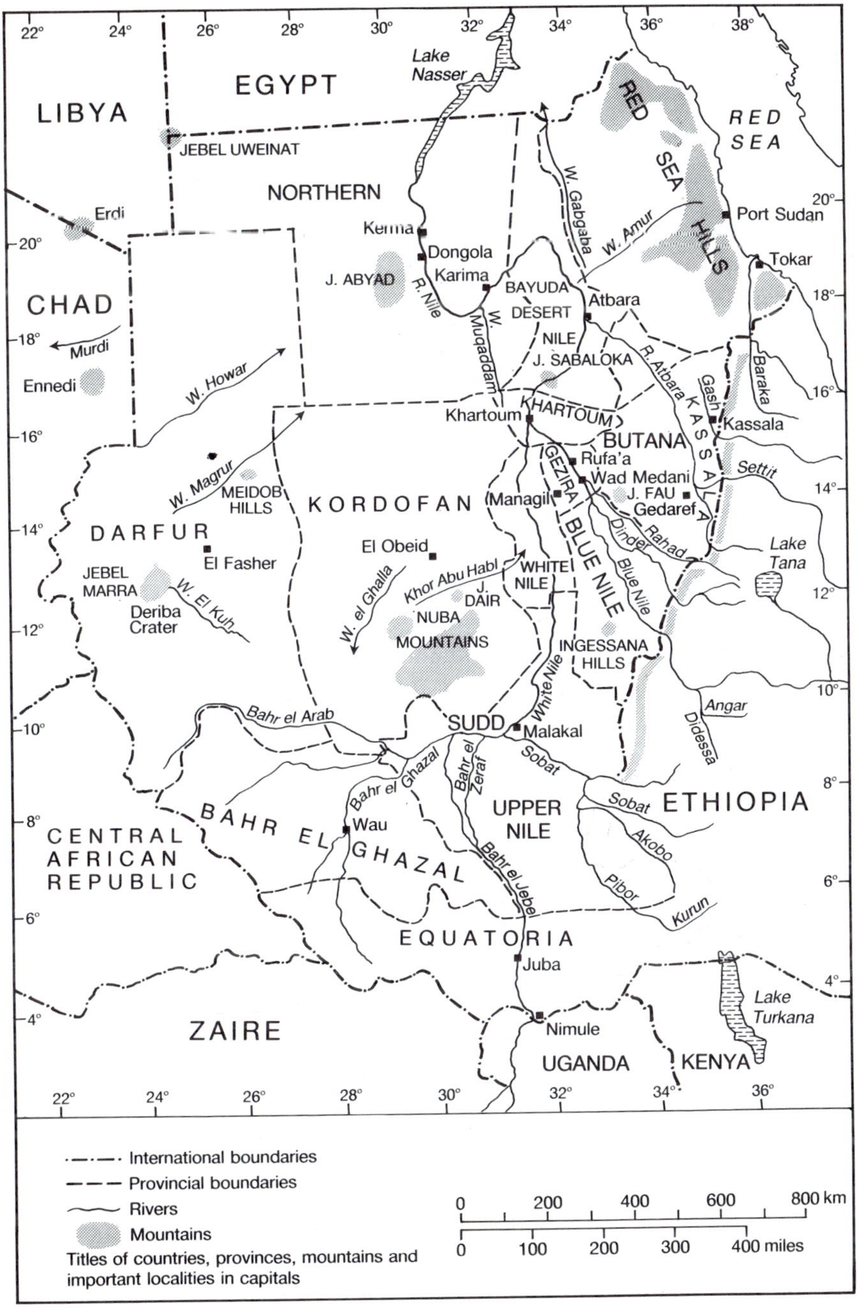

FIG. 1.2. Sudan: regions.

instance, consist of granite hills and gneissic ridges rising to about 600 m above the plain, culminating in Jebel Dair (1412 m). The Ingessana Hills in Blue Nile Province, by contrast, are formed of serpentinites, while the Nubian sandstone, which covers much of the country north of Khartoum, generally forms dissected horizontal plateaux, rising to about 100 m above the plain, with surrounding detrital aprons and tabular outliers.

The plains east and south of Khartoum, including the low grassy tracts around the Sudd, are dominantly formed of clay, hard in the dry season but a sea of mud after rains. In much of Kordofan and Darfur, however, these plains are sandy and gently undulating. All areas are drained by a network of watercourses which are larger and more densely spaced on the clays and towards the south.

Geological history

Precambrian, Palaeozoic, and Mesozoic

The geology of the Sudan has been summarized by Whiteman (1971) and Vail (1978). The country forms an integral part of the African Shield. This consists of a Basement Complex of metamorphosed and folded Precambrian schists and gneisses of sedimentary and volcanic origin as well as synorogenic or late orogenic granite emplacements and basic and ultrabasic bodies (Fig. 1.3). It also includes post-orogenic younger intrusives, many in the form of ring structures and linear dike swarms. Many igneous rocks occur which are of later age and some may be quite recent. The Basement underwent tectonism and igneous activity which lasted into Palaeozoic times. This elevated it into a land surface and was the last time the region experienced folding movements intense enough to form mountain ranges.

In the Palaeozoic period this landscape experienced a period of erosion and then a marine submergence leading to further deposition. Renewed uplift again led to emergence and a resumption of erosion. This removed most of the evidence for the deposition, but the Nawa formation is thought to belong to this period. It was not until the Cretaceous that an extensive cover of sediments was deposited. The chief representative is the mainly continental Nubian sandstone which now covers about one-quarter of the country. It was formed of sandstones and mudstones transported from Ethiopia towards the Mediterranean and deposited under sub-aqueous conditions probably in braided river systems debouching into delta fans or across flood plains (Vail 1978, p. 26). Many of the beds show features such as ripple marks and graded bedding, indicating a SE–NW direction of movement. They were uplifted soon after deposition and have been exposed to sub-aerial erosion ever since, so that extensive tracts have been swept clear of the deposit.

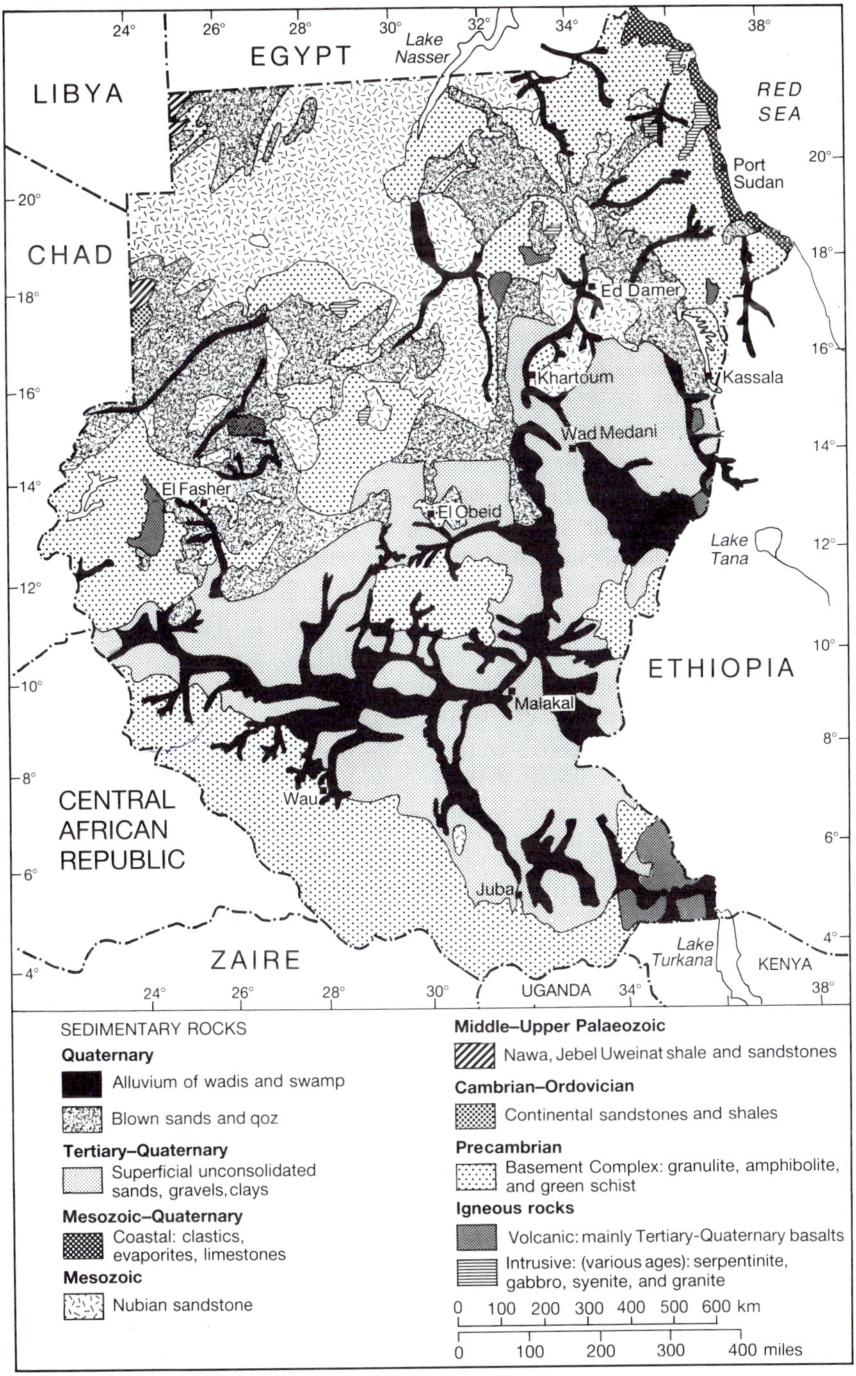

FIG. 1.3. Sudan: geology (after Vail 1978).

Tertiary

The Tertiary period saw the commencement of the widespread continental deposition, which continued through the Quaternary and today covers much of central Sudan. These periods were, however, strongly influenced by tectonic events. In the Cretaceous period, Africa had a different global position from today. The Equator ran through south-western Nigeria and north-western Sudan. A slight clockwise rotation in the Miocene and Pliocene brought it into contact with Europe. This had two important sets of consequences, one tectonic and the other climatic.

Tectonically, it caused the Red Sea Rift and uplifted the Red Sea Hills on one side of the continent and raised the Atlas Mountains on the other. Inland, there was both vulcanism and updoming in Hoggar, Tibesti, Air, and Jebel Marra. Thick volcanic flows covered much of Ethiopia, while more isolated and smaller volcanic fields developed in the Sudan—notably Jebel Marra, the Meidob Hills and in the Bayuda Desert. Jebel Marra with its fumaroles and hot springs is still potentially active. Downwarping formed the trough through Egypt and the Sudan along which the Nile now flows and a NE–SW depression in the Sudan near to the Ethiopian Highlands, which has subsequently received their detritus. This vulcanism and updoming also disturbed the drainage systems, for instance, blocking the outlet of the Wadis Howar and Magrur around the Jebel Abyad Plateau to the Nile near Dongola. Slight subsequent uplift in the east is shown by marine sediments now above water on the present coastal plain.

The climatic consequence of the rotation was a progressive southward shift of the equatorial forest belt and desiccation of those areas it had once covered. This was reinforced by the decrease in precipitation which resulted from the global reduction in evaporation, associated with the late Tertiary cooling of the oceans. The desiccation had dramatic effects in northern Sudan. Major rivers, which had previously eroded large valleys, dwindled and dried up, becoming choked with alluvium and colluvium washed off the previously vegetated slopes. Wind, no longer impeded by vegetation, formed dunes and exported silt- and clay-sized particles to be trapped as loess in the semi-arid fringes of the desert, leaving the surface to be armoured progressively by the concentration of residual stones and gravel (Williams 1982).

Quaternary

The Quaternary period saw an alteration of pluvial and interpluvial periods. The former were marked by accelerated erosion and sedimentation. The main feature of the pluvial periods was the formation of the wide clay plains in the south-eastern quadrant of the country, derived largely by fluvial deposition from the Ethiopian Highlands via the Blue Nile drainage system. There was also an extensive fixation of sand dunes and an increased development of soil horizons. During interpluvial phases, *wadis* dried up, vegetation retreated southwards and aeolian activity increased.

The Blue Nile drainage system included the proto-Blue Nile, Dinder, and Rahad Rivers. These were confined between the Managil Ridge and Jebel Fau and flowed in a north-westerly direction from north of Sennar towards the White Nile. The channels were sinuous and appear to have been active in the 12 000–11 000 BP period when higher floods supported the 'Early Khartoum' culture (Arkell 1953). By 7000–6000 BP, they had become incised in their present locations. Their former courses are now marked by sandier materials and a series of high-level meander cut-offs and ox-bows surrounded by clays deposited by overbank flooding. Evidence of such wetter conditions is provided by the presence of shells of the terrestrial snail (*Limicolaria flammata* Caill.), now living under 400–800 mm of rainfall in the Gedaref and Malakal areas, in the top 1.5 m of Gezira soil (Adamson *et al*. 1982).

The White Nile had a lower gradient, due partly to its lower source areas and partly to its constriction between the deltas of the Khor Abu Habl and the proto-Blue Nile south of Khartoum. Its course is marked by evaporite deposits and dunes which developed before about 12 000 BP when the wetter climate caused Lake Victoria to overflow into it. This extra discharge was probably further increased by intermittent overflows from Lake Turkana via the Pibor River between about 10 000 and 3000 BP (Harvey and Grove 1982). Much of the clay, however, owes its origin to *in situ* weathering of the Basement Complex rocks. Coarse detritus rich in alumino-silicates is eroded from hills on to the surrounding plains. This gives distinct light-coloured weathering aureoles around the hills, clearly visible from the air. Beyond these aureoles the seasonal alternation of inundation and desiccation appears to convert the original minerals into montmorillonitic clays. This process is evident in the Butana and the Gezira where soil clay percentages tend to increase southwards with the rainfall and in a direction opposite to that which would be expected if they were fining away from the Ethiopian source areas (Hunting Technical Services 1963).

The most extensive Pleistocene and Recent deposits west of the Nile are the *qozes*, which are gently rolling sand sheets and dunes with a general N–S trend and slopes rarely exceeding 10 per cent. They are composed of well-rounded quartz grains of Nubian sandstone origin and are now stabilized partly by special flora, partly by a thin surface crust of iron oxide and partly by fine silt and clay. The surface varies in colour from pale buff to deep red due to iron staining. They were formed mainly during the earlier drier phases of the Pleistocene, when their present location, which today has a rainfall of 200–700 mm, experienced the conditions which now occur some 6° further north where dunes are still actively forming. Wetter conditions from *c*. 10 000–7000 BP caused slope wash, terrace deposition, and soil formation. After this there was a second dry phase of dune formation before the onset of the present intermediate rainfall conditions (Warren 1970; Whiteman 1971).

Main geological formations

Basement Complex

The Basement Complex rocks outcrop over about one-quarter of the country. They consist mainly of undifferentiated gneisses and schists with a general NNE–SSW structural trend in most of the country curving around to NNW–SSE in the extreme south-east. Although data are still inadequate to give a comprehensive stratigraphic scheme, there seem to be three main metamorphic *facies*, dominated in chronological sequence by granulite, amphibolite, and green schist. The *granulite facies* occur mainly in the south where they correspond to similar rocks in Uganda and often form cratonic nuclei. The *amphibolite facies* are the most widespread, covering most of the country, while the green *schist facies* occur mainly in the east, especially in the Red Sea Hills and Ethiopia, but also in some places in the interior such as the eastern Nuba Mountains.

Palaeozoic

The Palaeozoic rocks consist mainly of coarse- to medium-grained arenaceous sediments, containing some siltstones and shales. There are three main localities, all in western Sudan and in adjacent parts of Chad, Libya, and Egypt. North and west of Wadi Howar large outcrops dip northwards and disappear beneath a cover of later deposits, re-emerging around the Jebel Uweinat inlier. The other significant occurrence is around the village of Nawa in Kordofan where mudstones, arkosic grits, and thin limestones overlie Basement rocks. These are, however, covered by Quaternary formations and are known only from boreholes. They are non-fossiliferous, but have been designated as Palaeozoic–Mesozoic on the basis of lithology and structure.

Mesozoic

The main Mesozoic formation in the Sudan is the Nubian sandstone, which covers about one-quarter of the country, mainly in the north. It extends under the Gezira and appears as isolated outliers in Bahr el Ghazal and Upper Nile Provinces. The thickness and wide extent of the beds indicates that it represents an extended temporal unit. Its identification depends largely on lithology because of a lack of fossils and marker beds. It consists usually of coarse, poorly sorted, pebbly sandstones, quartz pebble conglomerates, and arkoses with lesser shales and mudstones, becoming somewhat coarser towards its Ethiopian source. Calcareous and ferruginous horizons are common, especially the latter.

Tertiary

The most representative deposit is the Hudi chert, a fossiliferous opaline silica probably formed from volcanic material deposited in lakes in a number of locations in Northern Province not far from the Nile or its tributary *wadis*. At the surface it is characteristically weathered into ellipsoidal boulders up to 30 cm in diameter.

Other Tertiary deposits, thickening seawards, overlie the Basement Complex in the coastal strip along the Red Sea. They are of economic interest because they contain evaporites and, perhaps, gas and oil.

Quaternary

Quaternary formations blanket another quarter of the country. They consist mainly of water-laid clays, river alluvium, and wind-blown sands. Four names are given to the clay formations largely on the basis of geographical distribution: Umm Ruwaba, Gezira, El Atshan, and Butana.

The Umm Ruwaba formation extends westwards from the Blue Nile across Kordofan and into Darfur and covers extensive areas in the south around the major tributaries of the White Nile. Drilling has revealed nearly 300 m of unconsolidated gravelly sands, clayey sands, and green or grey clays.

The Gezira formation is the continuation of these deposits between the White and Blue Niles, varying in thickness from 1.5 m to 180 m with 70 m in the type locality of Ghubshan (14° 14′N, 33° 02′E). It consists of three main strata which are, in chronological order, the Mungata member of clayey sands, the lower sandy member of coarse gravelly sands, and the upper clay member—'Gezira clay', considered to range in date from late Tertiary to mid-Pleistocene. Since late Pleistocene lake margin features of the White Nile are cut into this deposit, its formation must have been completed before this.

The El Atshan formation is the continuation of the same deposits east of the Blue Nile, but is especially rich in basaltic materials and *kankars* (white to greyish calcium carbonate nodules) and, additionally, contains agate and ironstone nodules. The Butana formation is the northernmost tract of clay plain lying between the Nubian sandstone around Khartoum and the Atbara River. The clay is locally more than 12 m thick but thins out and becomes much coarser, eventually passing into clayey sands and gravels near the edges of pediments.

The mineral assemblage in sediments brought by the Blue Nile is somewhat different from that brought by the White Nile. Although both carry augite, hornblende, biotite, and garnet, the Blue Nile sediments contain more lava fragments and are higher in calcic plagioclase, iron ores, staurolite, kyanite, olivine, and apatite. The White Nile sediments are distinguished by a larger proportion of sillimanite, epidote, and sphene, and of these sillimanite is the most distinctive since it is absent from the Blue Nile sediments. However, since the White Nile receives volcanic material from the Ethiopian Plateau via the Sobat, the differences of its sediments from those of the Blue Nile are not always clear cut.

Recent deposits include wind-blown sands and valley alluvium. The former occur where the rainfall is less than about 500 mm and specifically north and west of a line roughly from the northern end of the Red Sea Hills through Khartoum and El Obeid to Jebel Marra. They are in dunes up to 50 m deep, which are aligned in a vast arc, flowing in a NW–SE direction in northern Egypt, sweeping around to become N–S along the Sudan's northern frontier, NE–SW along its western frontier, and ENE–WSW in Chad. This arc has the form of a clockwise spiral around a centre at about 28 °N, 18 °E in Libya, forming part of the broad pattern of sandflows established by Wilson (1971) for north Africa. The Nile and many of the smaller drainage systems have well developed alluvial deposits, some of which appear in Fig. 1.3. In the drier areas, in which flowing water is no longer found, the alluvial terraces testify to previously wetter conditions. The valleys of the northern *wadis* are generally sand-choked, often concealing the former drainage course, as illustrated by the Wadi Howar in Darfur Province and the Wadi Amur in Red Sea Province. Some of the larger rivers, especially those draining from Ethiopia and Jebel Marra, have developed large fan deltas on the plains.

In other areas, particularly over the Umm Ruwaba-type deposits of south-central Sudan, annual flooding by the main rivers causes seasonal overspill and large areas become inundated, spreading fine alluvium and forming swamps on low ground.

Present physiography of the Nile Valley

The Nile drainage derives from mainly outside the Sudan. The White Nile rises in Uganda, entering the Sudan as the Bahr el Jebel at Nimule, where it turns sharp left to follow the Aswa shear zone. At 65 km downstream, it swings north towards Juba, following the main Basement foliation. From then on for the next 1600 km it crosses unconsolidated Quaternary sands and silts until it meets the Blue Nile at Khartoum. Generally it flows northwards, except for one 200 km stretch, from the junction with the Bahr el Ghazal to Malakal whence it flows eastwards, possibly following an E–W fault alignment which parallels the Bahr el Arab and other streams and fractures in Bahr el Ghazal Province, Ethiopia, and the Central African Republic.

At Khartoum, the White Nile is joined by the Blue Nile, draining from Lake Tana through a gorge which cuts deeply through the younger rocks of the plateau. In the Sudan, its tributaries, the Atshan-Dinder and the Rahad have straight but converging courses, possibly reflecting faulting or warping of the underlying formations. The valleys of these rivers and that of the Atbara contain belts of recent alluvium, locally gullied into badlands, known as *kerrib*.

North of Khartoum, the combined Nile crosses the only outcrops of Basement gneisses within hundreds of kilometres and, even more extraordinarily, it cuts centrally through the only upstanding ground formed by the Sabaloka ring complex. It then crosses flat-lying sedimentary strata until, passing again over Basement Complex gneisses at Atbara, it follows almost exactly the changes in strike of the Basement rocks as far as Karima. This is surprising since, for much of its history, it must have flowed over the flat-lying sedimentary cover which is now being eroded off. Wherever it cuts across the grain of the country there are cataracts and rapids, the most important cataracts being the sixth at Sabaloka, the fifth at Atbara and the fourth at Karima.

Extensive river capture must have occurred during the evolution of this drainage system. For instance, the original Blue Nile might have continued north perhaps down the Wadi Muqaddam into the Dongola reach, while the Atbara might have flowed northwards past Atbara and on towards Wadi Gabgaba, or else westwards south of the Bayuda loop. These changes of flow might have been initiated by gentle updoming over the central Bayuda, perhaps due to volcanic activity of quite recent date (Vail 1978, p. 4).

The main Nile from Kerma to Wadi Halfa is again strongly influenced by the Basement rock structures and many bends and changes in direction closely reflect the fold trends of the underlying rocks, again forming cataracts where they are transgressed, notably the third at the southern and the second (now drowned by Lake Nasser) at the northern end of this reach. The first cataract is again on Basement at Aswan.

Soils

Sudan soils reflect, firstly, the broad climatic zonation of the country and, secondly, the modifying effect of local factors. The zonation is well marked because of the country's level character and wide extent, covering over 19° of latitude. It traverses regularly spaced rainfall belts from extreme desert in the north through steppe and savanna to woodland in the south. The soils are zonal where they reflect the climatic datum, azonal where local factors preclude its development, and intrazonal where they dominate it. Many soils, especially in the north of the country, show traces of formation during wetter periods of the Quaternary, notably in the survival of horizons showing rubefication or impregnation with lime, gypsum, or soluble salts. Vertisolization occurred in areas, for instance near Khartoum, where rainfall is inadequate to cause it today. Figure 1.4, based on the FAO/UNESCO (1974) *Soil map of the world*, shows the distribution of the main soil types.

Zonal soils

In the northern deserts, the scanty rainfall can only effect some local weathering of rock and shallow translocation of clay or salts and thus only forms faint and shallow horizons. As rainfall increases southwards, weathering accelerates, translocated layers intensify and deepen and organic matter increases.

Arid regions (no land use) Yermosols are the soils of the northern desert which have more horizonation than bare rock or mobile dunes. This generally takes the form of some clay, gypsum, or lime accumulation in the shallow subsoil and the absence of appreciable organic matter. They may have a *takyr* (cracked clay) surface. In waterless areas of northern Kordofan, Darfur, and neighbouring parts of Northern Province these and other soils, mainly of coarse texture, support a cover of juicy plants, known as *gizu*, suitable for camel grazing, the name sometimes being applied to the environment in general (Tothill 1948).

Semi-arid regions (extensive grazing only) Arenosols are coarse-textured soils with limited weathering of the parent material and some translocation of lime, gypsum, or clay particles. They occur typically on the *qozes* of Kordofan, Darfur, and neighbouring parts of Northern Province. They are characterized by 60–70 per cent coarse sand, 20–30 per cent fine sand and 5–10 per cent clay, with the proportion of clay increasing with depth and in the interdune hollows. The profile is generally structureless, compacted, buff to red, becoming paler with depth but with some mottling from lime, iron, and manganese concretions. The cation exchange capacity is low and there is occasionally high sodium saturation. Where *qozes* interdigitate locally with basin clays, the soils of the rises of the resulting soil catena are designated *atamur* and those of the basins, *naga'a*. Xerosols are developed on finer materials and resemble yermosols except that the somewhat higher rainfall gives more organic matter and deepens the horizons of accumulation.

Sub-humid regions (grazing and seasonal rain agriculture) Cambisols are developed on weakly-weathered rocks without migration of the weathering products. They are mainly associated with uplands and are widespread in

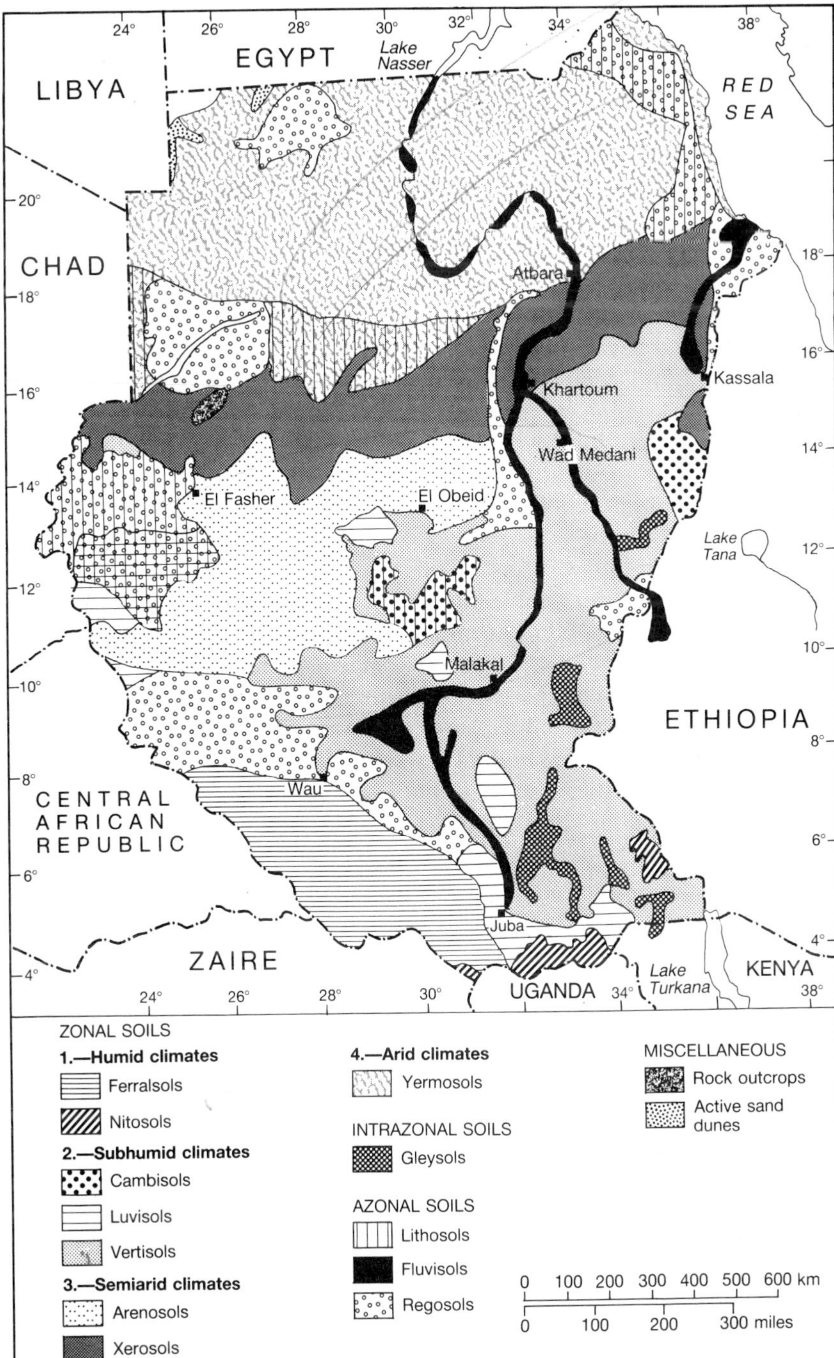

FIG. 1.4. Sudan: soils (after FAO/UNESCO 1973).

Ethiopia, the only significant occurrences in the Sudan being in the border areas around Gedaref and in the Nuba Mountains.

Luvisols are mainly distinguished by having an 'argillic' horizon of subsoil clay accumulation under rainfall which is adequate to translocate clays but inadequate to sustain the vegetation which would give an organic topsoil. They occur mainly in bushland in the southern Sudan and around hills in Kordofan and Darfur.

Vertisols are dark, cracking, montmorillonitic clays sometimes known as 'black cotton soils' and are widespread on the detrital plains derived from the Ethiopian uplands and Basement Complex outcrops in the Sudan. Their colloquial name is *badob*, but they are also called *fuda* where the surface is a granular mulch and, locally, on both sides of the Blue Nile, as *dahrat*. The non-cracking clays of pediments around the base of hills, which become very hard when dry, are known as *gardud*. In the south, vertisols have more organic matter and occur in and around the Sudd, notably on the seasonally flooded grazing lands along the watercourses draining into it known as *toiches*.

In the Gezira, the vertisols are deep, dark-coloured, low in organic matter, very slowly permeable when wet, and deeply cracked when dry. The clay content is 50–60 per cent and there is very little coarse material except for surface *kankars*. The top 60 cm is dark brown, and the second 60 cm grey, below which there is a considerable depth of yellowish brown soil. Calcium carbonate and gypsum concretions occur in the subsoil. The cation exchange capacity is high. They are generally fertile, although often restricted by alkalinity or salinity, both conditions being associated with greyer colours and less cracking, and increasing from south to north and towards the centre of depressions. In the central Gezira, lime and gypsum concretions and soluble salts reach a maximum at a depth of 1–2 m.

Humid regions: rain agriculture Nitosols are reddish-brown tropical soils with argillic horizons and some organic matter. They occur in the hilly areas with dense bushland in the south-eastern corner of the country, where they frequently form an upper member of soil catenas which extend downwards to hydromorphic soils in river valleys and basins.

Ferralsols are red tropical soils with organic topsoils over oxic subsoils. The latter are pedological horizons dominated by iron and aluminium oxides resulting from the almost total destruction of aluminosilicate minerals and the leaching of bases and silica. Clays are synthesized but relatively little is translocated. Ferralsols occur throughout the woodland zone on the Basement rocks of the Nile–Congo watershed where they are often dotted with termite mounds. As in the areas with nitosols, the undulating nature of the country often makes them the upper members of soil catenas which extend downwards to end in valleys and basins. Where ferralsol subsoils are exposed to the air they may become irreversibly hardened into a barren ferricrete shield.

Intrazonal soils

Intrazonal soils are due either to site conditions which concentrate water or else to the dominance of a particular parent rock. The only example which occurs in areas large enough to appear in Fig. 1.4 is the gleysol. Gleysols have profiles resulting from permanent or semi-permanent waterlogging, causing the exclusion of aerobic organisms. This, in turn, results in a peaty surface layer and a gleyed subsoil. Gleying is the reduction of oxygen-bearing compounds by anaerobic bacteria and is visible in the conversion of iron compounds from the reddish ferric to the bluish-grey ferrous state. Gleysols are most widespread in the extensive areas of marshland on the south-eastern clay plains. Where drained, they are often suitable for agriculture. Other intrazonal soils occurring in the Sudan have little agricultural potential. These include the solonchaks (salty soils) in and around irrigated areas and the rankers (soils with a shallow A horizon over non-calcareous rock) on hillslopes in the south.

Azonal soils

Fluvisols are soils on recent alluvium and support most of the irrigation outside the Gezira. They are located mainly along the Nile and its tributaries, along major seasonal wadis, and on inland deltas such as those of the Gash and Baraka. The sloping silty land in the bed of the Nile is known as *gerif* land, while the somewhat salty land along its bank just above high flood level is known as *qurer*. In the Gash Delta, the fine alluvial land is known as *lebbad* and the sandier land as *mashandow*.

Lithosols are very shallow soils on bare rock, unusable for agriculture. Regosols are other azonal soils on unconsolidated materials, some of which may be usable under irrigation.

Soil productivity

The overwhelming limitation on land use throughout the Sudan is drought. The dependence of rain-fed agriculture, forestry, and range on soil moisture is so complete that all other soil properties are secondary. Only under irrigation do they begin to play an important role. Thus, soil productivity in rain-fed and irrigated lands demand separate consideration.

Trees are an indication of potential agricultural productivity under rain-fed conditions. Smith (1949) showed that in the Sudan the distribution of tree species was almost wholly dependent on rainfall and site type and that these were interdependent in a fairly simple way. In general, sites increase in favourability with coarseness of soil texture, depression in the landscape, and the proximity and discharge of neighbouring watercourses. Clay soils require one and a half times as much rainfall to give the same soil moisture as sands. Because each tree species has a preferred range of soil moisture conditions, it is found on the same site types within one rainfall zone and on site types in other rainfall zones giving equivalent soil moisture conditions. For instance, *Acacia senegal* occurs on sites ranging from hillsides in Kordofan and Rufa'a (15 °N, 33 °E) where the rainfall is 300 mm, to sheet erosion slopes at Nimule (4 °N, 32 °E) and *Acacia tortilis* is found in seasonal runnels in the desert with a rainfall of 50 mm, on level sands at Khartoum with 150 mm, on clays in Kassala Province with 300 mm, and on erosion slopes in Butana (14–16 °N, 34–36 °E) with 500 mm. With no recorded exceptions, all species progress through their rainfall span via the same sequence of site types and, given the same rainfall, the site types fall into the same sequence of favourability for tree growth. The list of site types with their approximate rainfall equivalence is given on Table 1.1.

This scheme has some relevance to the quality of rangeland, but only a few of the site types are suitable for rain-fed agricultural crops because these require relatively level plots of substantial size with workable stone-free topsoils.

Productivity under irrigation reflects a wider range of soil properties

TABLE 1.1. Site types in the Sudan in order of favourability for tree growth

Site type description	Rainfall required by datum species (*Acacia seyal*)
(A) Hard surface slopes, i.e. sheet slopes	1000 mm
(B) High, old flood plain; if now flooded, only for days	
(C) Low flood plain, flooded for weeks at a time	800 mm
(D) Mounds, usually of termite origin, in swamp	
(E) Land-locked pools in clay areas	
(F) Clay plains, no runoff and no standing water	500 mm
(G) Mature sand plains, no runoff and no standing water	
(H) Immature sand, new dunes and old sandhills	300 mm
(I) Small well-watered and well-drained pockets in sands	
(K) Hills of rough rocky surface	150 mm
(L) *Wadis* or large seasonal watercourse flushing for one or two hours after rain	100 mm
(M) Hard plains of grit or rock	
(N) Seasonal runnels flushing for one or two hours after rain	
(O) Banks of permanent rivers and streams	

Source: Smith (1949).

than under rain-fed agriculture because the dominating drought constraint is removed. In general, in the Sudan, the main irrigated soils are of two types: the vertisols of the central clay plains and the fluvisols along river valleys. The vertisols, as previously indicated, are intrinsically highly fertile but suffer from the physical limitations of very slow permeability and high plasticity and stickiness when wet. They can yield good crops where salinity and alkalinity are controlled by maintaining a downward movement of soil water and a favourable Ca:Na ionic balance.

River alluvium contains a higher percentage of silt and sand and is, therefore, more permeable. Annual inundations have made some valleys the centres of irrigation for millennia. *Seluka* cultivation follows the annual retreat of the flood in the bed of the Nile, whose higher alluvial banks are irrigated by pumps or *saqqiyas* (waterwheels). Large annually flooded inland deltas, such as those of the Gash north of Kassala, the Baraka at Tokar, and the Wadi El Kuh south of El Fasher, provide important agricultural areas, while patches of irrigation are found in most smaller *wadis*.

References

Adamson, D. A., Williams, M. A. J., and Gillespie, R. (1982). Paleogeography of the Gezira and the lower Blue and White Nile valleys. In *A land between two Niles* (ed. M. A. J. Williams and D. A. Adamson), pp. 165–220. A. A. Balkema, Rotterdam.

Arkell, A. J. (1953). *Shaheinab*. Oxford University Press, London.

FAO/UNESCO (1974). *Soil map of the world, at a scale of 1:5 000 000*, Vol. 1 Legend, and Sheet VI-2 (1973) covering the Sudan. UNESCO Press, Paris.

Harvey, C. P. D. and Grove, A. T. (1982). A prehistoric source of the Nile. *Geographical Journal*, **148**, 327–36.

Hunting Technical Services Limited (1963). *Roseires soil survey, Report No. 1. Gezira extension area soil survey and land classification*, Vol. 1. Sir M. MacDonald and Partners, London.

Smith, J. (1949). *Distribution of tree species in the Sudan in relation to rainfall and soil texture*, Sudan Government Ministry of Agriculture Bulletin No. 4. Ministry of Agriculture, Khartoum.

Tothill, J. D. (ed.) (1948). *Agriculture in the Sudan*. Oxford University Press, London.

Vail, J. R. (1978). *Outline of the geology and mineral resources of the Democratic Republic of the Sudan and adjacent areas*, Overseas Geology and Mineral Resources Report No. 49. Institute of Geological Sciences, London.

Warren, A. (1970). Dune trends and their implications in the central Sudan. *Zeitschrift für Geomorphologie*, **10**, 154–80.

Whiteman, A. J. (1971). *The geology of the Sudan Republic*. Clarendon Press, London.

Williams, M. A. J. (1982). Quaternary environments in northern Africa. In *A land between two Niles* (ed. M. A. J. Williams and D. A. Adamson), pp. 13–22. A. A. Balkema, Rotterdam.

Wilson, I. G. (1971). Desert sandflow basins and a model for the development of ergs. *Geographical Journal* **137**, 180–99.

2

Climate, hydrology, and water resources

R. P. D. WALSH

Climate

Climatic regions of the Sudan and their meteorological controls

The whole of the Sudan, apart from the highest peaks of the Jebel Marra area, the Red Sea Hills, and the southern mountains, experiences a tropical climate, in that monthly mean temperatures always exceed 18 °C. There are, however, major contrasts in rainfall and humidity and several distinct climatic types, characterized by contrasting agricultural crops and systems, can be identified (Fig. 2.1 and Table 2.1). The climate of the Sudan ranges from Desert types in the north and Tropical Semi-Arid and Sub-Humid climates with a short summer rainy season in central Sudan, to Tropical Wet-Dry and Tropical Rainy climatic types with progressively longer wet seasons in the south. Summary characteristics of each climatic type are given in Table 2.1.

The climatic contrasts within the Sudan reflect the great N–S extent of the country (3–23 °N) and can be explained in terms of the distribution of surface and upper air pressure systems and the position of the migratory Intertropical Discontinuity (ITD) at different times of the year (Fig. 2.2). The ITD, which generally is aligned east–west along the migratory equatorial low pressure trough, is the boundary between two air masses of contrasting moisture and temperature characteristics: to the north is the dry, stable, descending (and hence rainless) air of the north-east trade winds, associated with the subtropical anticyclone situated over the Sahara in winter and the Mediterranean in summer; to the south of the ITD is the moist, somewhat cooler, south-westerly air derived from Equatorial Africa and the tropical Atlantic and Indian Oceans.

Rainfall in the Sudan and the distribution of the various climatic types are closely related to the seasonal migration of the ITD and the rain belt to the south of it. The ITD, in the Sudan, migrates from a latitude of 3 °N in January to its northernmost position of 17 °N in August (Zahran 1979). It advances more slowly than it retreats and hence its April position at 10 °N is considerably south of its October position at 13½ °N. The length of the

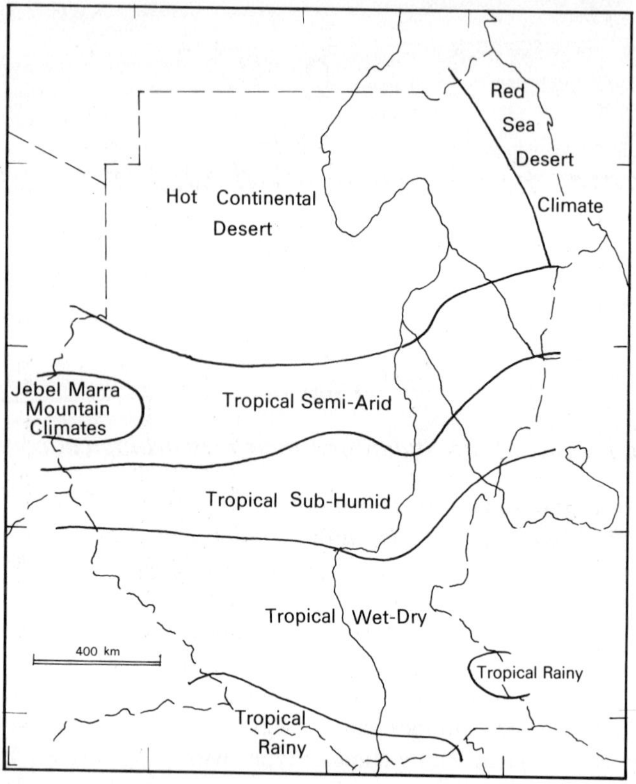

FIG. 2.1. Sudan: climatic regions.

wet season increases southwards as the length of time that the ITD lies northward of locations increases; hence the progression from Desert, through Semi-Arid, Sub-Humid, Wet-Dry to Rainy climatic types from north to south in the Sudan. In January, the whole of the country lies under the influence of the dry NE trades; the only region to experience significant rain from November to January is the Red Sea coast, which is affected by occasional depressions from the Eastern Mediterranean that regenerate in the Red Sea. It is this winter rainfall, together with the high humidity of a maritime location, that makes the Red Sea Desert climate a distinct type. By April, the Saharan anticyclone has weakened and has started to migrate northwards to the Mediterranean. A low pressure belt has established itself over central-south Sudan and the ITD lies just north of Malakal at 10 °N. By this time the wet season has commenced at locations south of 7–8 °N. By July, the ITD has advanced to about 17 °N which is close to its northernmost position in August and rainy conditions affect the whole of the Sudan south of 15–16 °N; however, further north, NE trades persist and account

TABLE 2.1. Sudan: climate classification

Climatic type	Annual rainfall (mm)	Wet months	Seasonality index	Other climatic characteristics	Agriculture
Tropical Continental Desert	<200	<0.5	⩾1.25	Relative humidity always <40%. Very short summer wet season in south. Large diurnal and annual temperature range	Nomadic pastoralism. Cultivation only via irrigation along Nile and Atbara
Red Sea Desert	<200	<0.5	1.10–1.25	Relative humidity >40% throughout year. Winter rainfall maximum. Smaller temperature ranges	Nomadic pastoralism
Tropical Semi-Arid	200–500	0.5–1.9	1.10–1.25	Highly variable annual rainfall	Widespread cultivation by dry farming, but frequent crop failures
Jebel Marra mountain climates	>500	⩾2.0	1.10–1.25	Mean annual temperature <25 °C. High diurnal range. Increase in rainfall with altitude	As Sub-Humid, but also citrus fruits, vegetables and potatoes
Tropical Sub-Humid	500–800	2.0–3.9	0.90–1.15	More reliable annual rainfall for dry farming crops	Widespread cultivation; more reliable crop production. Zone of major mechanized farming
Tropical Wet-Dry	800–1200	4.0–5.9	0.60–0.90	Long dry season and long wet season	Cattle rearing dominant. Some millet cultivation
Tropical Rainy	>1200	⩾6.0	<0.60	Short dry season. Warmest months are in 'winter' dry season. Very low annual temperature range	Maize and eleusine. Coffee, tea plantations. Some fruits of the wet tropics in mountains

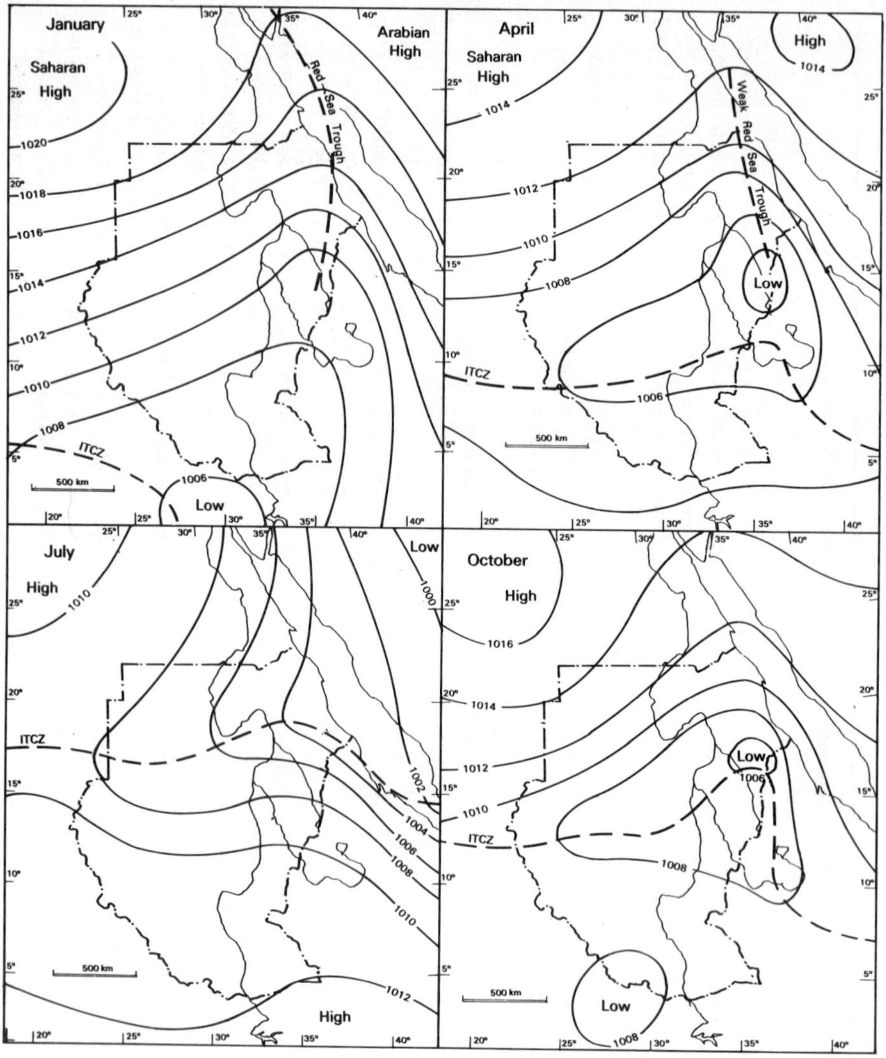

FIG. 2.2. Seasonal changes in surface pressure (millibars) and the position of the Intertropical Discontinuity (ITD or ITCZ) in the Sudan region. Source: Bhalotra (1963).

for the desert climates of northern Sudan and the Red Sea coast. By October, the Saharan High has started to re-establish itself and the low pressure belt and ITD have begun to retreat southwards; the rain belt has likewise retreated south and is largely confined to the Tropical Wet-Dry and Rainy zones.

The climatic characteristics of the various zones of the Sudan are now examined in more detail by considering each of the principal climatic elements.

Rainfall

Annual rainfall Mean annual rainfall in the Sudan over the period 1951–80 (Fig. 2.3(a)) ranged from zero in the extreme north-west, where it rarely rains, to over 1400 mm in the mountains of the Tropical Rainy zone of the extreme south. Annual rainfall thus declines steadily away from the equator reflecting a parallel decline in the period of influence of the rainy zone of the moist south-westerlies south of the ITD. A few anomalies from this essentially latitudinal pattern are apparent. The influence of relief in

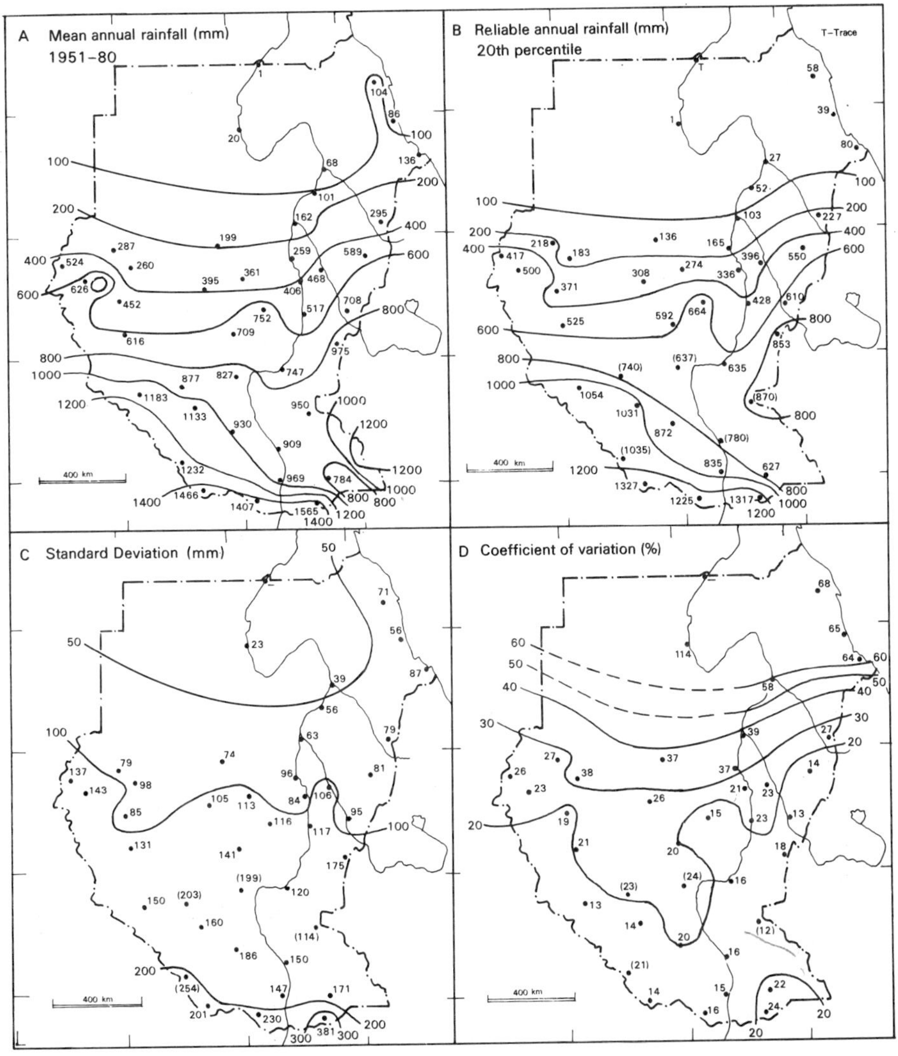

FIG. 2.3. Sudan: annual rainfall and its reliability and variability, 1951–80.

increasing rainfall is clear over the Red Sea Hills, the Jebel Marra region, the Nuba Mountains, close to the Ethiopian border, in the southern mountains, and on the Ironstone Plateau of the south-west borderlands of the Sudan. The rather higher rainfall of the Red Sea Desert region compared with the Tropical Continental Desert climatic area to the west, reflects the winter rainfall the Red Sea coast receives from temperate depressions.

Figure 2.3(b), showing reliable annual rainfall, maps the annual rainfall that was equalled or exceeded in 80 per cent of years over the period 1951–80. Year-to-year variability is summarized by the maps of standard deviation (Fig. 2.3(c)) and coefficient of variation (Fig. 2.3(d)), where the latter is the ratio of standard deviation to mean annual rainfall. The two parameters assess absolute and relative variability, respectively. It is interesting to note the anomalously low variability (relative to annual rainfall) of the sub-humid and semi-arid parts of eastern Sudan; at Kassala, Gedaref, and Er Roseires, the coefficients of variation are only 27, 14 and 13 per cent, respectively.

Rainfall regimes As shown in Fig. 2.4, all regions in the Sudan, except the Red Sea Desert experience a summer rainfall maximum with August the wettest month, when the ITD has reached its most northerly position.

The degree of concentration of a rainfall regime can be expressed quantitatively using the Seasonality Index (Walsh and Lawler 1981). Values of the index can range from zero for a regime with equal rainfall in each month to 1.83 for a regime where all the rainfall occurs in a single month. In the Sudan, values of the index increase with latitude as rainfall becomes concentrated into shorter rainy seasons. The range of values associated with each climatic type is given in Table 2.1. It is interesting to note that where annual rainfall is increased by relief, the Seasonality Index is unaffected. Thus, in the Jebel Marra area of Darfur, Zalingei (annual rainfall 626 mm) has an index value of 1.16, very similar to the 1.14 value at Nyala (annual rainfall 452 mm) at lower altitude to the south east.

Daily rainfall magnitude-frequency For detailed consideration of rainfall–agriculture relationships, daily rainfall data are important (Fig. 2.5). The number of rain days with larger daily falls (≥10 mm) is significant for recharging soil moisture, generating *wadi* flows and recharging shallow and deep groundwater systems (Hulme and Walsh, in press). Table 2.2 compares the annual frequencies of small (1.0–9.9 mm) and large (≥10.00 mm) daily falls at selected locations. The ratio of small to large falls declines with increasing annual rainfall, but the high proportion of annual rainfall accounted for by large falls is clear in all zones; in the Semi-Arid zone a few large falls account for over 50 per cent of the annual rainfall. It is interesting to note the higher frequency of small falls and high ratios of small to large falls at western, as compared with eastern, Sudanese locations at similar annual rainfalls in the Semi-Arid zone.

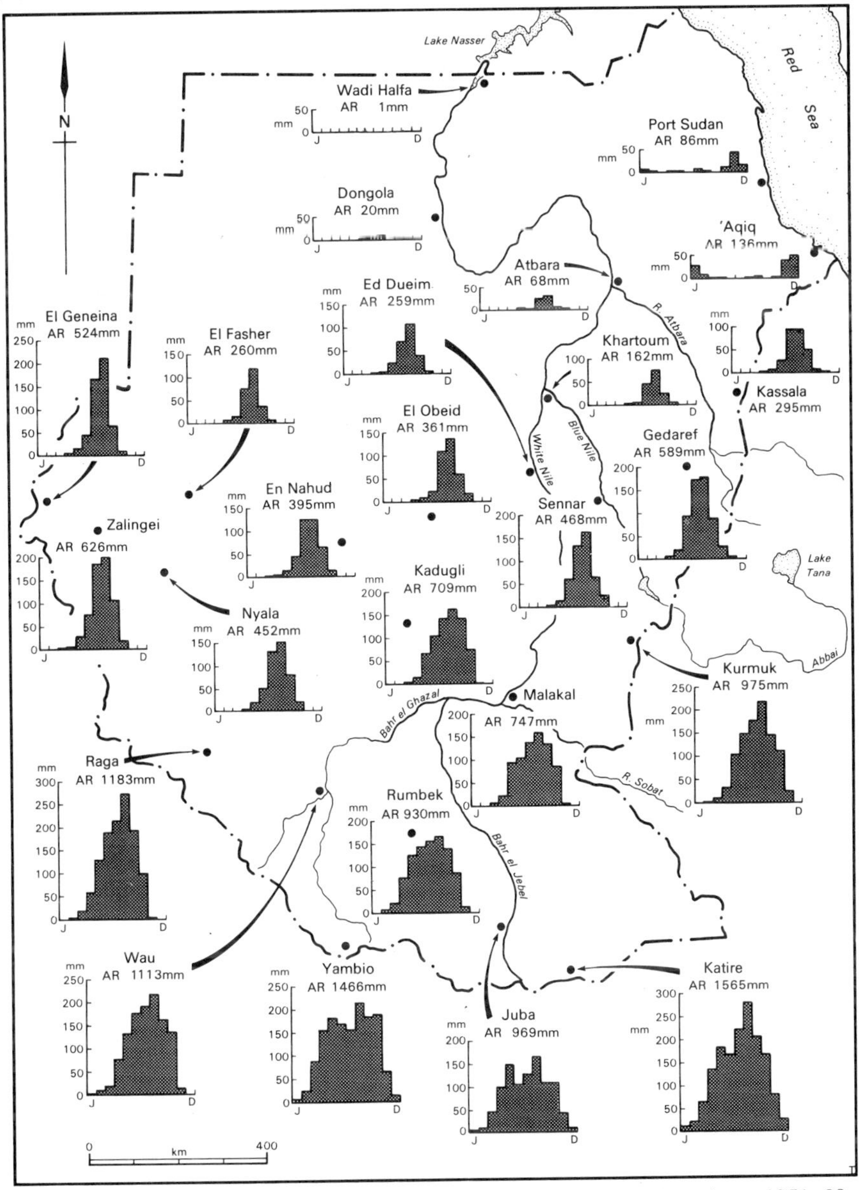

FIG. 2.4. Sudan: rainfall regimes at twenty-five representative locations, 1951–80. AR = mean annual rainfall

Of significance to erosion potential and peak *wadi* or river flows are mean annual maximum daily falls shown in Fig. 2.5(c). Highest recorded falls (Fig. 2.5(d)) are relatively modest by tropical standards, reflecting the 'continental' situation of the Sudan and the lack of a nearby ocean moisture

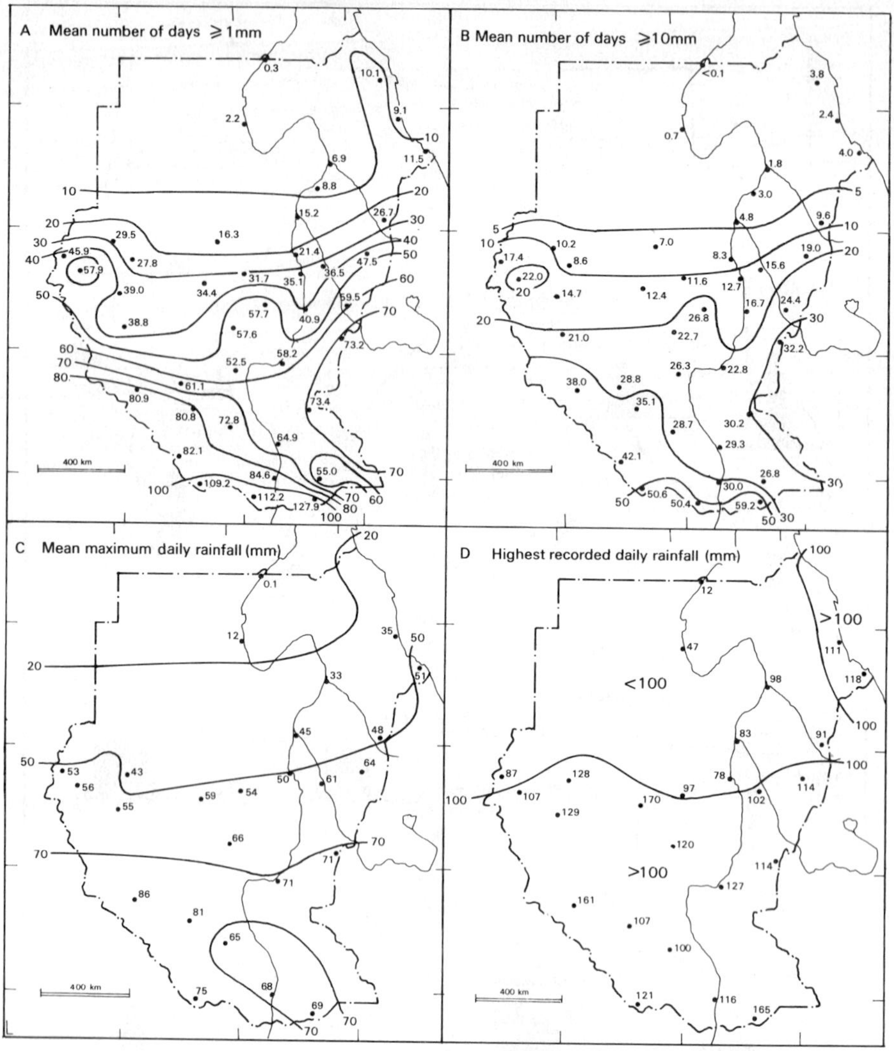

FIG. 2.5. Sudan: aspects of daily rainfall magnitude–frequency, 1951–80.

source that would be necessary to sustain high intensity rainfall and produce larger daily totals.

Sunshine and solar radiation

Monthly and annual means of duration of bright sunshine and solar radiation received at the earth's surface are given in Tables 2.3 and 2.4 for five locations in the Sudan. The natural tendency for both variables to follow a

TABLE 2.2. Mean annual frequency of different size categories of daily rainfall at Sudanese locations, 1951–80

Location	Climatic zone	Mean annual rainfall (mm)	Mean annual frequency of			Ratio of 1–9.9 : ≥10.0
			Rain-days	1.0–9.9 mm	≥10.0 mm	
Atbara	Continental Desert	68	6.9	5.1	1.8	2.8
'Aqiq	Red Sea Desert	136	11.5	7.5	4.0	1.9
Khartoum	Continental Desert	162	15.2	10.4	4.8	2.2
Ed Dueim	Semi-Arid	259	21.4	13.1	8.3	1.6
El Fasher	Semi-Arid	260	27.8	19.2	8.6	2.2
Nyala	Semi-Arid	452	39.0	24.3	14.7	1.7
Sennar	Semi-Arid	468	36.5	20.9	15.6	1.3
Zalingei	Jebel Marra	626	57.9	35.9	22.0	1.6
Kadugli	Sub-Humid	709	57.6	34.9	22.7	1.5
Juba	Wet-Dry	969	85	55	30	1.8
Wau	Wet-Dry	1133	81	46	35	1.3
Yambio	Rainy	1466	109	58	51	1.1
Katire	Rainy	1565	128	69	59	1.2

TABLE 2.3. Mean daily duration of bright sunshine at locations in the Sudan

Location	Period of record		Jan	Feb	Mar	Apr	May	Jun	Jul	Aug	Sep	Oct	Nov	Dec	Year
Wadi Halfa	1951–61	h day^{-1}	10.1	10.3	10.3	10.7	11.3	11.8	11.7	11.1	10.7	10.6	10.4	9.9	10.7
		%[a]	92	90	85	85	86	88	88	87	87	91	94	92	88
Port Sudan	1956–61	h day^{-1}	7.0	7.9	8.5	10.2	10.8	10.2	9.7	9.7	10.0	9.9	8.3	6.7	9.1
		%	62	69	71	81	84	87	74	77	80	84	74	61	75
El Fasher	1956–61	h day^{-1}	10.3	10.8	10.0	10.2	10.2	9.1	7.5	7.6	8.8	10.2	10.9	9.8	9.6
		%	88	92	82	82	80	71	58	60	72	86	95	90	80
Malakal	1952–61	h day^{-1}	9.9	9.2	8.9	8.1	7.8	5.2	5.0	5.4	5.9	7.3	9.5	9.8	7.7
		%	86	79	74	66	62	41	40	43	48	61	81	84	64
Juba	1957–61	h day^{-1}	9.2	8.1	7.3	6.7	7.8	7.4	6.1	6.7	7.8	7.8	8.0	8.9	7.7
		%	77	67	60	55	63	60	49	55	64	65	67	75	63

[a] Percentage of possible sunshine.
Source: Bhalotra (1963).

TABLE 2.4. Monthly and annual means of solar radiation (cal cm^{-2} d^{-1}) at the earth's surface at locations in the Sudan (years not specified)

Location	Jan	Feb	Mar	Apr	May	Jun	Jul	Aug	Sep	Oct	Nov	Dec	Year
Wadi Halfa	457	524	588	639	667	655	638	609	582	525	461	426	564
Port Sudan	354	449	539	612	625	577	564	552	547	486	383	340	502
El Fasher	457	534	579	594	592	550	532	534	563	529	486	456	534
Malakal	471	521	527	536	510	448	448	486	484	487	487	476	490
Juba	458	464	462	454	479	459	415	466	509	480	457	445	462

Source: Shahin (1985).

seasonal rhythm of increasing in summer and decreasing in winter in response to the migration of the overhead sun is interrupted to varying extents by the parallel migration of the intertropical cloud and rain belt and (in the case of the Red Sea region) by winter cloud and rain associated with temperate depressions. The sunshine and solar radiation regimes thus tend to be mirror images of the rainfall regimes.

Temperature and relative humidity

Very high temperatures and low relative humidities prevail for much of the year over the Sudan. Monthly means of air temperature and relative humidity are shown graphically in Fig. 2.6, together with a map of annual temperature range. Mean annual temperature varies little over the Sudan,

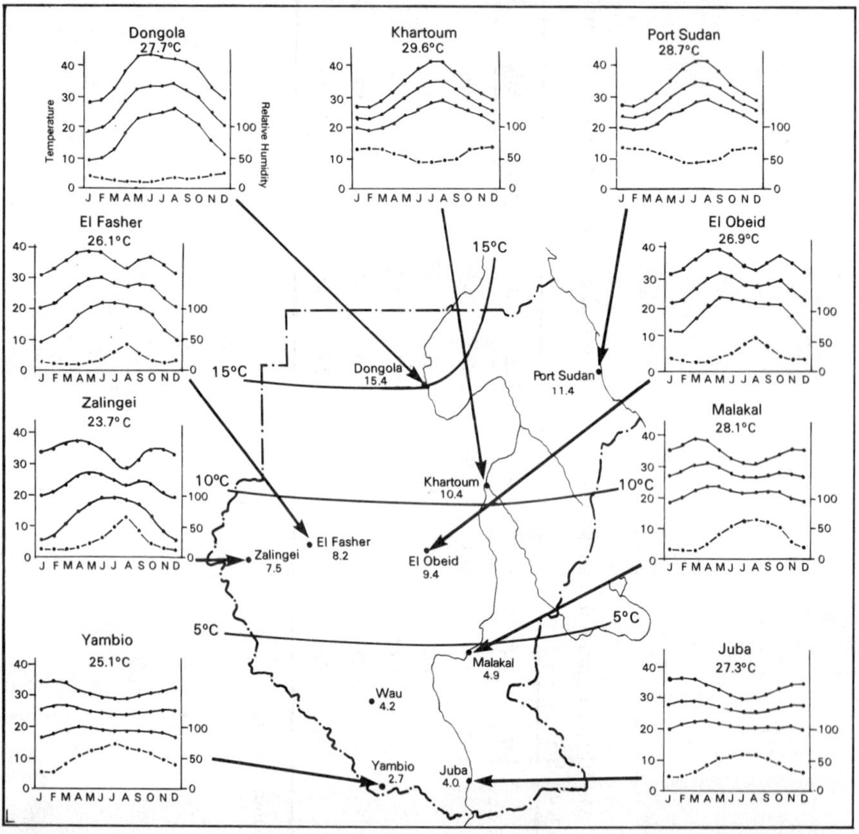

FIG. 2.6. Sudan: air temperature and relative humidity at nine stations, 1921–50. Continuous lines on graphs are monthly means of maximum, minimum, and mean air temperature; discontinuous lines are mean relative humidity at 12 noon; the figure above each graph is mean annual temperature; the map shows annual temperature range.

ranging from 29.6 °C at Khartoum to 27.3 °C at Juba, but rather lower at altitude, as at Zalingei (23.7 °C) and Yambio (25.1 °C) in the Jebel Marra and southern mountains, respectively. Annual temperature range increases poleward from 2.7 °C at Yambio to 15.4 °C at Dongola, but the maritime influence of the Red Sea reduces the range to 11.4 °C at Port Sudan.

As with sunshine and solar radiation, the impact of the rainy season and the incursion of humid south-westerly air behind the ITD in summer is very clear. Dongola and Port Sudan, which receive virtually no summer rainfall, have a summer maximum and winter minimum in temperature. At Khartoum, in the south of the Tropical Continental Desert zone, even the very short rainy season in July and August produces a sharp drop in temperature, with the mean maximum falling from 41.6 °C in June to 36.1 °C in August before rising again to 39.6 °C in October. The impact of the rainy season on temperature regimes increases southward. At El Fasher and El Obeid in the Semi-Arid zone, temperatures are highest in May/June; at Malakal in the Sub-Humid zone where the ITD arrives earlier, the temperature peak is in April; while in the extreme south at Juba and Yambio, where the dry season is short, temperatures vary little during the year and the 'winter' months are, in fact, the warmest.

Diurnal ranges of temperature are higher in the cloud-free dry season than in the wet season at all stations and are higher in the Tropical Continental Desert and Tropical Semi-Arid zones than in the rest of the country. At Dongola, the mean diurnal range exceeds 16 °C throughout the year. At El Fasher, the diurnal range from November to April exceeds 20 °C, but drops to 12.3 °C in August during the short wet season. The maritime influence reduces diurnal ranges along the Red Sea coast, where in the cloudy, rain-prone winter months diurnal ranges dip to 7.2 °C. In contrast, altitude increases diurnal ranges in the dry season at Zalingei, where the mean January diurnal range is 28.4 °C. In southern Sudan, diurnal ranges are considerably lower than in the north, averaging 13.7 °C at Malakal, 12.7 °C at Juba and 13.1 °C at Yambio, but less than 10 °C in the wettest months.

The lowest temperatures recorded at the stations in Fig. 2.6 were 2.8 °C at Dongola and 0.0 °C at Zalingei. Frosts are unknown on the plains, but do occur at higher altitudes in the Jebel Marra.

Relative humidity is very low (10–30 per cent) in the NE trades, north of the ITD, but considerably higher (>50 per cent) in the cloudy, rainy zone of the south-westerlies to the south of the ITD. During the year, relative humidity follows rainfall regimes at most Sudan stations, the duration of higher relative humidity declining from nine months in the Tropical Rainy zone to two to three months in the Tropical Semi-Arid zone and to zero months at Dongola in the Tropical Continental Desert zone. The only exception to this latitudinal pattern is the Red Sea climatic region, where the maritime influence and depressional activity in winter result in relative humidities over 40 per cent throughout the year and over 60 per cent in the winter months.

Wind speed

The direction of surface winds over the Sudan at different times of the year is indicated by the isobar patterns in Fig. 2.2. North-east or northerly trade winds prevail north of the ITD and cover the whole country in winter; south-westerly and variable winds occur south of the ITD and these reach their maximum extent in August. On average, wind speeds are significantly higher in northern than in southern Sudan (Table 2.5). The consistency and strength of the NE trades, allied with their extreme dryness, increases evaporation potential considerably during the long dry season in northern and central Sudan. Dust storms (*haboobs*), which affect northern and central Sudan (Bhalotra 1963, p. 56) can strip topsoil, damage crops, and silt up irrigation canals.

Evaporation and evapotranspiration

A detailed discussion of these topics is beyond the scope of this chapter and the interested reader is referred to recent comprehensive reviews by Shahin (1985) and Awadulla (1977; 1981; 1983).

The only long-term countrywide evaporation measurements available for the Sudan are those obtained from Piche evaporimeters. These tend to overestimate free water evaporation because of the 'island effect' of the small evaporation surfaces, but a correction factor of 0.5 has been found to be reasonable for converting Piche measurements to estimates of free water evaporation and potential evapotranspiration (PE) (Bhalotra 1963, p. 43) and has been used in the construction of PE graphs from long-term Piche records for representative stations in the Sudan (Fig. 2.7). Mean PE is highest in the Desert and Semi-Arid areas, averaging 7.7 mm day^{-1} at Khartoum and 6.6 mm day^{-1} at El Obeid, but somewhat lower (5.2 mm day^{-1}) at Port Sudan on the Red Sea coast, where higher relative humidities reduce the evaporation rate. PE declines southwards in the wetter zones in response to higher relative humidity, lower wind speeds, lower solar radiation and sunshine, and lower temperatures. Thus, mean PE is 4.8 mm day^{-1} at Malakal, 3.3 mm day^{-1} at Juba, and only 1.8 mm day^{-1} at Yambio. Though PE is undoubtedly less than that measured further north, there is some evidence that correction factors in the humid south should be higher (Shahin 1985, p. 185), as Penman estimates exceed Piche-derived PE values (Table 2.6). However, Penman equations have themselves been found to underestimate evaporation by 10 per cent in experiments at Khartoum (Shahin 1985, p. 206). In view of this, evaporation estimates should probably be regarded only as rough guides to potential evaporation levels.

Evapotranspiration of irrigated crops and other vegetation in the Sudan has been estimated using a wide variety of formulae, such as those of Olivier (1961), Hargreaves (1968), Blaney and Criddle (1966), and Penman

TABLE 2.5. Mean surface wind speed (knots) and annual frequency of duststorm days at locations in the Sudan (years of record not specified)

Location	Jan	Feb	Mar	Apr	May	Jun	Jul	Aug	Sep	Oct	Nov	Dec	Duststorm days
Wadi Halfa	7.5	8.5	8.7	9.2	8.6	8.0	6.5	6.9	8.2	8.0	7.0	6.5	7.4
El Obeid	8.0	8.7	7.1	6.3	6.0	7.1	7.0	5.9	4.9	5.3	6.6	6.4	6.4
Malakal	7.8	8.5	5.8	5.1	5.5	4.9	2.4	4.2	3.1	3.1	3.7	5.8	nd
Juba	2.8	2.7	3.3	3.3	2.8	2.1	1.9	2.0	1.9	2.0	2.2	2.3	nd

Source: Bhalotra (1963).

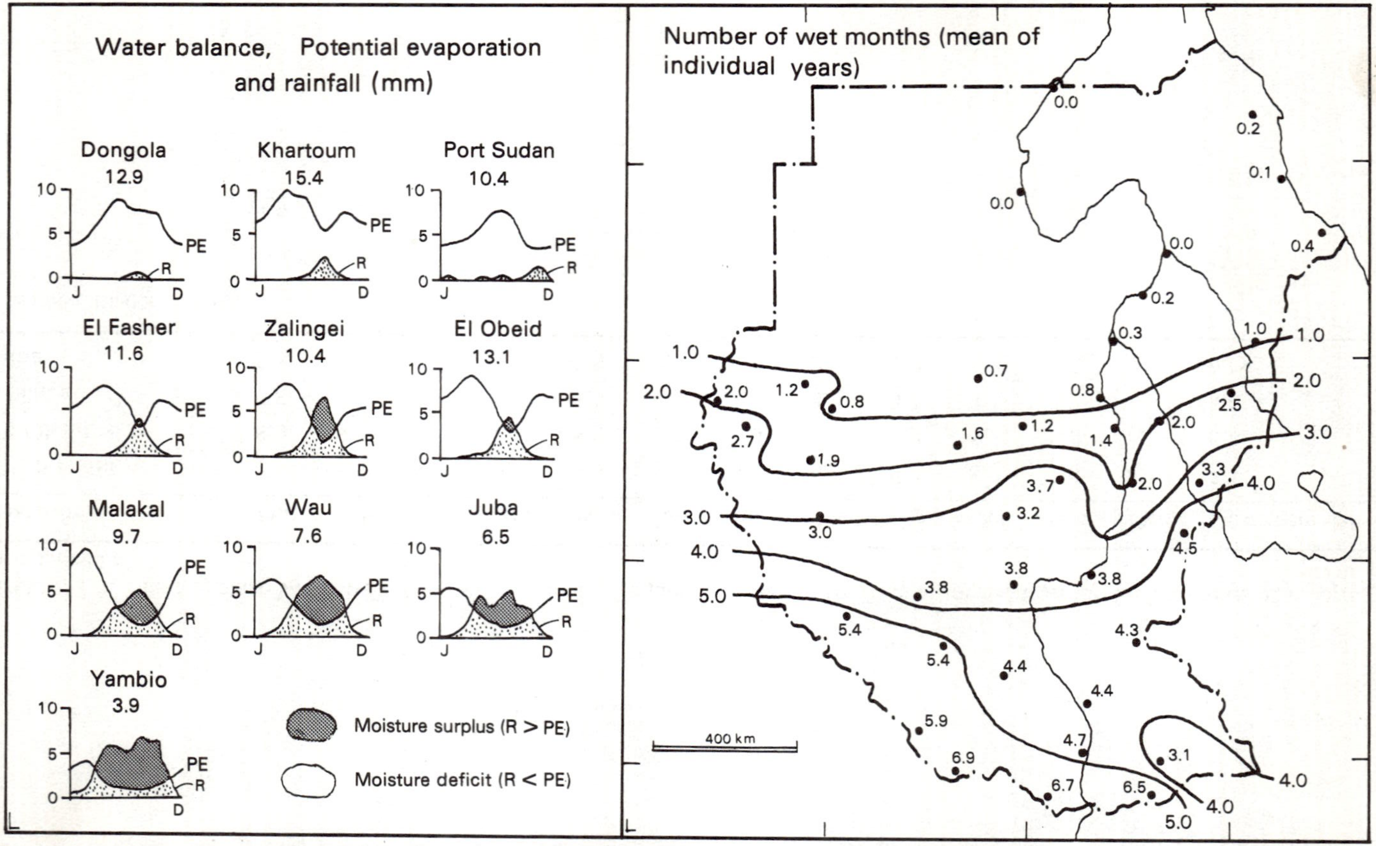

FIG. 2.7. Sudan: mean length of the wet season. LEFT: Water balance graphs using monthly means of rainfall and PE (where PE = Piche evaporation × 0.5). Figures beneath station names are unconverted annual means of Piche evaporation. All data in mm d^{-1}. RIGHT: mean number of wet (≥ 100 mm) months per annum, 1951–80.

TABLE 2.6. Comparison of monthly means of open water evaporation as estimated by Penman and Piche (× 0.5) methods (all figures in mm d^{-1}) (years not specified)

Station	Method	Jan	Feb	Mar	Apr	May	Jun	Jul	Aug	Sep	Oct	Nov	Dec	Year
Juba	Penman	6.4	6.9	6.6	5.8	5.2	4.7	4.1	4.5	4.9	5.2	5.3	5.7	5.5
	Piche × 0.5	6.2	6.0	4.9	3.7	2.4	2.2	1.7	1.7	2.2	2.7	3.6	4.6	3.5
Malakal	Penman	5.5	6.6	6.8	6.7	5.8	4.5	4.0	4.1	4.8	5.0	5.2	5.8	5.4
	Piche × 0.5	8.3	9.1	7.7	5.4	3.5	2.2	1.5	1.3	1.5	1.8	4.1	6.8	4.5
Khartoum	Penman	7.2	7.8	9.3	10.5	10.4	10.1	8.4	6.8	7.1	8.9	8.2	7.1	8.5
	Piche × 0.5	6.5	8.0	9.5	10.0	9.5	9.0	7.0	5.5	6.0	7.5	7.0	6.5	7.5

Source: Shahin (1985).

TABLE 2.7. Estimates of water use[a] by different crops at Wad Medani using the Hargreaves and Blaney-Criddle formulae (years not specified)

Month	Cotton		Wheat		Dura		Lubia	
	H	B-C	H	B-C	H	B-C	H	B-C
January	9.5	4.7	6.7	5.4	—	—	4.5	2.7
February	9.1	3.8	10.4	4.4	—	—	—	—
March	7.6	1.9	10.4	2.2	—	—	—	—
April	6.5	1.3	—	—	—	—	—	—
May	—	—	—	—	—	—	—	—
June	—	—	—	—	—	—	—	—
July	—	—	—	—	3.8	1.8	—	—
August	1.4	1.0	—	—	5.9	5.1	—	—
September	3.3	3.5	—	—	7.4	4.8	2.3	0.9
October	7.4	5.4	—	—	7.5	1.3	6.1	3.4
November	10.2	5.8	2.5	1.9	—	—	11.6	5.4
December	10.2	5.3	4.4	3.6	—	—	8.9	3.9
Year/Season	1842	964	885	480	544	347	943	461

[a] Monthly data in mm d^{-1}; year/season data in mm.
Source: Shahin (1985).

(1950). However, there have been few measurements of actual water loss using lysimeters against which to assess the accuracy of these methods; indeed, the methods themselves when applied in the Sudan give disturbingly different results (Table 2.7). Until more measurements of evapotranspiration are available, all methods and formulae, however complex, should be treated with scepticism.

Wet season length and structure

Monthly rainfall and PE are often used to derive monthly water balances and delimit the wet season by identifying months with a water surplus. In Fig. 2.7, the length of the wet season, defined by the mean number of wet ($\geq$100 mm) months per annum, ranges from zero months in the extreme north to over six months in the Tropical Rainy zone. This representation implies that water deficits occur when monthly rainfall is less than 100 mm and assumes that PE is 100 mm at each station throughout the year which, as the previous section on evaporation demonstrated, is not the case. In the water balance graphs (Fig. 2.7), rainfall and PE regimes are plotted together so that periods of moisture surplus (R>PE) and moisture deficit (R<PE) can be identified graphically. This representation allows for variations in PE by using the Piche-derived estimates of PE from the previous section. The increase in length of wet season from zero months at

the desert stations and a single month at the semi-arid station of El Fasher to seven months at Yambio in the Tropical Rainy south is clear.

Detailed analysis of wet seasons, to delimit the onset and end of a wet season and any significant breaks in the wet season, requires the use of daily climatic data. A recent approach developed by Hulme (1985, p. 131; 1987; Walsh *et al.* 1988) for central Sudan has used a simple water balance model based on daily rainfall and daily Penman PE data. The model is aimed specifically at identifying the onset and end of the wet season and distinguishing them from false starts and breaks within the wet season. Figure 2.8 shows how the wet season lengthens southwards from Khartoum

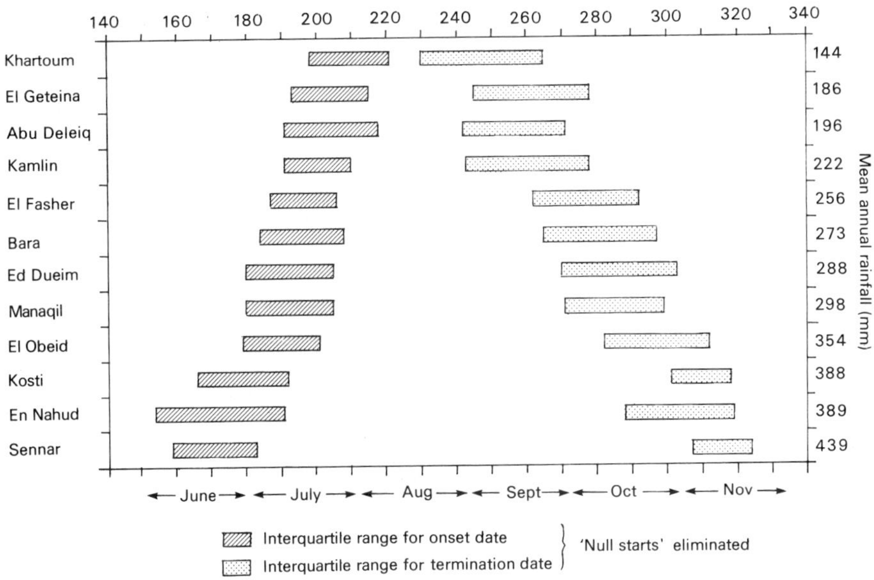

FIG. 2.8. Variability in the dates of onset and termination of the wet season at stations in central Sudan. Source: Hulme (1985).

to Sennar. As the interquartile ranges show, the onset varies rather less from year to year than the end of the wet season. Figure 2.9 shows the structure of the wet season, including mid-season breaks, for each year of record for three stations in central Sudan: Khartoum, which has an extremely variable and often absent wet season; and Ed Dueim and Kosti, which have longer wet seasons but quite frequent false starts.

Recent climatic change

In the Sudan, some climatic variables, particularly rainfall, have been subject to major change both during the current century and in the longer

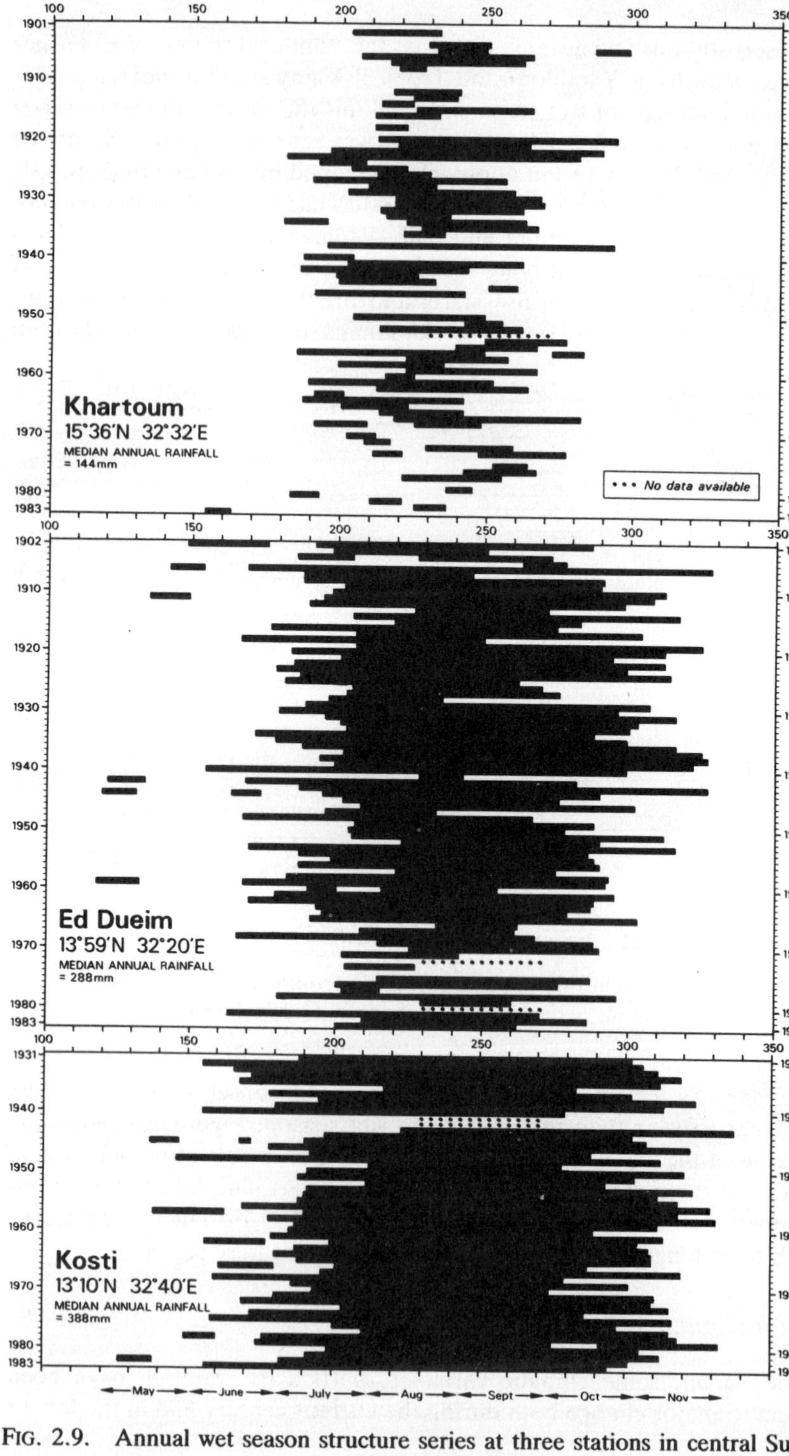

FIG. 2.9. Annual wet season structure series at three stations in central Sudan. Source: Hulme (1985).

term. In common with the rest of the Sahelian zone, there has been a major recent decline in annual rainfall in Semi-Arid and Sub-Humid Sudan, particularly since 1965 (Trilsbach and Hulme 1984), and this decline has continued and intensified in the 1980s (Walsh *et al.* 1988). Figure 2.10 shows deviations of annual rainfall from the 1921–80 mean at Ed Dueim (White Nile Province), El Fasher (Northern Darfur), and central Sudan (12–16°N) as a whole. Three rainfall epochs are apparent: a rather dry period during the first two decades of this century; a wet epoch from around 1920 to the early 1960s, with particularly wet conditions in the 1920s and 1930s, and a very dry period from 1965 to date. The recent decline is particularly striking. At Ed Dueim, mean rainfall in 1965–84 was 214 mm, 40 per cent lower than the 1920–39 mean (356 mm) and 27 per cent below the 1921–80 mean (295 mm). At El Fasher, 1965–84 rainfall at 202 mm was 38 per cent below that received in 1920–39 (325 mm), and 29 per cent below the 1921–80 mean. The graph for stations in central Sudan as a whole shows a similar pattern. When annual rainfall maps for the wet 1920–39 and very dry 1965–84 periods are compared (Fig. 2.11), the 200 mm and 300 mm isohyets can be seen to have retreated southwards some 120 km. In terms of the climatic classification presented earlier, it is clear that the Tropical Continental Desert zone has advanced and the Tropical Semi-Arid and Sub-Humid belts have retreated southwards correspondingly. The boundaries depicted in Fig. 2.1, based on 1951–80 averages, are intermediate between the wet mid-century and current very dry epochs and are clearly dynamic even in the short-term. Furthermore, historical accounts point to a series of dry epochs, of approximately equal severity to the current one, in the 1680s, in 1740–60, 1820–40 and 1895–1920 and a long and severe dry period in AD 1150–1500 (Nicholson 1983). Indeed, it appears that it was the mid-twentieth century wet period which was anomalous rather than the current dry epoch, which is much more typical of the climate of Semi-Arid and Sub-Humid Sudan over the past thousand years.

Parallel to the marked decline in annual rainfall have been changes in the length, structure, and reliability of the wet season. Marked contractions in the wet season have been noted (Hulme 1987; Walsh *et al.* 1988) at Khartoum, El Geteina, and Ed Dueim in central Sudan (Fig. 2.12) and, in each case, more of the contraction has occurred at the end rather than at the start of the wet season. Furthermore, the reliability of rainfall in mid- and late-season has also declined. Six mid-season breaks have occurred since 1968 at Ed Dueim compared with only six in the preceding sixty-six years.

Hydrology and water resources

River and groundwater systems of the Sudan

The whole of the Sudan, except the eastern slopes of the Red Sea Hills, lies within the Nile Basin, yet the contribution of rivers rising in the Sudan to

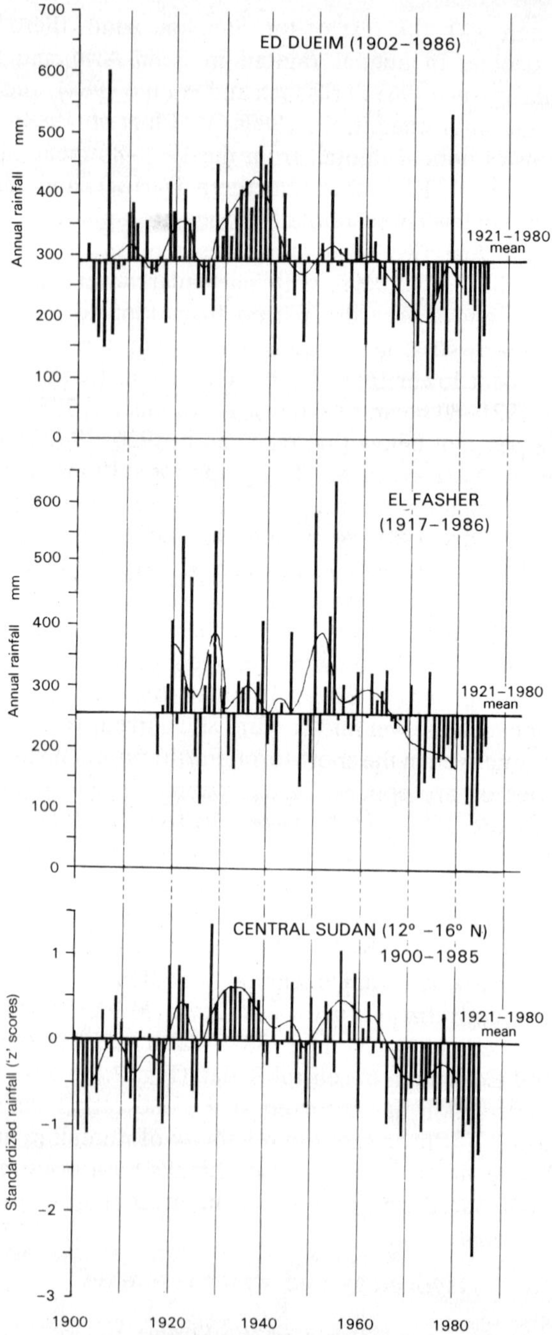

FIG. 2.10. Deviations of annual rainfall from the 1921–80 mean at Ed Dueim (White Nile), El Fasher (Northern Darfur), and central Sudan (12–16 °N). Source: Walsh *et al.* (1988).

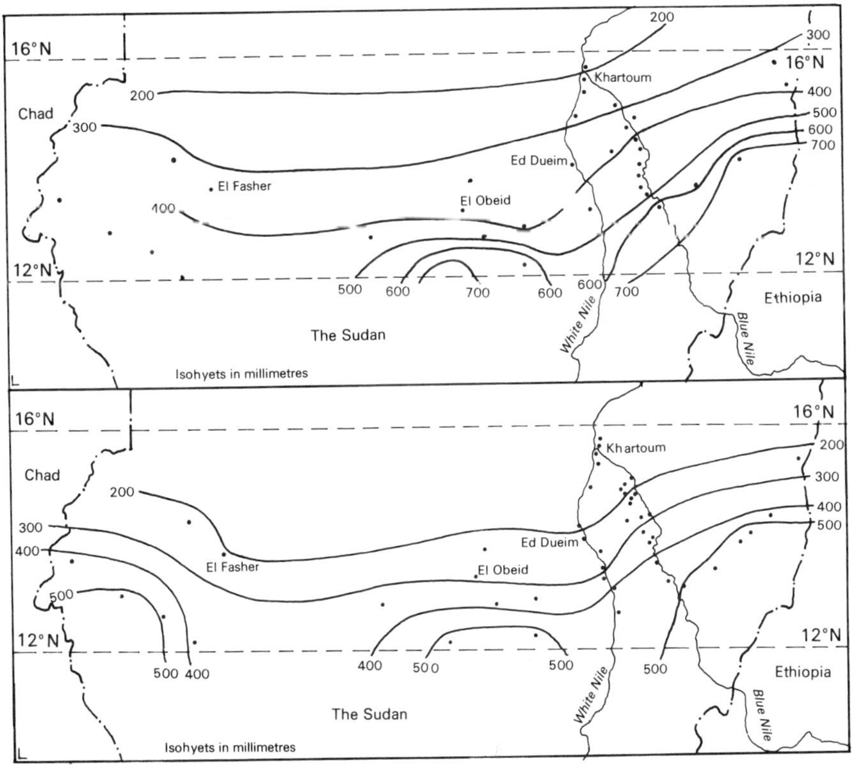

FIG. 2.11. Changes in mean annual rainfall between the periods 1920–39 (top) and between 1965–84 (bottom) in central Sudan. Source: Walsh *et al.* (1988).

the flow of the Nile is extremely small. The Nile derives almost all its flow from the Equatorial Lakes region via the White Nile (Bahr el Jebel) and from the Ethiopian Highlands via the Blue Nile, Atbara, and Sobat. Mean annual discharges and flow regimes over the period 1912–73 at various points in the Nile system within the Sudan are shown in Fig. 2.13. At Mongalla, shortly after entering the Sudan from the Equatorial Lakes region, the White Nile has a mean annual discharge of 31.4 milliards (where 1 milliard = 1000 million m^3 (or 995 $m^3\ s^{-1}$). The fairly even regime reflects the lack of marked dry seasons and the variety of rainfall regimes over the various tributary basins of the Equatorial Nile. The White Nile loses much of its flow in the Sudd marshes and, despite additional flows from the Sobat and Bahr el Ghazal totalling 13.6 milliards, the flow of the White Nile at Malakal (29.4 milliards) is less than it is at Mongalla. The Bahr el Ghazal and associated rivers draining the whole of the south-west of the Sudan (a catchment area of 528 000 km^2) also lose most of their flow in the Sudd as they encounter low gradients and relatively permeable

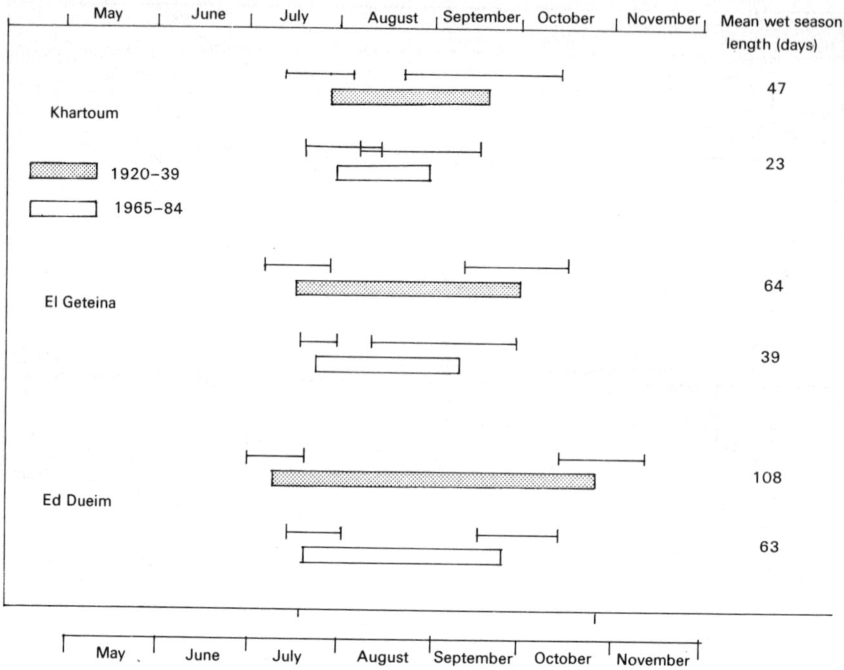

FIG. 2.12. Changes in the duration of the wet season between 1920–39 and 1965–84 at Khartoum, El Geteina, and Ed Dueim. The bars indicate the interquartile ranges of wet season onset and termination dates. Source: Walsh *et al.* (1988).

Umm Ruwaba sediments on leaving the impermeable Precambrian Basement Complex rocks of their upper catchments. Shahin (1985, p. 365) estimates that of the 14 milliards of flow entering the Sudd swamps from the Bahr el Ghazal catchment, only 0.7 milliards arrive at the White Nile. Annual evapotranspiration of river floodwaters from the swamp areas of the Bahr el Ghazal and Bahr el Jebel is estimated at 1880 mm and 1728 mm, respectively, to which evapotranspiration of rainwater from the swamps, which receive an annual rainfall of around 900 mm, should be added.

The flow of the Sobat ranges from less than 10 million m^3 day^{-1} from February to April to over 60 million m^3 day^{-1} in October/November after the late summer rainfall maximum over southern Ethiopia. The Blue Nile and the Atbara are progressively more seasonal in their flow regimes with discharge peaking earlier (in August) than in the Sobat, reflecting the shorter rainy season of northern Ethiopia. The mean flow of the Blue Nile (49 milliards per annum) is substantially higher than the White Nile (26 milliards) at the confluence at Khartoum and it is Blue Nile water that is largely used in the major irrigation projects.

The contribution of the western Sudan Sub-Humid and Semi-Arid zones

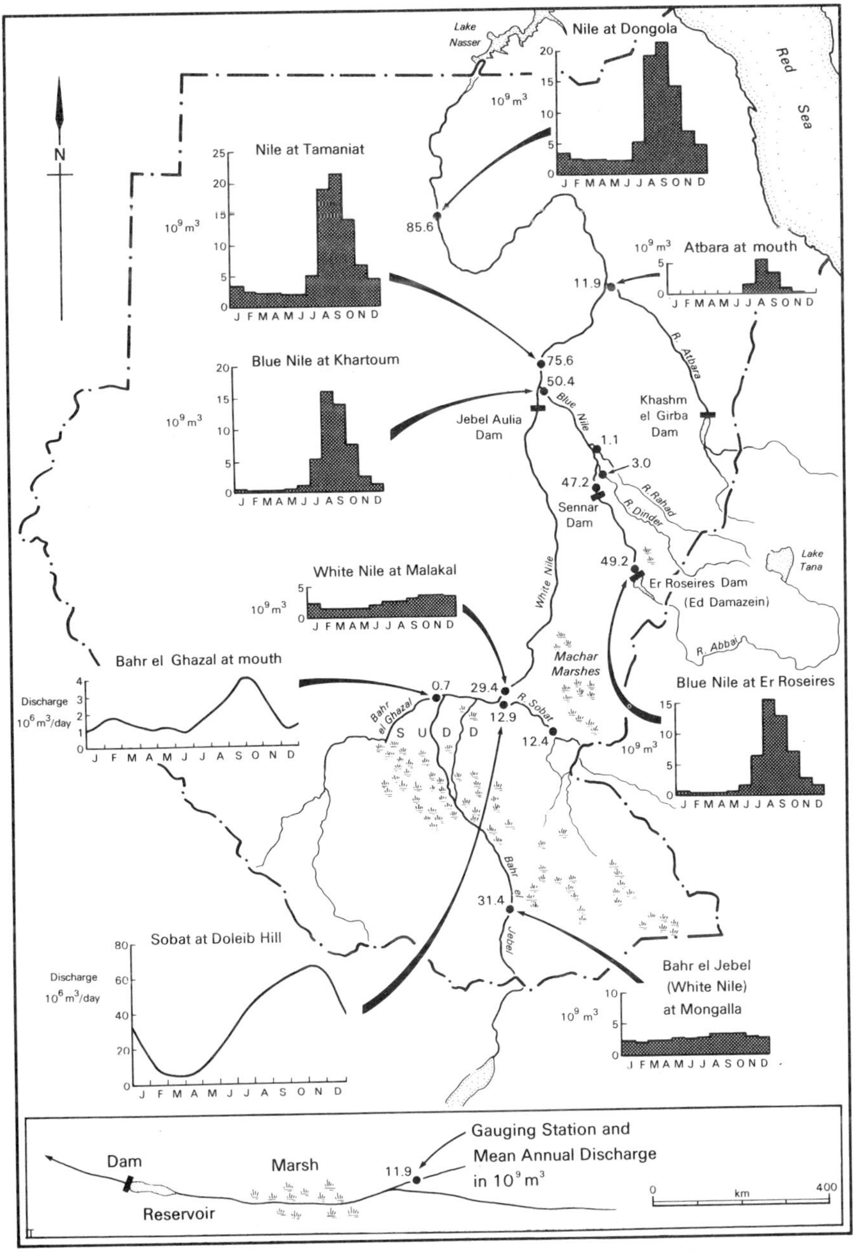

FIG. 2.13. Annual discharges and flow regimes of the Nile river system in the Sudan. All data are for the period 1912–73 except for the Bahr el Ghazal (1923–73) and were derived from Shahin (1985).

to the flow of the Nile is negligible. Although there are important local and regional ephemeral river systems and considerable surface flow following heavy storms in the rainy season, the rivers fail to reach the Nile, losing their flow through seepage and evaporation.

The flow of the Nile as it enters Lake Aswan (85 milliards per annum) is around 20 per cent less than the combined flows of the White Nile, Sobat, Atbara, and Blue Nile (around 105 milliards) as they enter the Sudan; it is also over 6 milliards less than the flows of the White Nile at Malakal, the Blue Nile at Khartoum, and the Atbara at its Nile confluence, showing the significance of transmission losses due to seepage and evaporation *en route*.

Flows display considerable variability from year to year (Table 2.8), though the flow of the White Nile is regulated significantly by the Sudd,

TABLE 2.8. Flow variability of the Nile river system in the Sudan

River and location	Mean annual flow (milliards)	Standard deviation (milliards)	Standard deviation (% of mean)
Bahr el Jebel at Mongalla	31.4	12.6	40.1
White Nile at Malakal	29.4	5.7	19.4
Blue Nile at Khartoum	50.4	10.3	20.4
Atbara at Nile confluence	11.9	3.9	32.8
Nile at Dongola	85.6	12.9	15.1

Source: derived from data in Shahin (1985).

with flow variability declining from 40.1 to 19.4 per cent between Mongalla and Malakal. This is due mainly to increased evapotranspiration losses in the Sudd in flood years. The flow of the Atbara is more variable than that of the Blue Nile, reflecting the lower and more variable rainfall of the Atbara's more northerly catchment.

Two types of groundwater system are important in the Sudan: deep, regional scale aquifer systems that characterize the sedimentary Nubian sandstone and Umm Ruwaba rock areas (Fig. 2.14); and localized, shallow, often perched aquifers that occur throughout the country. The latter are found particularly where *qoz* sand overlies clay or igneous rock in central Sudan, in sand and silt deposits of ephemeral watercourses, and in colluvial and alluvial deposits around inselbergs. Whereas the shallow, perched aquifers are small, have high annual recharge/storage ratios and are very reliant on the amount of recharge each wet season, the deep, regional systems have immense storages, low annual recharge/storage ratios and are hence independent of short-term rainfall and recharge fluctuations. In the northern Desert and Semi-Arid areas, radiocarbon and tritium isotope techniques have dated some regional aquifer water as

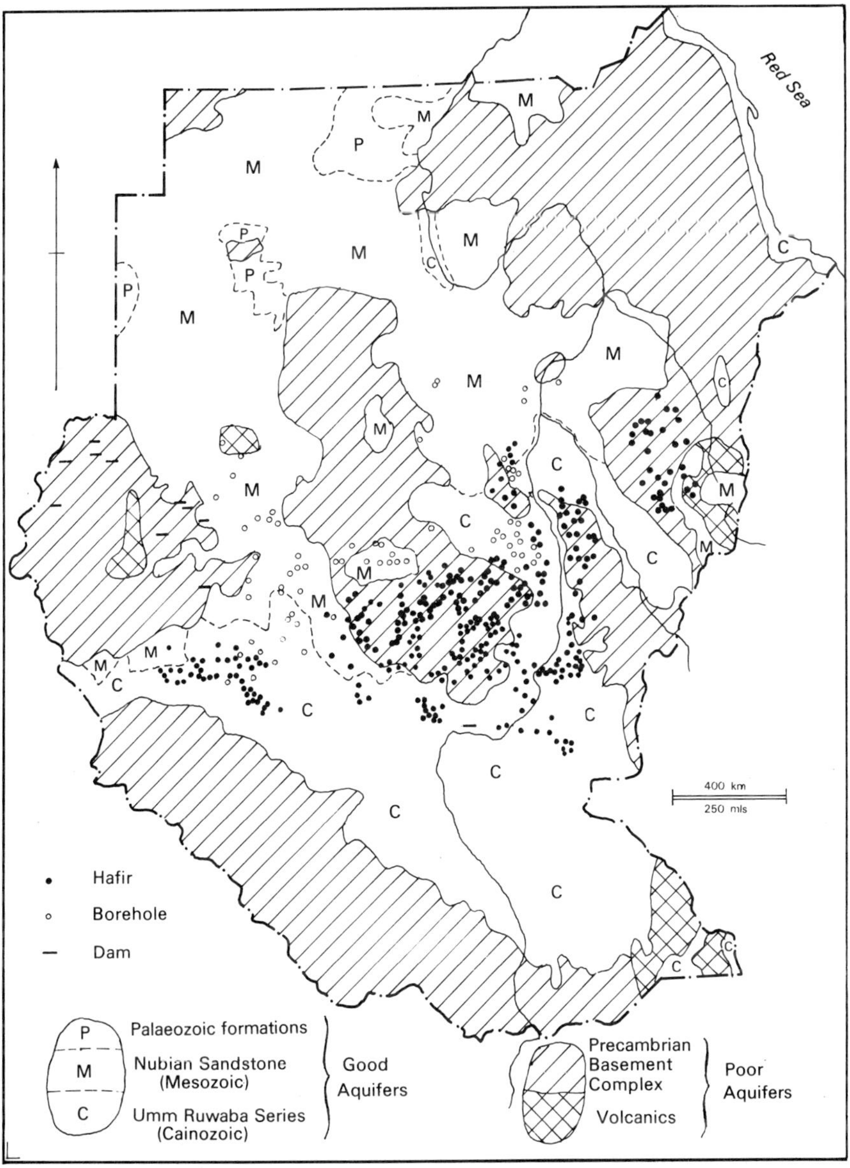

FIG. 2.14. Distribution of aquifers, mechanized boreholes, and *hafirs* in the Sudan. Source: Barbour (1961); Rodis *et al.* (1968); Trilsbach (1983); Al Awad *et al.* (1985); Hulme (1985); Mohammed (1986).

Pleistocene, in age around 36 000 BP, and present recharge in these areas is considered negligible (Salam unpublished paper, 1982; Rodis *et al.* 1968). Aquifers close to the Nile in these areas are recharged from seepage from the Nile itself. Further south in the Tropical Sub-Humid and Wet-Dry climatic zones, rainfall is higher and recharge, aided by the widespread flooding of the Bahr el Ghazal, Bahr el Jebel, and Sobat systems, becomes much more important. In Kordofan, annual recharge has been estimated by Rodis *et al.* (1968) to be zero in the Semi-Arid and Desert north, 5 per cent to the north of the Nuba Mountains, but 20 per cent south of the Nuba Mountains.

Recharge of the shallow, local aquifers occurs partly through infiltration of rainfall on slopes of moderate-to-high permeability (as on the *qoz* and bouldery inselberg slopes) but, more important is the concentrated recharge that occurs through infiltration of ephemeral river flow in localized sandy deposits of *wadi* beds and of overland flow in colluvial debris at the base of inselberg slopes. Significant recharge by these latter methods occurs only as a result of rainfalls large enough to produce large volumes of overland flow and high river flows; field evidence suggests storm thresholds of 20 mm for minor and 40 mm for major surface flow events and thus it is the frequency of large rainfalls rather than annual rainfall that is important.

Water resource development of the Nile system in the Sudan

The perennial, dependable flow of the Nile has for millennia acted as a magnet for settlement and civilization in the Sudan, providing a reliable supply of water for domestic use, livestock, and small-scale 'garden' irrigation, but major schemes to harness the Nile waters for agriculture were not initiated until the Anglo-Egyptian Condominium. After the old Aswan Dam had been constructed in 1898–1902 to extend perennial irrigation in Egypt, the Sennar Dam was built in 1925 on the Blue Nile (Fig. 2.13) to store water for the Sudan primarily for cotton irrigation in the Gezira Scheme south of Khartoum on the Nile Plain between the two Niles. The irrigation water flows by gravity along the Gezira Canal and then to the fields via a hierarchy of feeder canals with associated sluice gates. Supply of irrigation water to the Gezira commences at the end of July and continues to the end of March; the canal then supplies water at a reduced rate for domestic purposes alone. The Sennar Dam is also used to generate hydroelectric power.

The construction of the Sennar Dam and the use of Nile water for irrigation in the Gezira alarmed Egypt and led to the First Nile Waters Agreement of 1929 between Egypt and the Sudan. This Agreement which stated that '. . . no works should be constructed or measures taken on the Nile or its branches or on the lakes from which it flows, in the Sudan or in the territories under British administration, which would affect the flow of the river in such a way as to cause prejudice to the interests of Egypt . . .', clearly favoured Egypt and effectively prevented further major irrigation schemes in the Sudan from being initiated. The Sudan was restricted to the annual use of 4 milliards compared with 48 milliards for Egypt. The

Second Nile Waters Agreement of 1959 redressed the balance somewhat; under its terms, the average annual flow of the Nile is shared, so that the Sudan may take 18.5 milliards and Egypt 55.5 milliards, as measured at Aswan. For the Sudan, it is more convenient to refer to the volume of water at Sennar, where, allowing for transit evaporation and seepage losses, the equivalent volume it may use is estimated at 20.55 milliards. The Second Agreement also states that, should the average Nile flow increase, additional water would be divided equally between the two countries.

This later agreement has permitted greatly increased irrigation in the Sudan, particularly along the Blue Nile, its tributary the Rahad, and on the Atbara. The irrigation has been facilitated by the construction of two dams, the Khashm el Girba Dam in 1965 on the Atbara and the Er Roseires Dam in 1966 on the Blue Nile; both also are used to generate hydroelectric power. A major problem with the Khashm el Girba Dam has been siltation; as a result of accelerated soil erosion in Ethiopia and high suspended sediment transport by the highly seasonal Atbara, the reservoir's original 1.3 milliard capacity was reduced to 0.95 milliards by 1971 (Shahin 1985, p. 431) and approximately 0.65 milliards by 1979 (Ministry of Irrigation 1979). Irrigation from the Khashm el Girba reservoir to the New Halfa Irrigation Scheme is by gravity flow canals for nine months of the year but the decline in water level in the reservoir necessitates pumping in the three months prior to the rains. The Er Roseires Dam has permitted major expansions to the Gezira Scheme by providing 3 milliards storage additional to the 0.8 milliards of the Sennar Dam.

The 1959 Agreement has also allowed considerable irrigation along the White Nile using water stored behind the Jebel Aulia Dam, which had originally been constructed in 1937 to provide additional storage to the old Aswan Dam for Egypt, but which has been rendered superfluous by the construction of the High Aswan Dam in 1970. The reservoir behind the Jebel Aulia Dam had an original capacity of 3.5 milliards, since reduced by siltation to 2.2 milliards, and at times of maximum storage extends 314 km upstream. This has facilitated a large number of public and private pump irrigation schemes to be developed covering about 500 000 feddans in 1979 (Ministry of Irrigation 1979). There are plans to decrease the amount stored and to lower White Nile levels in order to reduce the considerable evaporation losses from the reservoir and use the conserved water for additional irrigation schemes, but new pumps would have to be installed by existing schemes to reach the lower Nile levels.

By 1979, the area irrigated by Nile waters in the Sudan totalled 3 948 000 feddans, of which 68 per cent was fed by the Blue Nile, 13 per cent by the White Nile (between Malakal and Khartoum), 10 per cent by the Atbara, 8 per cent by the main Nile below Khartoum, and only 1 per cent in southern Sudan at the Mongalla Sugar Scheme on the Bahr el Jebel (Ministry of Irrigation 1979).

The schemes to date have been supplied within the 20.55 milliard Sudan allocation under the terms of the 1959 Agreement, but it has been predicted (Ministry of Irrigation 1979) that total abstractions will have reached this limit by 1990. Further irrigation and domestic/industrial water supply needs will depend on extra water being made available through water conservation projects. Various schemes, brought together in the form of a Nile Waters Master Plan (Ministry of Irrigation 1979) have been proposed for the period 1980–2000. Central to the strategy is the construction of the Jonglei Canal, so as to divert part of the flow of the Bahr el Jebel to the east of the Sudd, thereby reducing flooding and evaporation loss over the Sudd and increasing the flow of the White Nile at Malakal by 4.4 milliards; this would mean increased water availability for the Sudan of 2.13 milliards. Construction of the canal commenced in 1980 but mechanical problems and civil war in the south have prevented its completion. Further conservation schemes that would yield 6.87 milliards in extra water resources for the Sudan have also been proposed: a second Jonglei Canal and the construction of channels to increase the flow from the Bahr el Ghazal through the Sudd and from the Machar Marshes north of the Sobat. These water conservation schemes would then allow increased storage (at a heightened Er Roseires Dam on the Blue Nile and at a new dam at Rumela close to the Ethiopian border on the Atbara) and a further expansion of irrigation both along those rivers and along the White Nile and Main Nile. A number of HEP run-of-the-river projects are also envisaged, notably at Merowe on the Main Nile and on the Bahr el Jebel.

Water supply in rain-fed areas away from the Nile

Away from the Nile and its major tributaries, stream water acts as a perennial water source only in the tropical rainy zone of the extreme south; elsewhere watercourses are seasonal or ephemeral and are dry throughout the long dry season. In these areas, water resource developments have been geared to improving the availability and reliability of water supply for people and livestock rather than provision of water for irrigation.

Sources of rural water supply can be classified as traditional or modern and surface or groundwater. For further details see Hulme (1985), Barbour (1961), Trilsbach (1983), and Al-Awad *et al.* (1985).

Surface sources Use of stream water is confined to the wet season. In the dry season, the Baobab or Tebeldi tree (*Adansonia digitata*) is an important small-scale source; the hollow interior of a typical tree can store 1000–2000 litres of water and large ones can hold up to 9000 litres (Idris 1960). More important, however, are surface pools located in minor topographic depressions and along watercourses. Important, particularly for livestock watering, they can be improved by diversion of wet-season runoff using minor earthen banks and by enlargement through excavation, but they suffer from high evaporation, dependence on runoff of the previous wet season and hence early exhaustion during the dry season.

The main modern development of surface sources has been the construction of *hafirs* (excavated reservoirs). A major *hafir* building programme commenced under the British in 1947. The programme concentrated on the impermeable igneous Precambrian Basement Complex rock areas where there are few groundwater resources to develop, notably in the Nuba Mountains area of Southern Kordofan and the Butana Land of Eastern Sudan (Fig. 2.14). Mechanically dug, the *hafirs* were generally of 10–15 000 m^3 capacity and were located on clay soils so as to increase runoff inputs and decrease seepage losses. A total of 361 were built between 1947 and 1955 and there were 592 by the end of 1966 (Oliver 1968). Some were fenced and their water use officially controlled. The main problems with *hafirs* have been the following:

(1) high evaporation losses, with usually only about 50 per cent of the water stored being usable (Lebon 1965);
(2) reductions in capacity due to silting and bank erosion by livestock;
(3) that they rarely last through the dry season, particularly following a poor wet season;
(4) their role in spreading malaria;
(5) inadequate management.

Groundwater sources The traditional method, still prevalent throughout rural Sudan, involves the dug well. It has been estimated that 80 per cent of the water used for domestic and livestock purposes in rural Sudan currently comes from this source (Salam 1982). The dug wells can be permanent or ephemeral (i.e. freshly located and dug each dry season) and vary from a few metres to over a hundred metres deep, though most are shallow. The deeper ones tend to be more reliable, as they tend to tap the larger regional aquifers of the Nubian sandstone and the Umm Ruwaba Series. The main problems of the shallow wells are their unreliability, resulting from their dependence on the rainfall and recharge of the previous wet season, and their propensity to pollution, but they have the advantages of cheapness and relative ubiquity. Recharge of *wadi* bed wellfields can often be increased by constructing earthen dams across the *wadi* at strategic points (Hulme 1985, p. 223).

Developments in the modern era have sought to remedy the deficiencies of the dug well. To help protect against pollution, lining of wells with stone or brick has been encouraged. In the Kadugli area of Southern Kordofan, a UNESCO programme of installing permanent hand-pump units at shallow well sites has been initiated, with the well covered by concrete and with culverts to direct water away from the pump for livestock use (Walsh

1983). However, the main development has been the mechanized borehole, the history of development of which has been reviewed by Hulme (1985, p. 26). Boreholes were first developed for the water supply of towns, but the 1950s saw an expansion of the rural drilling programme, which was concentrated in the Nubian sandstone and Umm Ruwaba rock areas of the Semi-Arid and Sub-Humid zones of Darfur, Kordofan, and White Nile (Fig. 2.14). By the mid-1960s, the number of boreholes in Kordofan had reached 175 (Rodis *et al.* 1968) and a renewed expansion occurred from the early 1970s; the total number in the Sudan now runs into thousands. Associated with the organized borehole sites are wateryards, which are fenced areas within which are located from two to five wells, a large elevated steel storage tank and separate watering facilities for domestic use and livestock. Theoretically, such borehole sites should result in larger and more reliable water supplies since they are tapping large aquifers with low recharge/storage ratios but, in practice, there have been major problems with many boreholes, notably fuel shortage, mechanical breakdown, lack of spare parts and maintenance expertise, breakdown of the allocation system during water shortage periods, and excess demand caused by large-scale migration to these new watering points (Trilsbach 1983; Mohammed 1986). Despite these developments towards the provision of more reliable water supplies, most people away from the Nile still depend on shallow wells and surface pools, which are the most susceptible to drought.

Recent hydrological changes of significance to agriculture

The flows of different parts of the Nile system have changed significantly in recent years (Table 2.9). The flow of the Bahr el Jebel increased suddenly in 1961 in response to a large rise in rainfall levels over parts of the Equatorial Lakes region; the mean flow of 1961–73 was 64 per cent above the long-term mean. This increase was reduced somewhat by increased evaporation losses of the floodwaters in the Sudd, and the White Nile at Malakal showed an increase of just 25 per cent. In contrast, the flows of the Sahelian belt Blue Nile and Atbara have declined from 1965, with mean flows for 1965–73 some 22–29 per cent below their long-term means and the decline has intensified in the 1980s. The net effect of these contrasting changes has been a decline of 11 per cent in the flow of the main Nile at Dongola and, if sustained, this must have implications for the amount of irrigation water available to the Sudan under the terms of the 1959 Agreement.

The recent decline in rainfall in Semi-Arid and Sub-Humid Sudan has had marked hydrological consequences. In north-west White Nile (Walsh *et al.* 1988) these have included: reduced frequencies of *wadi* flows and overland flow; a major decline in shallow aquifer recharge in *wadi* bed and inselberg front locations, but little impact on deep regional aquifer levels;

TABLE 2.9. Recent changes in annual discharge of the Nile river system in the Sudan

River and location	Annual discharge (milliards)			% Change
	Long-term mean (1912–73)	1961–73	1965–73	
Bahr el Jebel at Mongalla	31.4	51.5	—	+64
White Nile at Malakal	29.4	36.8	—	+25
Blue Nile at Er Roseires	49.2	—	38.5	−22
Atbara at Nile confluence	11.9	—	8.4	−29
Main Nile at Dongola	85.6	82.5	76.3	−11

and earlier exhaustion in the dry season of shallow wells, surface pools and *hafirs*. Major reductions in the cultivated area have been noted in both north-west White Nile and Northern Darfur and Kordofan and this is related, at least in part, to the lower rainfall, associated crop failures, and inadequate water supplies of the recent very dry epoch.

It is clear that neither the climatic nor the hydrological environment in the Sudan is static; agricultural planning in the Sudan needs to take this into account in the future.

Acknowledgements

Thanks are due to many people both in the Sudan and in Great Britain for their help in enabling me to write this chapter. Special thanks are due to Dr Michael Hulme of the Department of Geography, Salford University and Mr H. R. J. Davies of the Department of Geography, University College of Swansea for their invaluable advice and encouragement; and to Mr Guy Lewis and Mr Alan Cutliffe for drawing and photo-reducing the figures, respectively.

References

Al-Awad, A. A., Mohammed, Y. A., and El-Tayeb, S. A. (1985). The impact of improved rural water supplies on the environment: the case of East Kordofan District. In *Natural resources and rural development in arid lands: case studies from Sudan* (ed. H. R. J. Davies), pp. 16–25. United Nations University, Tokyo.

Awadulla, S. A. (1977). Water balance in the Sudan. Unpublished M.Sc. thesis. University of Sheffield.

Awadulla, S. A. (1981). On a climatic classification for the Sudan and short-term soil moisture conditions in the Central Region. Unpublished Ph.D. thesis. University of Sheffield.

Awadulla, S. A. (1983). *Potential evapotranspiration over the Sudan: an application of Penman's model over a dense spatial pattern*, Sudan Meteorological Service Scientific Note No. 3. Sudan Meteorological Service, Khartoum.

Barbour, K. M. (1961). *The Republic of the Sudan*. University of London Press, London.

Bhalotra, Y. P. R. (1963). *Meteorology of Sudan*, Sudan Meteorological Service Memoir No. 6. Sudan Meteorological Service, Khartoum.

Blaney, H. F. and Criddle, W. D. (1966). *Determining consumptive use for planning water developments*. American Society of Civil Engineers, Las Vegas.

Hargreaves, G. H. (1968). Consumptive use computation from evaporation. *Journal of Irrigation and Drainage Division, American Society of Civil Engineers*, **94**, Paper No. 5863.

Hulme, M. (1985). Secular climatic and hydrological change in central Sudan. Unpublished Ph.D. thesis. University of Wales, Swansea.

Hulme, M. (1987). Secular changes in wet season structure in central Sudan. *Journal of Arid Environments*, **13**, 31–46.

Hulme, M. and Walsh, R. P. D. (in press). Hydrological consequences of recent climatic change in west-central Sudan and some suggestions for future monitoring and assessment. In *Proceedings of the workshop on monitoring and controlling desertification in Sudan: 20–24 February 1983, Khartoum* (ed. D. Johnson), Clark University Press, Worcester, Massachussetts.

Idris, H. A. (1960). White Nile pump schemes. Unpublished B.A. thesis. University of Khartoum.

Lebon, J. H. G. (1965). *Land use in Sudan*, World Land Use Survey Regional Monograph No. 4. Geographical Publications Ltd., London.

Ministry of Irrigation (1979). *Nile waters master plan summary*. Ministry of Irrigation, Khartoum.

Mohammed, Y. A. (1986). Local administration, water supply and community participation in White Nile Province. In *Rural development in White Nile Province: a study of interaction between man and natural resources* (ed. H. R. J. Davies), pp. 108–13. United Nations University, Tokyo.

Nicholson, S. E. (1983). Sub-Saharan rainfall in the years 1976–80: evidence of continued drought. *Monthly Weather Review*, **111**, 1646–54.

Oliver, J. (1968). Problems of the arid lands: the example of Sudan. In *Land use and resources*, Institute of British Geographers Special Publication No. 1, pp. 219–39.

Olivier, H. (1961). *Irrigation and climate*. Edward Arnold, London.

Penman, H. L. (1950). Evaporation over the British Isles. *Quarterly Journal of the Royal Meteorological Society*, **76**, 372–83.

Rodis, J., Hassan, A., and Wahadan, L. (1968). *Ground-water geology of Kordofan Province, Sudan*, United States Geological Survey Water-Supply Paper No. 1757-J. United States Geological Survey, Washington D.C.

Salam, Y. A. (1982). Groundwater resources of Sudan. Unpublished paper presented at the seminar, Water for human needs, 6–26 November 1982. Institute of Environmental Studies, University of Khartoum.

Shahin, M. (1985). *Hydrology of the Nile Basin*. Elsevier, Amsterdam.

Trilsbach, A. (1983). Hydrology and water supply in the White Nile Province of the Sudan: modifications and problems. *Swansea Geographer*, **20,** 58–65.

Trilsbach, A. and Hulme, M. (1984). Recent rainfall changes in central Sudan and their physical and human implications. *Transactions of the Institute of British Geographers, New Series*, **9**, 280–98.

Walsh, R. P. D. (1983). Observations on hydrological problems of the Nuba Mountains area of the Sudan. *Swansea Geographer*, **20**, 16–25.

Walsh, R. P. D. and Lawler, D. M. (1981). Rainfall seasonality: description, spatial patterns, and change through time. *Weather*, **36**, 201–8.

Walsh, R. P. D., Hulme, M., and Campbell, M. D. (1988). Recent rainfall changes and their impact on hydrology and water supply in the semi-arid zone of the Sudan. *Geographical Journal*, **154**, 181–98.

Zahran, A. M. B. (1979). Sudan rainfall variability and tropospheric circulation. Unpublished M.A. thesis. University of Khartoum.

3

Natural vegetation

G. E. WICKENS

Historical review

The first major attempt to produce a description and map of the vegetation of the Sudan was in 1948 by the Sudan Government's Senior Economic Botanist, Dr F. W. Andrews, in a publication that had similar coverage and objectives to the present volume (Andrews 1948). The same classification with slight modifications was followed by Dr John Smith, the then Director of Agriculture and Forests, and a former Chief Conservator of Forests (Smith 1949). A decade later, M. N. Harrison, Pasture Research Officer in the Ministry of Animal Production and J. K. Jackson, Silviculturist in the Forests Department, each working independently, produced somewhat similar classifications of the vegetation, the former placing greater emphasis on grazing, the latter on tree resources. The two works were combined (Harrison and Jackson 1958) and this, together with the accompanying vegetation map at a scale of 1:4 000 000 has become the model for all future studies. The vegetation boundaries, based largely on field traverses, agree remarkably well with later surveys based on photo-interpretation. The present author, working in Kordofan Province, found that the informal classification of Harrison and Jackson (1958) required reorganizing into an hierarchial system in order to meet the requirements of more detailed mapping at a scale of 1:250 000 (Hunting Technical Services 1964). This modification was developed further for use in a number of FAO and UK (Overseas Development Administration (ODA)) projects in the Sudan, including the United Nations Development Program (UNDP) Jebel Marra Project (FAO 1968) and the Southern Darfur Project (Hunting Technical Services 1974). These studies resulted in a revision of the vegetation map of the Sudan (Wickens 1977), which was revised again in Parry and Wickens (1981) (Fig. 3.1).

A detailed description of the major vegetation communities of the Sudan that can be recognized on a pan-African basis, together with vegetation maps at a scale of 1:5 000 000 is included in the vegetation survey of Africa by White (1983). Unfortunately, the maps which were printed in 1981 were

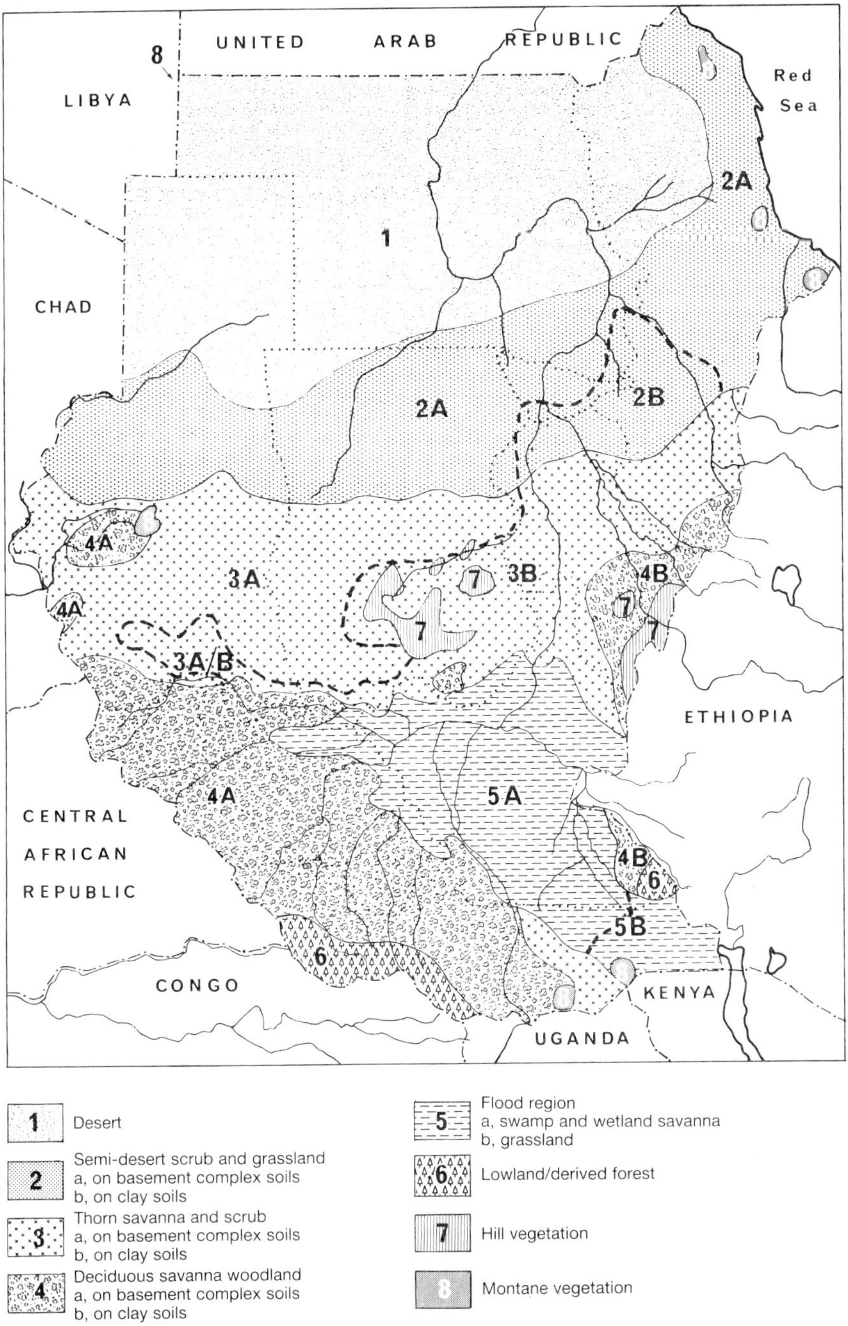

FIG. 3.1. Sudan: vegetation (reproduced from Parry and Wickens (1981) by kind permission of the Royal Geographical Society).

prepared before the latest published information (Parry and Wickens 1981) on the north-eastern salient of Sudanian woodland in south-western Darfur, formed by the Qozes Dango and Salsilgo, could be included.

Vegetation changes

The evidence for changes in the distribution of the vegetation in the Sudan as a result of long-term fluctuations of the climate during the Pleistocene and Holocene up to the present are discussed by Jackson (1957) and Wickens (1975*a*). Such studies are of value in determining the status and origins of the flora, such as the probable native status of the oil palm (*Elaeis guineensis*) (Wickens 1975*b*), or the origins of the boreal, Afro-montane and Saharo-montane elements in the flora of Jebel Marra, an isolated massif in western Darfur (Wickens 1976).

Evidence for short-term climatic fluctuations have been noted from both historical and meteorological records. The earliest records are those of the Pharaohs from the Nilometer near Aswan, replaced in AD 641 by the Nilometer at Roda Island, Cairo. This has provided a more-or-less continuous line of records since AD 840. More orthodox records are discussed by Grove (1978, 1985). Periods of drought occur at relatively regular intervals and the period of below normal rainfall (Sahelian drought), which began in 1968, is still continuing. Its effect upon the vegetation has been devastating. Drought is not, however, the prime cause of this devastation, it has merely accelerated a process of deterioration that started in the 1930s with increases in human and livestock populations following the introduction of prophylactic human and veterinary medicines. In the early 1940s the Soil Conservation Committee became aware of the problem and carried out some protective measures around some of the major townships (Soil Conservation Committee 1951). Constraints imposed by the Second World War and Independence on 1 January 1956 presented little opportunity for the necessary introduction of appropriate conservation measures for the peasant farming communities. The 1950s were a period of above average rainfall, which tended to obscure the insiduous increase of desertification. Settlements around boreholes in the drier parts of the Sudan increased, accompanied by rapidly increasing areas of deforestation and overgrazing. As already stated, the present drought has merely accelerated the process.

In 1971, the author revisited parts of Kordofan Province surveyed in 1962–3 (Hunting Technical Services 1964). Some areas were unrecognizable due to the disappearance of both vegetation and topsoil; in other areas the ground vegetation was much reduced. The shallow sandy soils overlying an indurated former land surface to the south of the deep sands of the *qoz* proper had been denuded of their sand and the trees were either dead or dying of drought. The sand had been sufficient to retain the rainfall and allow it time to percolate into the subsoil; its removal had exposed the surface to sheet wash, and had also allowed what little rainfall was available to evaporate. Such areas are unlikely to recover their former vegetation without remedial measures using expensive heavy equipment to break up the hard surface and allow the rainfall to percolate. Furthermore, the deep sands of the *qoz* proper had become very sparsely vegetated. Formerly, these areas carried a more or less contiguous cover of bunch grasses during the summer and these burned readily during the dry season. The damage caused by grass fires was extensive and bush fires were an unwelcome sight; now they would be greeted as a sign that the vegetation had returned to its pre-drought cover. It must be emphasized, however, that no recovery will be possible unless both the human and livestock populations are maintained below the actual carrying capacity of the land. Detailed studies in northern Senegal have suggested that some thirty years of total protection are required for the vegetation to return to pre-drought condition (Le Houérou and Grenot, in press). Current anthropozoic pressure makes any such recovery unlikely.

In 1977 the author returned to Darfur to report on vegetation changes that had taken place since the survey of the Jebel Marra area twelve years earlier (Hunting Technical Services 1968). The lowland area, especially in the Wadi Azum Basin, had become badly degraded, due largely to excessive numbers of livestock using an area which was formerly well-watered and had excellent winter grazing and browse. Severe gullying was observed on the western ash piedmont where a government mechanized farming scheme was established. In an earlier soil survey report (White 1966) the area was considered unsuitable for mechanized agriculture. There was evidence of some slight deterioration on the Jebel Marra but, as there had been no new movement of people or livestock into the massif and since the upland Fur practise both terrace and basin cultivation techniques, soil erosion is minimized and maximum use had been made of the existing rainfall (Hunting Technical Services 1977, Appendix E).

Major vegetation communities

The major vegetation communities recognized are shown in Fig 3.1. The following descriptions of these communities are based largely on the author's own detailed reconnaissance surveys in Kordofan and Darfur (Hunting Technical Services 1964, 1968, 1974, 1977), as well as journeys eastwards to the Red Sea. Because of the political unrest in southern Sudan, there has been no major ecological survey there in recent years apart from in the Jonglei Canal area (MFEP 1985), an ODA forestry project in the Imatong Mountains, and an independent study by the University of Copenhagen (Friis and Rasmussen 1981). Descriptions of the communities in southern Sudan are based, therefore, on those of Harrison and Jackson (1958). Undoubtedly, there has been considerable deterioration in the state of the vegetation since then due to deforestation and overcultivation, as well as a decrease in the annual precipitation.

Desert

Desert extends over vast areas of northern Sudan, occupying approximately 725 200 km^2 or 28.94 per cent of the country. The average rainfall is less than 75 mm and vegetation is virtually absent, except along watercourses, and consists mainly of ephemeral herbs and grasses known as *gizu*, that occur sporadically during the dry winter months following rare rain showers. The *gizu* occurs along the southern fringe of the Libyan Desert in approximately two years out of five, but was absent during the drought years of 1965–75. In 1975, however, a small area of *gizu* was recorded (Wilson 1978). The value of the *gizu* lies in its ability to support domestic livestock without their access to potable water during the dry season months of November to February.

According to Harrison (1955) the camels and sheep obtain all their moisture requirements from the vegetation during the three or four months that they are grazing the *gizu*. Herdsmen, sometimes with their families, depend mainly on their livestock for meat and milk, with occasional trips for water. The *gizu* plays an important role in the economy of several pastoral tribes but has received scant attention by the social anthropologists and ecologists. The present state of knowledge is summarized by Wilson (1978).

In years past, vegetation along the watercourses of the Atbai and Bayuda Deserts was sufficient to support a few hardy nomads and their livestock (Harrison and Jackson 1958). Sadly, the carrying capacity is now much reduced.

Semi-desert scrub and grassland

This occupies an area of approximately 492 100 km^2 or 19.64 per cent of the Sudan, traversing the country in a wide belt from the boundary with Chad in Northern Darfur to Kassala Province and the Red Sea coast. The average annual rainfall ranges from about 75 to 300 mm.

Harrison and Jackson (1958) recognize five subdivisions of this vegetation type and these are considered here as forming two major plant community groups, those on Basement Complex soils and those on clay soils, and a third minor group, the coastal halophytic vegetation.

Harrison (1955) estimated the carrying capacities of the two major community groups to be 26 livestock units per km^2 on Basement Complex soils compared with 18 livestock units on the clay soils of the Butana. Ten years later the effects of overgrazing, accelerated by the increased number of unregulated stock watering points (*hafirs* and boreholes) had greatly reduced these stocking rates. The current drought has brought a further and possibly permanent reduction in carrying capacity.

The distinction between scrub and grassland is somewhat arbitrary, depending on the density of trees and shrubs, primarily *Acacia tortilis* subsp. *tortilis*, *Leptadenia pyrotechnica, Salvadora persica, Maerua crassifolia*, and *Capparis decidua*. The major grasses include *Aristida funiculata, A. mutabilis, Stipagrostis plumosa*, and *Cenchrus setigerus*, with the perennial, sand-binding clump-grass *Panicum turgidum* often present on loose sand dunes.

The clay soils are generally devoid of trees and shrubs except along the watercourses where *Acacia mellifera, Capparis decidua, Ziziphus spina-christi*, and *Balanites aegyptiaca*, predominate. The Butana grasslands are typical on clay soils with *Cymbopogon nervatus, Sorghum purpureo-sericeum, Sehima ischaemoides*, and *Schoenefeldia gracilis* as the dominant grasses.

The grasslands over a considerable area of the Butana form a characteristic arceate pattern with the bare ground between the arcs forming the catchment area for the runoff rainfall to feed the vegetation arc immediately downslope. Such patterned ground represents a very delicate balance between rainfall and vegetation, which would be extremely unlikely to recover if destroyed. In June 1969, there was evidence already of overgrazing, sheet and wind erosion, leading to the formation of a clay desert (Wickens and Collier 1971).

Blepharis edulis is the outstanding pasture herb for camels and sheep during the dry season (Harrison 1955); cattle dislike the herb on account of its spiky seed heads. Within 20–30 km of watering points grazing by camels and sheep has greatly reduced or eliminated this herb whose seeds are retained within the seed head throughout the dry season and are, consequently, eaten. This is where an understanding of the biology of a species is essential for range management. *B. edulis* is highly nutritious (8.3 per cent true protein per seed head) but its low density would appear to give only one plant per mouthful.

Harrison (1955) argues that camels, which can survive 4–6 days without water and range over 50–70 km, make the most efficient use of the grazing and are the safest animals for the Butana. By comparison, cattle require daily watering, drink twice as much as camels, range over only 20–30 km and are more destructive to the vegetation.

Low-rainfall halophytic vegetation occurs along the 15 to 20 km wide Red Sea coastal plain. The littoral salt marshes are characterized by *Arthrocnemum glaucum, Halocnemum strobilaceum, Zygophyllum album, Nitraria retusa*, and *Suaeda monoica*, the latter being locally dominant over quite extensive areas, especially around the Tokar Delta. The grasses, *Aeloropus lagopoides* and *Sporobolus spicatus*, fringe the upper edge of the salt marshes and groves of the mangrove *Avicennia marina* occur along the seaward margin (Harrison and Jackson 1958).

Thorn savanna and scrub

Thorn savanna and scrub occupy 420 990 km^2 or 16.94 per cent of the Sudan in an E–W belt to the south of the semi-desert scrub and grassland zone in areas receiving between 250–650 mm annual rainfall. The vegetation is dominated by thorn (*Acacia* spp.) communities on either sandy soils derived from the Basement Complex or on clay soils. The vegetation is described in detail in Hunting Technical Services (1964) and summarized in White (1983).

Harrison (1955) estimated carrying capacities for communities on sandy soils to be 78 livestock units per km^2, on clay soils 31 units per km^2, and in the Gezira, where supplementary fodder is available, 195 livestock units per km^2. Hunting Technical Services (1964) recommended an average carrying capacity not exceeding 50 livestock units per km^2. Census data would suggest that overstocking has occurred in Kordofan Province over many years and with the onset of the drought years, deterioration in carrying capacity was inevitable.

Sandy soils are characterized by *Acacia tortilis* and *Leptadenia pyrotechnica* in the northern, drier regions merging more or less imperceptibly with *Acacia senegal* which dominates the major area of dunes and sand sheets. The perennial *Aristida sieberana* is often the dominant grass; *Eragrostis tremula, Aristida funiculata, A. stipoides, Anthephora* spp., *Cenchrus biflorus*, and *Cymbopogon schoenanthus* are associated grasses, the latter often forming pure stands on poorly drained soils.

Until the mid-1960s, annual grass fires swept through the vegetation of these sandy soils during the dry season. Fires are unusual now because the vegetation has been so severely devastated by the combined effects of overgrazing and drought.

Acacia senegal (*hashab*) produces commercial gum arabic and is an important export crop for the Sudan. Between two and three decades ago the *hashab* formed the major tree constituent of the bush fallow period and coppice growth was encouraged as part of the traditional gum cultivation cycle. Changes in the cultivation pattern, resulting largely from urbanization and the concentration of arable areas, led to a reduction or total loss of the tree/bush fallow period, accompanied by a loss of fertility and an increase in soil erosion and *Calotropis procera*, a characteristic species of exhausted lands. As a result, tappers have had to travel further during the dry season in order to tap the trees for their gum. The tapping process is described in detail in Hunting Technical Services (1964). Various aid agencies are currently encouraging the planting of gum trees, but the commercial effectiveness of these new plantations is decreased by an inadequate knowledge of the physiology of gum production and the factors which affect it. This, in turn, has a bearing on the selection and multiplication of high yielding strains.

Dense, impenetrable thickets of *Acacia mellifera* dominate the clay soils, which receive up to 450–500 mm annual rainfall; *Schoenefeldia gracilis* is often the dominant grass. Higher rainfall areas are dominated by open stands of *Acacia seyal*, with var. *fistula* in the depressions. *Balanites aegyptiaca* may be locally dominant on clay soils, especially along the margins. Grasses associated with *Acacia seyal* include *Hyparrhenia* spp., *Andropogon gayanus*, and the unpalatable *Cymbopogon nervatus*. The vegetation of the Gezira Plain, the riverine vegetation of the Blue and White Niles are described by Mubarak *et al.* (1982).

Included in the thorn savanna zone are inland delta systems draining from the Jebel Marra and the Nuba Mountains to the Bahr el Arab. They are known as the Baggara and Raqaba Catenas, respectively (Harrison and Jackson 1958). Although Harrison and Jackson regard them as two distinct catena communities and describe them in detail, the present author regards them as a single unit where any former distinguishing variations in vegetation have become blurred by overgrazing.

The Baggara Catena is characterized by areas of sandy clay loam known as *naga'a* which are generally bare of grasses, alternating with slightly higher areas of lighter soils with species including *Acacia senegal, Albizia amara*, and *Sclerocarya birrea. Acacia mellifera* and *A. seyal* are found at water receiving sites on the *naga'a* (Hunting Technical Services 1974). The Raqaba Catena was reputed to differ due to the presence of seasonally flooded water channels and swamp grasses including *Echinochloa pyramidalis, E. scabra*, and *Vetiveria nigritana*. Such channels also occur in the Baggara Catena.

The catena communities form an important part of the dry season grazing within the traditional annual migration of the Baggara cattle nomads. However, there is some evidence that the shortening of the rainy season and the reduction in precipitation have curtailed this migration pattern (Wilson and Clarke 1975).

Deciduous savanna woodland

This occupies some 511 410 km^2 or 20.25 per cent of the Sudan and occurs on Basement Complex and volcanic soils of the Sudan receiving between 450 and 1300 mm annual rainfall. In the north, savanna woodland merges with thorn savanna communities on sandy soils and *Combretum glutinosum, Albizia amara* subsp. *sericocephala, Sclerocarya birrea*, and *Terminalia brownii* dominate. Outliers of the latter species form arceate patterns near En Nahud and linear patterns in association with *Albizia amara* near Abu Zabad (Wickens and Collier 1971). Over considerable areas where the relatively thin veneer of sand overlying an indurated sandy clay B horizon has been removed by sheet erosion, the trees have died from drought during the past two decades. It is unlikely that such areas will ever recover.

The *Anogeissus leiocarpus* savanna woodlands on Basement Complex soils in Darfur are described in detail by Hunting Technical Services (1968, 1977) and Wickens (1977). The hitherto unknown Qozes Salsilgo and Dango are described in Hunting Technical Services (1977) and Parry and Wickens (1981). Further south *Khaya senegalensis, Isoberlinia doka*, and *Butyrospermum paradoxum* subsp. *niloticum* become increasingly important. The complex catenas and mosaics of these woodlands are described by Morrison *et al.* (1948).

Due to poor communications and internal strife, the present state of this vegetation type is not known fully. Overgrazing resulting from an influx of livestock from the drought-stricken areas to the north has certainly devastated certain areas, especially the fertile Wadi Azum Basin of western Darfur.

On the sloping clay soils bordering Ethiopia, *Anogeissus leiocarpus* and *Combretum hartmannianum* with some *Acacia seyal* alternates with belts of *A. seyal*. The major grasses include *Hyparrhenia* spp., *Andropogon gayanus*, and *Setaria incrassata*.

Flood Region vegetation

Three major ecological communities are recognized in the Flood Region, swamp and wetland savanna in areas with an average annual rainfall of 800–1000 mm and grassland in areas with average rainfall ranging from 600 to over 1000 mm.

The swamp region or Sudd has been studied intensively by the Jonglei Investigation Team (1954) and, more recently, by the National Council for the Development of the Jonglei Canal Area (MFEP 1985) and summarized by Mubarak *et al.* (1982). The Sudd is fed by the White Nile and its tributary the Bahr el Ghazal, has an area of 150 000 km^2 and is the largest swamp in the world (White 1983). The permanent swamp is dominated by the perennial *Cyperus papyrus*, often forming pure stands and associated species include *Typha latifolia, Phragmites australis, Echinochloa pyramidalis, E. scrabra*, and *Vossia cuspidata*. The recently discovered, rare and imperfectly known endemic grass *Suddia sagittifolia* also occurs in the Sudd (Renvoize *et al.* 1984).

Normally, the major reeds and grasses are rooted in the underlying soil but some species, especially *Vossia cuspidata* and *Cyperus papyrus*, extend outwards to form floating mats which frequently become detached to form floating islands. The deeper waters support both free-floating and submerged communities. Of the former the introduced water hyacinth, *Eichhornia crassipes*, is the most troublesome. Its rapid and so far uncontrollable expansion has seriously interfered with navigation and threatens irrigation canals. Some possibilities for utilizing the enormous biomass productivity of this troublesome weed have been discussed (National Academy of Sciences 1976).

Bordering the permanent swamps of the Sudd are the seasonally inundated plains known as the *toich*. The low lying areas where the water-table is too high for tree growth are dominated by *Hyparrhenia rufa* or *Setaria incrassata* grasslands, with small areas of *Acacia seyal–Balanites aegyptiaca* woodland savanna. The grasslands, which are burned seasonally, are the winter grazing for the cattle of the Dinka, Shilluk, and Nuer tribes, the cattle and their herdsmen move to higher ground, described below, during the rainy season.

On the relatively high ground that is not seasonally flooded are island groves of dom palm (*Hyphaene thebaica*), with the doleib palm (*Borassus aethiopicum*) locally dominant, mixed thorn savanna with *Acacia sieberana* and *A. polyacantha* subsp. *camplyacantha*, or mixed broad-leaved savanna with *Combretum* spp. Such island areas in the Flood Region provide dry land for man and stock during the rainy season and also land for crop cultivation.

The edaphic grasslands of south-western Sudan known as the Toposa area have been described in detail by Peers (1955). *Hyparrhenia* spp., *Setaria* spp., *Chrysopogon plumulosus*, and *Bothriochloa insculpta*, are among the more important grasses present and thickets of *Acacia mellifera* are also present. It is mainly a pastoral area with some subsistence cultivation. The presence of limited sources of potable water has meant that the Toposa area, unlike the remainder of the Sudan, has been understocked. The present situation is not known.

Lowland forest

Small relic areas of lowland forest—Lotti, Laboni, Talanga, and Azza, occur in an area with an average rainfall of over 1300 mm along the Uganda–Zaire border and on the Aloma Plateau. Characteristic species include *Celtis zenkeri* and *Chryosophyllum albidum* and these are regarded as representatives of former rain forest. Gallery forests, representing reduced forms of rain forest, extend along the river valleys into the surrounding deciduous savanna woodland. The characteristic species include *Khaya grandifolia, Erythrophleum guineense, Cola cordifolia*, and *Syzgium guineense* (Harrison and Jackson 1958). The apparently wild groves of oil palm (*Elaeis guineensis*) in the Yei area probably represent relic forest outliers (Wickens 1975*b*). The successful establishment of oil palm plantations in Equatoria is a reflection of its natural distribution.

Lowland rain forest occupies some 36 260 km^2 and there is an urgent need to conserve these forests from exploitation before they are permanently destroyed. The Talanga forest, for example, is already threatened due to the exploitation of its valuable timber.

Hill vegetation

Isolated areas of savanna and woodland vegetation survive on hill slopes, often in areas where similar communities are no longer found in the surrounding lowlands; this hill vegetation is best represented by the Nuba Mountains.

A general description of hill vegetation is given in Harrison and Jackson (1958). Their Darfur Hill Catena to the north and east of Jebel Marra consists mainly of *Acacia mellifera* thornland, with *Anogeissus leiocarpus* savanna woodlands to the south and west, both on Basement Complex soils, the intervening areas of inselbergs, rock outcrops and volcanic plugs, being too small for inclusion in Fig. 3.1, see Quézel (1969) for a detailed description of the area to the north of Jebel Marra.

The relatively rich vegetation of the Red Sea Hills, described by Kassas (1956), has been destroyed through overgrazing and its recovery is regarded as extremely unlikely. In all areas of hill vegetation, the surrounding piedmont plain provides fertile soils for cultivation.

Montane vegetation

Upland areas, often with tropical and temperate species known only from similar mountain areas in Africa, occupy about 6475 km^2 or 0.26 per cent of the Sudan. The depauperate Saharo-montane vegetation of Jebel 'Unweinat has been studied by Léonard (1969) and Osborn and Krombein (1969). It is an area of insignificant development potential.

The Jebel Marra itself has been described by Hunting Technical Services (1968), FAO (1968), Wickens (1977), and Miehe (1988), and summarized by White (1983). The role of *Acacia albida* for agro-forestry in the area is described by Miehe (1986). Three communities are recognized on the massif; hill savanna, gallery forest, and montane grassland. Representative samples of these communities have been recommended as conservation areas (FAO 1968) and more recently 2283 hectares of fallow land have been declared a forest reserve (Miehe 1986). The hill savanna and the montane grassland have potential for forestry development provided the problems of accessibility and transport can be solved. The fertile southern and western ash piedmont plains have good potential for non-mechanized agriculture and forestry, but the highly erodable nature of the ash soils require that any development be safeguarded by appropriate soil conservation measures.

The vegetation of the Imatong Mountains has recently been studied by Friis and Rasmussen (1981), who recognize lowland, transition, low montane, higher montane, and Ericaceous zones. The flora is extremely rich, containing representatives of one-third of the Sudan's total flora, including some 200 species not found elsewhere in the Sudan. The Imatongs are believed to have been settled only some 250 years ago and initially to have been relatively free of forest fires and cultivation. During the Second World War there was some felling of *Podocarpus* and some replanting with *Pinus* and *Cupressus* species. During the 1960s there was no further development due to political disturbances but interest in forestry development has revived recently with proposals for large-scale plantation development, including the exploitation of the Talanga forest. Proposals for conservation reserves have been proposed by Friis and Rasmussen (1981) but have yet to be implemented.

The neighbouring Dongotona and Didinga Mountains have been described briefly by Jackson (1950, 1951) and have a vegetation somewhat similar to the Imatongs.

The wetter parts of the Red Sea Hills on the Eritrean border contain pure stands of *Juniperus procera*, a characteristic Afromontane species (Harrison and Jackson 1958); such hills are relic outliers of the Ethiopian montane vegetation.

Conclusions

In the space allocated, it is possible to give only a very superficial outline of Sudan's vegetation. The general impression is of a rapid deterioration during the past twenty years, a situation which is familiar elsewhere in Africa. In some of the sand dune areas which have been destroyed through desertification, attempts are being made to reafforest with *Acacia senegal*. Unfortunately, insufficient attention is being given to water harvesting techniques as a means of re-establishing the vegetation and also as a means of increasing crop yields. There was once a hope that Sudan would become the 'bread basket' of Africa; regrettably, it is more likely to become the 'dust bowl'.

References

Andrews, F. W. (1948). The vegetation of the Sudan. In *Agriculture in the Sudan* (ed. J. D. Tothill), pp. 32–61. Oxford University Press, London.

FAO (1968). *Land and water resources survey in the Jebel Marra area, the Sudan: final report*. FAO/SF:48/SUD-17, FAO, Rome.

Friis, I. and Rasmussen, J. B. (1981). *Flora and fauna of the Imatong Mountains, Southern Sudan*. University of Copenhagen (mimeo).

Grove, A. T. (1978). Geographical introduction to the Sahel. *Geographical Journal*, **144**, 407–15.

Grove, A. T. (1985). The arid environment. In *Plants for arid lands* (ed. G. E. Wickens, J. R. Goddin, and D. V. Field), pp. 9–18. Allen & Unwin, London.

Harrison, M. N. (1955). *A report on a survey of the grazing areas of the Sudan*. Sudan Government, Department of Animal Production (mimeo).

Harrison, M. N. and Jackson, J. K. (1958). *Ecological classification of the vegetation of the Sudan*. Forests Bulletin No. 2 (New Series), Forests Department, Khartoum.

Hunting Technical Services (1964). *Land and water survey in Kordofan Province of the Republic of the Sudan. Report on the survey of geology, geomorphology and soils, vegetation, and present land use*. DOX-SUD-A-26, Doxiades Association, Athens.

Hunting Technical Services (1968). *Land and water resources survey of the Jebel Marra area: reconnaissance vegetation survey*. LA/SF/SUD/17, FAO, Rome (mimeo).

Hunting Technical Services (1974). *Land use planning survey in Southern Darfur. Annex 1: soil and vegetation resources*. Hunting Technical Services, Boreham Wood.

Hunting Technical Services (1977). *Agricultural development in the Jebel Marra area. Annex 1: land resources*. Hunting Technical Services, Boreham Wood.

Jackson, J. K. (1950) *The Dongotona Hills, Sudan*. Memoirs of Forestry Division, No. 1. Sudan Government.

Jackson, J. K. (1951). *Mount Lotuke, Didinga Hills, Sudan*. Memoirs of Forestry Division, No. **3**. Sudan Government.

Jackson, J. K. (1957). Changes in the climate and vegetation of the Sudan. *Sudan Notes and Records*, **38**, 47–66.

Jonglei Investigation Team (1954). *The Equatorial Nile project and its effect in the Anglo-Egyptian Sudan*. Government Printer, Khartoum.

Kassas, M. (1956). The mist oasis of Erkowit. *Journal of Ecology*, **44**, 180–94.

Le Houérou, H. N. and Grenot, C. J. (in press). The grasslands of the Sahel. In *Natural Grasslands* (ed. R. T. Coupland).

Léonard, J. (1969). Expédition scientifique Belge dans le desert de Libye. *Africa-Tervuren*, **15**, 101–36.

Miehe, S. (1986). *Acacia albida* and other multipurpose trees on the Fur farmlands in the Jebel Marra highlands, Western Darfur, Sudan. *Agroforestry Systems*, **4**, 89–119.

Miehe, S. (1988). Vegetation ecology of the Jebel Marra Massif in the semi-arid Sudan. *Dissertationes Botanicae*, **113**, 1–171.

MFEP (Ministry of Finance and Economic Planning) (1983). *Development studies in the Jonglei Canal area: final report. Volume 3, Limnological and plant studies, Volume 4. Vegetation studies, Volume 10. Range ecology survey, livestock investigations and water supply*. Mefit-Babtie Srl., Glasgow.

Morrison, C. G. T., Hoyle, A. C., and Hope-Simpson, J. F. (1948). Tropical soil–vegetation catenas and mosaics. A study in the south-western part of the Anglo-Egyptian Sudan. *Journal of Ecology*, **36**, 1–84.

Mubarak, M. O., Bari, E. A., Wickens, G. E., and Williams, M. A. J. (1982). The vegetation of the central Sudan. In *A land between two Niles* (ed. M. A. J. Williams and D. A. Adamson), pp. 143–64. Balkema, Rotterdam.

National Academy of Sciences (1976). *Making aquatic weeds useful: some per-*

spectives for developing countries. National Academy of Sciences, Washington DC.

Osborn, D. J. and Krombein, K. V. (1969). *Habitats, flora, mammals and wasps of Gebel 'Unweinat, Libyan desert*. Smithsonian Contribution to Zoology No. 11.

Parry, D. E. and Wickens, G. E. (1981). The *qozes* of Southern Darfur, Sudan Republic. *Geographical Journal*, **147**, 307–20.

Peers, A. W. (1955). Report of a grazing survey of Eastern District, Equatoria Province. In *A report on a survey of the grazing areas of the Sudan* (ed. M. N. Harrison), Appendix 7. Sudan Government, Department of Animal Production (mimeo).

Quézel, P. (1969). *Flore et végétation des plateaux du Darfur Nord-occidental et du Jebel Gourgeil*. Dossier 5 de la Recherche Cooperative sur Programme No. 45. Centre National de la Recherche Scientifique, Marseille.

Renvoize, S. A., Lock, J. M., and Denny, P. (1984). A remarkable new grass genus from southern Sudan. *Kew Bulletin*, **39**, 455–61.

Smith, J. (1949). *Distribution of tree species in the Sudan in relation to rainfall and soil texture*. Ministry of Agriculture Bulletin No. 4, Khartoum.

Soil Conservation Committee (1951). Management and conservation in the Anglo-Egyptian Sudan (extract from the Report of the Soil Conservation Committee (1944)). In *Management and conservation of vegetation in Africa*, Bulletin No. 41. Commonwealth Bureau of Pastures and Field Crops, Aberystwyth.

White, F. (1983). *The vegetation of Africa. A descriptive memoir to accompany the Unesco/AETFAT/UNSO vegetation map of Africa*. Unesco, Paris.

White, L. P. (1966). *Land and water survey of the Jebel Marra area. Reconnaissance and semi-detailed soil survey*. FAO LA:SF/SUD/17, Rome (mimeo).

Wickens, G. E. (1975*a*). Changes in the climate and vegetation of the Sudan since 20 000 BP. *Boissiera*, **24**, 43–65.

Wickens, G. E. (1975*b*). A preliminary note on the status of the oil palm (*Elaeis guineensis*) in the Sudan. *Sudan Notes and Records*, **55**, 185–8.

Wickens, G. E. (1976). Speculations in long-distance dispersal and the flora of Jebel Marra, Sudan Republic. *Kew Bulletin*, **31**, 105–50.

Wickens, G. E. (1977). *The flora of Jebel Marra (Sudan Republic) and its geographical affinities*. Kew Bulletin Additional Series **V**, HMSO, London.

Wickens, G. E. and Collier, F. W. (1971). Some vegetation patterns in the Republic of the Sudan. *Geoderma*, **6**, 43–59.

Wilson, R. T. (1978). The 'gizu': winter grazing in the South Libyan Desert. *Journal of Arid Environments*, **1**, 325–42.

Wilson, R. T. and Clarke, S. E. (1975). Studies on the livestock of Southern Darfur, Sudan. 1. The ecology and livestock resources of the area. *Tropical Animal Health and Production*, **7**, 165–87.

4

Indigenous animals

R. T. WILSON

The Sudan is an enormous country falling entirely within the northern tropics. Almost all of the described tropical ecozones are found within its boundaries and on its shores. The main vegetation zones run west to east and vary from arid desert in the north to humid rain forest in the south. In the east the Red Sea coast is fringed with coral reefs along much of its length: there is also a small number of inshore islands. The major vegetation zones are interrupted in places by hills and mountain massifs such as the Daju Hills in Northern Darfur, the Jebel Marra in western Darfur and the Imatong Mountains in Equatoria. The major rivers and a number of permanent and temporary lakes and swamps including the Sudd, diversify even further the range of habitats.

As a result of the wide ecological spectrum, the variety of animal life is extremely diverse. Representatives of most of the major taxa are to be found and previously unrecorded species are constantly being added to the Sudan list. The discovery of entirely new species and sub-species has not been an uncommon occurrence on zoological expeditions to Sudan in the recent past, and it can be expected that animals as yet unknown to science will be described in the future.

The organization responsible for the control, conservation, and exploitation of indigenous animals is the Sudan Wildlife Administration with its headquarters at Khartoum Zoo. There is also an active Department of Zoology in the Faculty of Science at the University of Khartoum which maintains a natural history museum. A number of firms in Sudan and some in Kenya and elsewhere specialize in hunting and photographic safaris, but diverse problems in recent years have rendered this sector of the economy almost completely inactive. The high potential for tourism and game fishing in the Red Sea is almost completely neglected.

The two major wildlife reserves in Sudan are the Dinder and Boma National Parks. These are not easily accessible even in the dry season and it is not possible to get to them at all in the rainy season. The Sudan Wildlife Administration has few resources and has difficulty in maintaining the parks to International Union for the Conservation of Nature (IUCN) standards. Traditional hunting is tolerated although the increasing use of firearms means that efficiency is greatly increased, much to the detriment of the game. There is severe competition from domestic stock for grazing and for water during the dry season. A number of other reserves, mostly small and usually created to conserve or protect a particular species, are scattered throughout the country. Plans to create new reserves, including a large and potentially important one at Kafia Kinji in Southern Darfur, appear to have little hope of being implemented.

There are rich literature sources to Sudan's indigenous animals. These are not only confined to scientific publications on mammals, birds, fish, insects, and even smaller organisms. The diaries and records of early travellers (Browne 1799; Burckhardt 1819; Nachtigal 1889) and the memoirs of members of the Sudan Political Service (Ffrench-Comyn 1911; Chapman 1921; Dugmore 1924; Brocklehurst 1931) also provide information on the Sudan fauna as do the archival monthly reports of various district and provincial offices, these last being particularly useful in relation to species of economic importance.

Mammals

The standard work on Sudan mammals (Setzer 1956) is now somewhat out-of-date. There are seven common species of domestic herbivores including three ruminants (cattle, goats, and sheep), the camel (best described as a pseudo-ruminant), and three equines (donkey, horse, and mule). The water buffalo (*Bubalus bubalis*) has been imported at various times and the domestic pig is found in small numbers in the larger towns and, more commonly, in the southern region. Between sixty and seventy species of larger wild mammals (i.e. bigger than the size of a duiker) have been recorded.

There are records relating to more than 100 smaller mammals belonging mainly to the orders Lagomorpha (hares), Chiroptera (bats), Insectivora (hedgehogs, shrews, and their allies), and Rodentia (rats and mice). Other small mammals belong to the orders Carnivora (predators), Hyracoidea (hyraxes or cavies), and Primates (apes and monkeys but specifically the galago or bush-baby in relation to small mammals). The numbers of species of small mammals should not, however, be regarded as very useful as the taxonomy of the Insectivora and Rodentia, in particular, is confused (numerous revisions also render it very fluid) and there are undoubtedly numerous new species to be discovered.

The intentional and unintentional activities of man have generally had deleterious effects on most larger wild mammals. Some species have disappeared altogether and there is evidence that many others have been reduced in number with their former ranges being so compressed that now they survive only as relict populations. The principal findings of recent aerial surveys, specifically designed to count animals and map their distribution (Watson *et al.* 1977), are shown in Table 4.1. In addition to seven species of domestic animals (excluding buffalo, dogs, and cats), only thirty-one species of large herbivores (twenty-six of the order Artiodactyla, Family Bovidae, four of the order Perissodactyla, including three equines and the white rhinoceros; and one Proboscidea, the elephant) were seen. Three generalist Artiodactyla (the giant forest hog, the bush pig, and the warthog of the Family Suidae), four primates, four carnivores, and one rodent (the porcupine) were also seen. Major game concentrations are now found principally in Equatoria and, to a lesser extent, in Bahr el Ghazal and Southern Darfur.

In the accounts which follow, domestic animals are grouped together and wild game within the major orders, although the latter do not necessarily follow the strict taxonomical system.

Domestic animals

The three true ruminants are the most important species of indigenous animals in Sudan. Cattle are numerically and economically paramount. The major concentrations are in Southern Darfur and Southern Kordofan where they are owned mainly by the Baggara (from *baqr* = cattle) tribes. These cattle are known as Western Baggara and belong to the major grouping classed as North Sudan Zebu (Mason and Maule 1960). Large numbers of the Kenana breed are found in White Nile and Gezira Provinces and of the Butana breed in Kassala and Khartoum Provinces. The Kenana and Butana are considered to have potential for improved milk production and measures are currently being taken to conserve this genetic resource. The Nilotic tribes of the south also keep large herds of cattle of the Sanga type. The Dinka and the Shilluk have most cattle with concentrations in Bahr el Ghazal, Jonglei, and Upper Nile Provinces.

Most sheep are found in the northern provinces. These are mainly of the Sudan Desert type. The Kababish have large flocks and, as a result, Northern Kordofan carries more sheep than any other province. However, Northern and Southern Darfur, Kassala, Gezira, Blue Nile, and White Nile also each have more than one million sheep. In the south, sheep are of the Nilotic or Dwarf type, large numbers being found in Upper Nile Province. Goat types and distribution are similar to those described for sheep.

Camels are generally described as baggage or riding types. The latter are classified into breeds of which the Anafi and the Bishari are best known. Northern Kordofan and Kassala have most camels but there are large populations in Northern and Southern Darfur, Northern, Gezira, and Red Sea Provinces.

TABLE 4.1. Numbers, overall density, and group sizes of the larger herbivorous mammals in the Sudan

Species: Common name	Latin name	Estimated total ('000)	Density (No km^{-2})	Group size (No)
Domestic stock				
Cattle	*Bos indicus*	15367.2	6.16	46.1
Sheep	*Ovis aries*	12222.1	6.50	58.7
Goats	*Capra hirca*	11300.1	4.53	20.7
Camels	*Camelus dromedarius*	2360.9	0.95	16.7
Equines	*Equus caballus*	882.1	0.36	2.5
	Equus asinus			
	Equus caballus × *E. asinus*			
Wildlife				
Buffalo	*Syncerus caffer*	320.1	0.13	27.7
Bushbuck	*Tragelaphus scriptus*	12.5	0.01	1.6
Bongo	*Boocercus euryceros*	0.9	—	1.4
Elephant	*Loxodonta africana*	133.8	0.05	7.4
Eland	*Taurotragus oryx*	10.0	—	17.3
Giant Eland	*Taurotragus derbianus*	17.9	0.01	10.2
Yellow-backed Duiker	*Cephalophus silvicultor*	1.4	—	1.5
Grey Duiker	*Sylvicapra grimmia*	46.8	0.02	1.5
Red Duiker	*Cephalophus natalensis*	1.2	—	1.0
Giraffe	*Giraffa camelopardalis*	179.9	0.07	7.6
Gazelle	*Gazella dorcas*	45.1	0.02	3.4
Gazelle	*Gazella rufifrons*	326.7	0.13	13.1
Gazelle	*Gazella dama*	4.9	—	1.7
Hippopotamus	*Hippopotamus amphibius*	15.9	0.01	4.2
Hartebeest	*Alcelaphus bucelaphus*	157.5	0.06	4.5
Uganda Kob	*Kobus kob*	10.1	0.50	4.5
White-eared Kob	*Kobus kob*	1255.6	—	29.2
Greater Kudu	*Tragelaphus strepsiceros*	5.3	—	2.2
Lesser Kudu	*Tragelaphus australis*	1.4	—	6.0
Nile Lechwe	*Kobus megaceros*	40.3	0.02	12.6
Oribi	*Ourebia ourebia*	37.9	0.02	2.5
Oryx	*Oryx beisa*	14.1	0.01	17.6
Reedbuck	*Redunca redunca*	80.0	0.03	2.2
White Rhinoceros	*Ceratotherium simum*	0.2	—	1.0
Roan Antelope	*Hippotragus equinus*	55.4	0.02	5.2
Sitatunga	*Tragelaphus spekei*	0.3	—	1.4
Tiang (Korrigum)	*Damaliscus korrigum*	714.2	0.29	18.9
Waterbuck	*Kobus defassa*	35.2	0.01	3.2
Burchell's Zebu	*Equus burchelli*	86.8	0.03	49.3
Grevy's Zebra	*Equus greyvi*	11.5	0.01	35.0
Nubian Ass	*Equus asinus*	1.6	—	2.3

Source: Watson *et al.* (1977).

Gezira and Southern Darfur Provinces are the principal areas of donkey concentration: they are used mainly for transportation of crops. In Northern Darfur and Northern Kordofan donkeys are primarily used to carry water. Horses are much less numerous than donkeys, are now mainly considered as a prestige animal, and about two-thirds of the country's total are found in the two Darfur Provinces.

Proboscidea

About sixty-six per cent of elephants are found in Western and Eastern Equatoria with most of the rest in Bahr el Ghazal Province. Elephants were much more numerous and widespread in earlier times and even up to the middle of the present century. Ivory was an important export from Darfur before its incorporation into Sudan in 1916. Increases in the price of ivory during the 1940s led to an increase in the number of tusks exported from Southern Darfur, from 15 in 1942 to 72 in 1944 and 78 in 1945, when a quota of six elephants per year was imposed. During the 1940s and 1950s crop raiding by elephants was still a problem in Darfur, but the last confirmed sightings on Jebel Marra were in 1966 and one was seen in central Southern Darfur in 1972 (Wilson 1979*a*). Sudan is one of the few countries where elephant can still be hunted on licence but 'crop protection', poaching, and civil unrest in the south are factors that will lead to a continued decline in numbers.

Carnivora

In the aerial surveys of 1975–7 only four species of larger carnivora were seen. Estimates of the number of lion (*Panthera leo*), for the whole country, were 1610 with more than fifty per cent of these in Southern Darfur and twenty-five per cent in neighbouring Bahr el Ghazal. Most jackal (*Canis aureus*) were seen in Northern and Southern Darfur with an estimate of 2000 out of the national total of 3029. Of 1250 hyaenas, about two-thirds were in White Nile Province: two species occur, the striped (*Hyaena hyaena*) and the spotted (*Crocuta crocuta*), but no distinction was made between these. The estimate for cheetah (*Acinonyx jubata*) was 144 animals, all these being in Western Equatoria.

All these estimates are likely to be much lower than the true populations, but give some indication of the distribution and relative abundance of these four species. The larger predators have been particularly persecuted in Sudan, not for trophies but in response to their nuisance value and their depredations on domestic stock. Trapping, shooting, and poisoning have been used in control measures and these resulted in the legal (and reported) deaths of 76 lion, 20 leopard, 688 hyaena, and 656 jackals and foxes in Darfur in the six-year period 1947–52 (Wilson 1980*a*). This did not prevent lion killing 113 cattle, 81 donkeys, 56 camels, 25 horses, and large numbers of goats and sheep in one small area in 1955. Larger carnivores are not common in Boma Park in spite of the potential prey: this is because the prey are migratory while the predators are territorial (Fryxell 1983).

Other large species still extant include the leopard (*Panthera pardus*), most common in the southern forested areas although it has been seen as far north as 16° (Maydon 1923), almost as far into the desert as the cheetah recorded at El 'Atrun oasis at 18°N (Newbold and Shaw 1928). The African hunting dog (*Lycaeon pictus*) was commonest in the past in the desert, a pack killing 33 cattle and 25 donkeys at 16°N during 1946. The fennec fox (*Fennecus zerda*), and Ruppell's fox (*Vulpes ruppellii*), are also recorded from the northern areas (Shaw 1936). The sand fox (*V. pallida*), has been reported from Jebel Marra (Thomas and Hinton 1923), but has not been seen for many years and is possibly extinct there.

Smaller cats that probably still occur include the serval (*Felis serval*), and the African wild cat (*F. libyca*), although in the latter case confusion with feral domestic cats or with crosses of wild and domestic forms might account for some reported sightings. Various members of the families Mustelidae and Viverridae are not uncommon: these include the zorilla (*Ictonyx striatus*), ratel (*Mellivora capensis*), slender mongoose (*Herpestes sanguineus*), genet (*Genetta genetta*), and white-tailed (*Ichneumia albicauda*), Egyptian (*Herpestes ichneumon*), and banded (*Mungos mungo*), mongooses. All these can be pests of small-scale poultry enterprises. The civet (*Viverra civetta*), which can still be seen fairly frequently, used to be kept as a domestic animal and 'milked' for its perfume.

Primates

In comparison to some African countries, Sudan has a small number of primates. Nonetheless, there are records of seven species including one prosimian, five cercopithecoids, and one hominoid. The prosimian Senegal galago (*G. senegalensis*) is probably the most widespread and the hominoid chimpanzee (*Pan troglodytes*), the most restricted. The cercopithecoids are well distributed as a group but are generally found south of 16°N except for the Anubis baboon (*Papio doguera*), which has a population in the Erkowit Hills at 19°N. The other cercopithecoids include: the Grivet monkey (*Cercopithecus aethiops*) of which there are three sub-species, the nominate (*Cercopithecus aethiops aethiops*), *Cercopithecus aethiops arenarius* from the Imatong Mountains, and *Cercopithecus aethiops marrensis* on Jebel Marra; the blue monkey (*Cercopithecus mitis*), confined to a small area on the Ugandan border at about 4°N, 32°W (Butler 1966); the Patas or red Hussar (*Erythrocebus patas*); and the black and white Colobus (*Colobus polykomos*). Patas have been recorded as far north as 17° and as much as 480 km from open water (Thomas and Hinton 1923) but the northern limit is now about 13 °N.

The chimpanzee is limited in numbers in an area close to that where the blue monkey is found. The Colobus is probably still not uncommon in high forest in the Imatong and Didinga Mountains, but is heavily hunted for its skin. Patas and baboons are common and widespread and these and the Grivet can cause considerable damage to crops. The galago has an important role in pollinating the baobab tree (*Adansonia digitata*).

Group sizes of baboon determined in the aerial survey were 12.1, Patas troops averaged 2.0 animals and Grivet monkeys 14.0. In a series of ground surveys group sizes were slightly different being 14.0 for baboon and 3.9 for Patas (Wilson 1979*b*).

Artiodactyla

Even-toed ungulates comprise the most species and the greatest numbers of Sudan's large mammals. All are herbivores, being grazers (eating grass and herbs in the field layer), browsers (eating leaves, shoots, fruit, and twigs of shrubs or trees), or mixed feeders. Species or species groups complement each other in the utilization of the feed resource although there is some overlap among them in diet and there is often overlap and competition between wild and domesticated species.

The white-eared kob outnumbers all other species of large mammal. It is migratory, spending the dry season (October–June) in Boma Park and moving in an elliptical orbit south-westwards in the early rains (July) returning north-eastwards towards Boma later in the year (September). Almost the whole of the national herd moves as a single unit; at about one million animals, it is possibly the largest congregation of a single ungulate species in the world. This spectacular sight, although less well recorded than that of the wildebeest in the Serengeti/Mara ecosystem, equals it in splendour (Fryxell 1983). The white-eared kob is a major competitor with domestic animals for grass in the eastern part of the Southern Region. The range of the Uganda kob overlaps that of the white-eared in Jonglei and Eastern Equatoria Provinces but it is present in much smaller numbers.

The korrigum—also, and commonly in arabic, known as the *tiang*—and the lelwel hartebeest are rather similar species of the savanna grassland zone. They are, respectively, the next most numerous wild animals after the white-eared kob. The korrigum, in particular, was formerly much more widely distributed in Sudan, being common in the early twentieth century in Darfur as was the hartebeest. The Darfur korrigum is not the same subspecies as that of eastern and southern Sudan, being part of a West African population extending from Chad to Senegal. Most korrigum are now found in the Jonglei area while hartebeest are generally distributed in the southern provinces with a small remnant population extending into Darfur. The roan antelope has ecological requirements similar to those of the two previous species and has a similar current distribution. All three species can cause considerable damage to crops and this is a primary cause for persecution having reduced numbers and range. Rinderpest, which greatly reduced the numbers of susceptible species such as buffalo, eland, and giraffe in the late nineteenth and early twentieth centuries did not affect the populations of hartebeest and roan to any extent as they are virtually immune to the virus (Carmichael 1938). It is not known, however, whether they act as reservoirs for transmission of the disease to domestic stock.

The Tragelaphinae is a sub-family of antelopes in which the males have corkscrew shaped horns and the females are hornless, except for the eland in which both sexes have horns. They are usually browsers, are cryptic and shy and males are generally solitary except in the breeding season when they maintain small harems. They are of little agricultural importance although eland can be a nuisance in modern ranching operations as they are habitual breakers of fences. They are hunted for their horns and skins, the latter being attractively marked with stripes and spots. The common eland is presently restricted to Eastern Equatoria while the main concentration of the giant or Lord Derby's eland is in Bahr el Ghazal. The giant eland is a species of major international concern and the Sudanese population probably represents most of those left in the world: strict measures should be taken to conserve it. Bushbuck are rather widely but thinly distributed throughout the south. Most greater kudu are found in Eastern Equatoria although there is still a sizeable population on Jebel Marra. Lesser kudu tend to be found in slightly more arid areas than the greater kudu and while they have overlapping distributions in Equatoria and Darfur there is also a small number of lesser kudu in the Red Sea Province. There is possibly still a small population of bongo in Western Equatoria.

The buffalo, a characteristic species of the savanna-swamp mosaic, is present in relatively large numbers throughout the southern provinces with maximum numbers in Bahr el Ghazal although there are many in Eastern Equatoria and in Upper Nile Provinces. A small population still survives in White Nile and a few vagrants occasionally cross the Bahr el Ghazal River into Southern Kordofan and Southern Darfur. There are historical records of buffalo as far north as El Fasher and even until the mid-1940s buffalo were seen as far north as Kas (12 °N). Buffalo are potentially harmful to crops and can cause heavy damage to rice: they are one of the 'big five' trophy animals and are hunted professionally as well as for subsistence. Other typical species of the river and swamp complex are the Nile or Mrs Gray's lechwe, confined to southern Sudan and Uganda; the reedbuck, still reasonably common in the southern provinces and in Southern Darfur and Southern Kordofan; the waterbuck with a similar distribution to the reedbuck but more demanding in its requirements and rarely found more than 2 km from water; and the sitatunga, a highly specialized swamp-dwelling tragelaphid with elongated toes and probably rapidly approaching extinction in Sudan. The waterbuck, although not particularly persecuted, is an important reservoir of some livestock diseases because of the large tick burden it carries.

Giraffe are obligate browsers and are present in relatively large numbers in the southern provinces. Their scarcity in the almost ideal habitats of Southern Darfur and Southern Kordofan is, in part, due to disturbance by domestic livestock and in part to their being hunted from horseback by the Baggara (Cunnison 1956; Lampen 1933).

The hippopotamus, also a potential pest of crops close to any source of water, still survives along watercourses and in permanent pools throughout the basins of both Niles.

From several points of view, including aesthetics, zoogeography, physiology, and ecology the most interesting of Sudan's mammals are the addax (*Addax nasomaculatus*) and the scimitar-horned oryx (*Oryx dammah*). The latter has a slightly wider range than the former, being found as far south as 14 °N and occasionally down to 12 °N (Audas 1951). Traditional uses of oryx include the hide which is supposed to produce a superior quality of rope: addax horns are used for digging salt at El 'Atrun oasis. Numbers appear to have been maintained reasonably well until the mid-1950s, but sightings have been increasingly rare since then, the principal reasons for this near-demise being the use of the automatic rifle and motorized transport. The Barbary sheep (*Ammotragus lervia*) is another species which, if not already, is close to extinction: its former habitats included much of the rocky desert, particularly the Meidob Hills in Darfur and the Erkowit Hills in Red Sea Province. The Nubian ibex (*Capra ibex*) was present in Red Sea Province until early this century.

Of the gazelle, the most specialized is the Addra or dama. This species can still be seen on any lengthy journey along the desert fringes west of the Nile but is undoubtedly rare and confined almost entirely to Northern Darfur and Northern Kordofan. The Dorcas gazelle occurs in small numbers over most of the desert, being commonest again in the two northern provinces of Darfur and Kordofan. The red-fronted gazelle is still relatively numerous in savanna areas and especially in Jonglei Province and Grant's gazelle is found in Boma Park.

Smaller antelope have little effect on the agricultural or livestock economy. The three species of duiker (Table 4.1), as well as the oribi, are much more numerous than indicated by the aerial census. The dik-dik (*Madoqua saltiana*) is present in Boma National Park and on Jebel Marra (Cloudsley-Thompson 1965) but its status elsewhere is uncertain although probably rather rare.

Members of the pig family can cause considerable damage to crops by their rooting and burrowing habits. Warthog are widespread and numerous and particularly troublesome in groundnut crops while bush pig are restricted to denser savanna in the Equatoria provinces and a few giant forest hog also occur there.

Perrissodactyla

These odd-toed ungulates include the mainly browsing rhinoceros and members of the equine family.

Black rhinoceros were fairly numerous in Bahr el Ghazal and in Southern Darfur in the early years of the present century but there are no records since 1946, when one was seen near Wau (Blancou 1958). White rhino were limited to very small numbers in Western Equatoria in the 1970s but may now be extinct there (Hillman-Smith *et al.* 1986).

Burchell's and Grevy's zebras have overlapping ranges in Eastern Equatoria and while neither are present in very large numbers, there are large populations elsewhere in eastern Africa. The status of the Nubian wild ass is uncertain as it is often difficult to distinguish between the pure form and its crosses with feral domestic donkeys. Very small numbers remain in Red Sea and Kassala Provinces but its situation is undoubtedly precarious.

Rodents

The porcupine (*Hysterix* sp.), the largest of the rodents, is common and widespread although as it is largely nocturnal, it is not often seen. Of much more importance is the side-striped ground squirrel (*Xerus erythropus*), abundant everywhere and a notorious pest of groundnuts.

Smaller rodents, particularly the cyclically eruptive Muridae, are serious competitors with man for cereal grains at both pre- and post-harvest stages. The main pest is the multimammate rat (*Praomys natalensis*) with an eruptive cycle of five or six years. Gerbils and jerboas are interesting desert species with peculiar physiological and behavioural adaptations to life under conditions of extreme aridity.

Other mammal groups

Bats, especially the commensal *Tadarida* which invades any permanent structure, are the most prevalent of other mammalian groups along with insectivores (mainly shrews but also the hedgehog which is also to some extent commensal with man). Hares are also common. Hyraxes are only recorded from Jebel Marra. The armadillo-like pangolin and the aardvark are less common but are widespread and useful predators on ants and termites.

Marine mammals of tourist and scientific interest include the dugong—now very rare—and dolphins and porpoises.

Birds

The standard work on Sudan birds (Cave and McDonald 1955) is now somewhat out of date but will be replaced increasingly by more modern texts (Brown *et al.* 1982). Sudan is a major wintering and passage area for migrants from the Palaearctic regions and has a wide range of indigenous species but few endemic forms other than at the sub-species level.

The ostrich (*Struthio camelus*) is the largest living bird and has been of considerable economic importance to Sudanese communities. It is still fairly numerous to the south and in Southern Darfur and Southern Kordofan with smaller numbers in Red Sea and Northern Darfur Provinces. Eggs and feathers have long formed items of trade (Wilson 1976): continued demand for these and for the fancy skin contribute to a steady decline in the species. Bustards of the genus *Otis* are also subject to hunting pressure for food.

Sudan is home to about two-thirds of the world population of the rare and bizarre whale-headed or boat-billed stork (*Balaeniceps rex*). Although feared and normally protected by the Nilotic tribes of the southern flood zone, it has specialized requirements for feeding and nesting (Guillet 1978) which are being denied increasingly due to river development and the growth in human and livestock populations. Prognosis for its survival is little better in Uganda, the second of its known breeding areas, but at Bangwelu in Zambia it appears relatively safe.

Vultures are virtually absent over much of northern Sudan, presumably due, in part, to the poison campaigns directed at ground predators. In addition to direct and indirect human effects, climatic changes have affected a number of bird species (Wilson 1982).

Some birds are extremely destructive of grain crops. The best known of these is the Sudan dioch (*Quelea quelea*): vast sums have been spent by international organizations on attempts to control it, so far, it seems, with little success. Other weaver birds, the males of some of them having spectacular plumage, probably cause as much damage in the aggregate.

Beneficial birds, for example, those which eat termites or hunt rodents, include starlings, bee-eaters, rollers, swallows, swifts, hornbills (including the very large ground hornbill (*Bucorvus abyssinicus*)), Sudan's national emblem the secretary bird (*Sagittarius serpentarius*), and diurnal and nocturnal raptors. The helmeted guinea fowl (*Numida meleagris*) is probably neutral in its effects, taking some grain but eating many potential insect pests: it is hunted heavily and trapped but appears in no danger.

Fish

Sudan has icthyofauna in both fresh and salt waters. Comprehensive treatments for freshwater fish exist (Sandon 1950; Amirthalingham and El-Yassa Khalifa 1965) but there is no comparable text for sea fish. With the exception of small areas in the extreme west which drain into the Chad Basin, minor drainage into the Red Sea, and some internal ephemeral basins, all freshwater in Sudan forms part of the Nile system. Commercial fishing is not well developed either from salt or freshwater but subsistence fishing provides considerable protein of animal origin, particularly in the south.

Although superficially well-known, detailed studies of the distribution and abundance of freshwater fish still need to be undertaken. Recent studies in Darfur (Wilson 1980*b*) have shown, for example, that 27 genera of fish occur in the combined Nile and Chad systems, but that the Nile perch (*Lates niloticus*) is absent from the latter. A recent study of the fish in Dinder Park showed that 11 of the 14 families of the Blue Nile basin were present but these comprised only 21 of the 46 genera: however, a fish new to the Blue Nile, *Heterotis niloticus*, was recorded (Mahmoud 1986).

Freshwater fish of economic value include the piscivorous Nile perch and the tiger fish (*Hydrocyon lineatus*), the latter providing fine sport on the line and being excellent eating, although bony. The tiger fish is a ferocious predator and there is evidence that it reduces the overall diversification of species where it occurs. The Cichlidae is an important family of good eating fish including *Haplochromis* and *Tilapia*. Cichlids have a wide range of dietary tastes and can be used to assist control of phyto- and zooplankton.

The balance of aquatic micro-organisms has been greatly disturbed in the Nile due to flood control and irrigation works. As a result, chironomid insects, particularly *Simulium damnosum*, the black fly belonging to a family which causes river blindness, have become a pest, favoured by the ideal breeding habitat provided by fast flowing water at dam spillways. They have been a pest at Sennar since 1925, at Jebel Aulia (and in Khartoum where they swarm from December to April and are responsible for many allergic conditions) since the early 1930s, and at Wadi Halfa since 1938 when the Aswan Old Dam was heightened (Rzoska 1968). Cichlids can also be used to control bilharzia vectors, as in the case of *Hemichromus bimaculatus* in Ghana. Dams have helped some fish to exploit better their preferred habitats. For example, the inshore shallow water community on Lake Nubia is dominated by the phytoplankton feeder *T. nilotica* while the deeper water contains more zooplankton feeders of the Characidae such as *Alestes dentex, A. nurse*, and the tiger fish. In general, the economic potential of freshwater fish in Sudan is not exploited adequately.

Freshwater fish of general or scientific interest include relict species such as the African lungfish—two species occur in Sudan, the East African *Protopterus aethiopicus* and the West African *P. annectens*—and the polyp (*Polypterus senegalus*). Catfish of the family Mormyridae are endemic to Africa and particularly adapted to ephemeral muddy water. Catfish such as *Gymnarchus niloticus* and *Mormyrops boulengeri* set up weak electric fields and migrate backwards into their burrows as water dries up—in this way they survive the drought and it is an amazing sight to see thousands suddenly appear at the onset of the next rains. A particular use of *Clarias lazera* in Sudan is to keep shallow wells free of algae. Catfish and the rather similar barbels provide good eating if caught alive and kept to evacuate themselves overnight in clean water. An adaptation of some *Synodontis* spp., for example *S. batensoda* and *S. membranaceus* (catfish of the family Mochocidae) which feed on the surface is that they swim upside down and have developed counter-shaded patterns to aid concealment: bottom feeders such as *S. schall*, which squeaks when caught, feed the right way up and have normal shading.

In the Red Sea both elasmobranch (cartilaginous) and teleost (bony) fish occur. The former sub-class includes all the sharks. Many, including the hammerhead, are huge and extremely dangerous. Typical sharks such as the tiger shark (*Galeocerda cuvieri*), and porbeagles and mackerel sharks, Maho (*Isurus glaucus*) and White (*Carcharodon carcharias*) are also common. Other elasmobranchs include saw fish, skates, stingrays, electric rays, and the enormous manta ray (*Manta birostris*), which may be as much as 7 m across the 'wings' and weigh two tonnes.

The wide variety of bony fish include moray and conger eels, flatfish, left- and right-handed flounders, soles, razorfish, seahorses, and many brightly-coloured inhabitants of inshore coral reefs (the Sudanese coral reef is currently suffering severe degradation by the crown-of-thorns starfish (Ormond and Campbell 1971)). Sporting fish are usually fast-swimming surface predators and include tuna, wahoo, bonito, albacore, marlin, and sailfish.

As in the case of the freshwater resource, Red Sea fish are subject to very limited exploitation for food, tourism and sport.

Reptiles

The Nile crocodile (*Crocodylus niloticus*) is the major large reptile. It is still numerous in the basin and principal swamps of the White Nile and to a lesser extent of the Blue Nile. It has almost disappeared entirely from Bahr el Ghazal and the Chad catchment in Darfur (Wilson 1978) as a result of hunting for its skin and from killing to protect domestic stock.

Poisonous snakes of most varieties (mambas, cobras, spitting cobras (family Elapidae), and puff adders (family Viperidae)) are present in suitable areas. The python (*Pithon sebae*) is common from the savanna zone southwards.

Other reptiles of inland Sudan are: the Nile turtle (*Trionyx triunguis*) well adapted to an aquatic life with lungs having branchi furnished with hair-like processes to receive oxygenated blood from the carotid arteries; the desert or leopard tortoise (*Testudo pardalis*); and the grooved tortoise (*T. sulcata*), particularly well adapted to dry conditions as it lays its eggs in late autumn or winter and these hatch eight months later during the short rains. Lizards of many families (monitors, chamaeleons, agamas, skinks, and geckos) occur throughout, many having special adaptations to heat, aridity, or moisture.

In the sea there are probably five species of turtle. All are rare but one at least breeds on the inshore islands (Hirth and Abdel Latif 1980).

Insects and ticks

The orders Orthoptera (grasshoppers), Isoptera (termites), and Hymenoptera (bees and ants) are dominant in this group. The insect with potential for causing most economic loss is the desert locust (*Schistocerca gregaria*), which has a major outbreak locus in the Red Sea area: there have been four widespread plagues since 1910, not including the one reported in 1986. Grasshoppers provide a source of protein for many birds and mammals, including man.

Many termites have considerable economic effects and both Microtermes and Odontotermes are pests of cotton in the Gezira. *Hodotermes mossambicus* feeds on grass and cereal seeds. Macrotermes build large mounds, which can be a nuisance on mechanized farming schemes. Ants have various life styles from nomadic hunting (the so-called 'safari' ants of the family Dorylidae) to general food gathering ('harvester' ants), the latter affecting the human food supply in a manner similar to that of termites and grasshoppers.

The tsetse fly is the vector of trypanosomiasis in animals and sleeping sickness in man. It renders large areas of potential grazing unavailable and reduces the efficiency of man in areas where he has some tolerance. Other members of the family Tabanidae (biting and horse flies) cause consequential loss in livestock through the disturbance they create and direct loss in the case of camels as they are capable of mechanically transmitting trypanosomiasis to them.

Ticks, which are vectors of diseases such as heartwater and babesiosis, reduce livestock output. They can also cause direct losses when present in large numbers. Mites are responsible for mange in all domestic animals but are a particular problem on camels.

References

Amirthalingham, C. and El-Yassa Khalifa, E. (1965). *A guide to the common commercial freshwater fishes in the Sudan*. Government Printing Press, Khartoum.

Audas, R. S. (1951). Game in Northern Darfur. *Sudan Wildlife and Sport*, **2**, 11–14.

Blancou, L. (1958). Note sur le statut actuel des Ongules en Afrique Equatoriale Francaise. *Mammalia*, **22**, 399–405.

Brocklehurst, H. C. (1931). *Game animals of the Sudan. Their habits and distribution*. Gurney and Jackson, London.

Brown, L. H., Urban, E. K., and Newman, K. (1982). *The birds of Africa, Volume I*. Academic Press, London (and subsequent volumes in press or projected).

Browne, W. G. (1799). *Travels in Africa, Egypt, and Syria from the years 1792–1798*. London.

Burckhardt, J. L. (1819). *Travels in Nubia*. John Murray, London.

Butler, H. (1966). Some notes on the distribution of primates in the Sudan. *Folia Primatologica*, **4**, 416–23.

Carmichael, J. (1938). Rinderpest in African game. *Journal of Comparative Pathology*, **51**, 264–8.

Cave, F. O. and McDonald, J. D. (1955). *Birds of the Sudan*. Oliver and Boyd, Edinburgh.

Chapman, A. (1921). *Savage Sudan: its wild tribes, big-game, and bird life*. Gurney and Jackson, London.

Cloudsley-Thompson, J. L. (1965). The fauna of Jebel Marra, western Sudan. *Sudan Silva*, **2**, 24–8.

Cunnison, I. G. (1956). Giraffe hunting among the Humr tribe. *Sudan Notes and Records*, **37**, 49–60.

Dugmore, A. R. (1924). *The vast Sudan*. Arrowsmith, London.

Ffrench-Comyn, D. C. E. (1911). *Service and sport in the Sudan*. Bodley Head, London.

Fryxell, J. (1983). Wildlife migration in Boma National Park. *Swara*, **5** (5) 12–15.

Guillet, A. (1978). Distribution and conservation of the Shoebill (*Balaeniceps rex*) in the southern Sudan. *Biological Conservation*, **13**, 39–49.

Hillman-Smith, K., ma Oyisenzoo, M., and Smith, F. (1986). A last chance to save the northern white rhino? *Oryx*, **20**, 20–7.

Hirth, H. F. and Abdel Latif, E. M. (1980). A nesting colony of the Hawksbill turtle *Eretmochelys imbricata* on Seid Ada Kebir island, Suakin Archipelago, Sudan. *Biological Conservation*, **17**, 125–30.

Lampen, G. D. (1933). The Baqqara tribes of Darfur. *Sudan Notes and Records*, **16**, 97–118.

Mahmoud, Z. N. (1986). Ichthyofauna of the Dinder National Park, Sudan. *African Journal of Ecology*, **24**, 27–30.

Mason, I. L. and Maule, J. P. (1960). *The indigenous livestock of eastern and north-eastern Africa*, Technical Communication, No. 14 (revised). Commonwealth Agricultural Bureaux, Farnham Royal.

Maydon, H. C. (1923). North Kordofan to south Dongola. *Geographical Journal*, **61**, 34–41.

Nachtigal, G. (1889). *Sahara and Sudan, IV Wadai and Darfur*. (English edn. 1971, ed. A. G. B. and H. J. Fisher). University of Los Angeles Press, Berkeley.

Newbold, D. and Shaw, W. B. K. (1928). An expedition in the south Libyan desert. *Sudan Notes and Records*, **11**, 103–94.

Ormond, R. F. G. and Campbell, A. C. (1971). Observations on *Acanthaster planci* and other coral reef echinoderms in the Sudanese Red Sea. *Symposium of the Zoological Society of London*, **28**, 433–54.

Rzoska, J. (1968). Observations on zooplankton distribution in a tropical river dam-basin (Gebel Aulyia, White Nile, Sudan). *Journal of Animal Ecology*, **37**, 185–98.

Sandon, H. (1950). Illustrated guide to the freshwater fishes of the Sudan. *Sudan Notes and Records*, **25**, 1–61.

Setzer, H. W. (1956). Mammals of the Anglo-Egyptian Sudan. *Proceedings of the United States National Museum*, **106**, 447–587.

Shaw, W. B. K. (1936). An expedition to the southern Libyan desert. *Geographical Journal*, **87**, 193–221.

Thomas, O. and Hinton, M. A. C. (1923). On the mammals obtained in Darfur by the Lynes–Lowe expedition. *Proceedings of the Zoological Society of London*, 247–71.

Watson, R. M., Tippett, C. I., Rizk, F., Jolly, F., Beckett, J., Scholes, V., and Casbon, F. (1977). *Sudan National Livestock Census and Resource Inventory*. Resource Management and Research, Nairobi.

Wilson, R. T. (1976). The ostrich, *Struthio camelus*, in Darfur, Sudan. *Bulletin of the British Ornithologists Club*, **36**, 123–5.

Wilson, R. T. (1978). Notes on the crocodile, *Crocodylus niloticus*, in Darfur, western Sudan. *East African Wildlife Journal*, **16**, 65–7.

Wilson, R. T. (1979*a*). Wildlife in Southern Darfur, Sudan: distribution and status at present and in the recent past. *Mammalia*, **43**, 323–38.

Wilson, R. T. (1979*b*). The primates in Darfur, Republic of Sudan. *Folia Primatologica*, **31**, 219–26.

Wilson, R. T. (1980*a*). Wildlife in Northern Darfur, Sudan: a review of its distribution and status in the recent past and at present. *Biological Conservation*, **171**, 85–101.

Wilson, R. T. (1980*b*). The small scale fisheries industry in Southern Darfur, Sudan. *Bulletin de l'Institut Fondamental de l'Afrique Noire*, **42A**, 823–35.

Wilson, R. T. (1982). Environmental changes in western Darfur, Sudan, over half a century and their effects on selected bird species. *Malimbus*, **4**, 15–26.

5

History, ethnicity, and agriculture in the Sudan

L. KAPTEIJNS and J. SPAULDING

The Sudan is home to a human community with a long history and a rich and diverse culture. Not all the elements of the Sudanese experience are known, nor is everything known of immediate relevance to agriculture. The purpose of the present chapter is to delineate in broad terms some aspects of the Sudanese historical and cultural experience that do carry such relevance. It is well to conceptualize the conduct of agriculture as systemic activity, in which some components are fixed by natural determinants, others by technological factors and the whole given structure and put in motion by the social arrangements of its practitioners. As a consequence of the systemic character of these social arrangements it is possible to make meaningful distinctions between groups distributed culturally through space or historically through time; in many cases, though not all, these distinctions are also validated by the criterion of language, the repository of culture and often the delineator of ethnic identity.[1]

The earlier history of agriculture

The Sudan has a long pre-agricultural past, known today largely through archaeology and linguistic inference, in which people made their living by hunting and gathering. The legacy of the pre-agricultural age, otherwise beyond the scope of the present essay, is one of perspective, from which vantage point the initial introduction and subsequent development of agriculture in the Sudan (between the seventh and the fourth millennia BC and following) may be seen as a series of ingenious human responses to conditions of material scarcity. Each stage in this sequence of responses

[1] The modern 'tribes' which populate much European literature about the Sudan are largely the product of the historical experience to be outlined below.

entailed an intensification of labour, the more thorough exploitation of the natural environment, and increasing levels of social differentiation within an ever more populous human community (Kapteijns and Spaulding, 1985).

The great majority of modern Sudanese trace their ultimate linguistic ancestry to two important communities that flourished locally across the chronological boundary between the pre-agricultural and agricultural eras—the specialized aquatic hunting and fishing culture of the early Nilo-Saharans and the specialized seed-gathering culture of the speakers of northern Afroasiatic (Sutton 1974; Ehret 1979). The Kordofanian languages of the remaining ancient speech community, Congo-Kordofanian, appear to share a similar time depth of historical presence in the Sudan, but they have not yet been associated with an archaeologically-attested Mesolithic culture (Thelwall and Schadeberg 1983). A possible early community of Khoisan hunter–gatherers in the southern part of the country has long been absorbed by later food producers (Ehret 1982).

As the extended wet phase of the era 10 500–3000 BC drew to a close and drier conditions approximating the modern came to prevail (Harvey 1982), some broad features of subsequent Sudanese ethnic regional geography took shape. Afroasiatic speakers dominated the Nile banks in Egypt, but at Sudanese latitudes the forefathers of the modern Beja community were largely confined to the highlands east of the great river. Kordofanian speech survived in or near the Nuba Mountains, where a number of descendant languages (for example, Heiban, Talodi, Katla, Tumale) are found today. Most of the rest of the Sudan was settled by Nilo-Saharan speakers and one may infer that, in many cases, their communities were stable and long-lived, for over the next several millennia their ancestral tongue developed into many rather widely divergent languages. Some of the modern groups who trace their linguistic ancestry and approximate regional location to the Nilo-Saharans of this period include the Zaghawa, Berti, Masalit, and Fur of the far west, the Kunama and Berta of the Ethiopian borderlands, and the speakers of Komuz, a family of languages that includes Gumuz, Twampa (Uduk), Komo, and a number of smaller languages of the western Ethiopian borderlands along with Kadugli and its kindred languages, spoken in the southern Nuba Mountains (Bender 1983).

Future cultural dominance, however, lay with two Nilo-Saharan communities whose area of settlement included the basins of the great rivers. In the south, from the upper White Nile westward along the Bahr el-Ghazal and on toward Lake Chad lived the community ancestral to the speakers of the modern Central Sudanic languages, represented in or near the Sudan today by the Bongo, Moru, Kresh, Yulu, Mangbetu, Lendu, and others. The lands east of the upper White Nile, and the whole of the Nile Valley north as far as Egypt, formed the core of the homeland of the speakers of Eastern Sudanic languages. Several communities from this linguistic tradition (Meroitic, Nubian, Nilotic) figure prominently in the later history of the Sudan and will be introduced in various contexts below; here one may mention some less conspicuous Eastern Sudanic groups whose approximate present location may plausibly be traced back to the initial period of Nilo-Saharan settlement. Nera (Barya) is spoken along the Sudan border in lowland Eritrea, Gaam (Ingessana) in a range of hills west of the Blue Nile near modern Damazin, and Nyima and Temein in the Nuba Mountains. Daju languages are spoken in now-isolated communities extending at intervals from the Nuba Mountains westward to Dar Sila in Chad and beyond, while Mararit is spoken in parts of western Darfur and eastern Chad, and Tama further north. Along the Ethiopian borderlands south of the Sobat lived the ancestors of the modern Murle and Didinga and, on the plains beyond extending into Uganda, the forefathers of the Ik and their kin.

Scholarly reconstruction of the cultures of the Sudanese peoples of this remote period is still in its infancy, being confined to preliminary studies of a few communities. On an impressionistic basis, however, it would seem that most or all of the groups introduced above were food-producers who cultivated *Sorghum bicolor* and sometimes other grains, and herded cattle, sheep, and goats; hunting, gathering, and often fishing also remained important components of the economy. Most communities made pottery and their tool kits of stone, bone, and other materials were refined in design and specialized in function. The total population of the Sudan in this age was probably comparatively small and the scale of each individual social community was almost certainly very limited.

By about 3000 BC the very success of the food-producing way of life within the more fertile and favoured parts of the Sudan had given birth to a population large enough to exert pressure upon the available natural resources, a strain felt most keenly along the Nile corridor and, particularly, in the comparatively arid north. There followed a series of social and technological innovations, primarily among the centrally-located members of the Eastern Sudanic speaking community, that were to grant this group a long age of cultural hegemony. One important set of strategies entailed the development of new ways of livelihood that subordinated or even abandoned the cultivation of crops in favour of an emphasis upon the herding of livestock. In northern Sudan the adoption of specialized pastoral pursuits elevated the fortunes of the (Afroasiatic-speaking) Beja of the lands east of the Nile and of the (Eastern Sudanic-speaking) linguistic forefathers of the Nubians, Daju, and Tama of the broad plains of the west (Thelwall 1982; Adams 1982). From the region east of the great swamps of the White Nile, a land that probably included the home of the modern Burun, the ancestors of the modern Nilotic-speakers began a long age of pastoral expansion to the south and west; by the turn of the modern era Nilotic was spoken as far south as the Kenya Highlands and across the White Nile swamps westward around the southern flank of the Nuba Mountains. The heart of the southern Sudan was dominated by cattle-keeping peoples ancestral to the modern Dinka and Nuer, Anuak and Bari, Latuka and others, while Central Sudanic speech and culture were increasingly confined to tsetse-infested zones around the south-western perimeter of the Nilotic community where pastoral pursuits could not flourish. Moreover, during the first millennium BC, parts of this cattle-free south-western zone had been settled by speakers of Ubangian languages of the Congo-Kordofanian family, at the eastern edge of their historic zone of predominance. Both early settlers such as the ancestors of the Ndogo and the later Azande wielded a system of forest agriculture based on root crops that permitted them to prosper in an environment deemed marginal at best by the Central Sudanic peoples, who cultivated cereals and kept some livestock (Saxon 1982).

The adoption of a predominantly pastoral mode of livelihood was not a simple and final decision. Rather, as the years lengthened, a long series of alternative pastoral strategies became available, each of which bestowed certain benefits upon those communities who adopted it, but which also imposed new requirements. Some new strategies were made possible by technological improvements such as the introduction of ironworking, the camel, disease-resistant Zebu cattle, and the local domestication of rugged *Sorghum caudatum* that proved unusually well-suited to a supportive secondary role in livestock-oriented economies (Stemler *et al.* 1975),[1] as did the recent American import, maize. The rise to prominence and geographical expansion of the Jur, Shilluk, Acholi, and their kin within the previously established Nilotic-speaking world after AD 1000 may perhaps be best understood in this light. Other new strategies depended primarily upon reformulations of the prevailing constitutions of social organization; the emergence of a culturally-distinct Nuer movement from the midst of the Dinka-speaking community after AD 1500 may serve as an example (Newcomer 1972).

Most present scholarly interpretations of the history of the Sudan between the adoption of food production and the rise of the state would probably accept the broad paradigm offered here; pastoral livelihood strategies, often couched in Eastern Sudanic languages, expanded at the expense of systems based upon sedentary mixed agriculture, and within the community of pastoral emphasis, a further series of innovative refinements gave new advantages to some groups at the expense of others. However, there are numerous specific points of controversy and much remains to be learned about the nuances and the chronology of these developments.

[1] The Chari-Nile linguistic community has since been disaggregated by scholars but the authors' ethnobotanical evidence retains its validity.

For limited illustrative purposes, it may be justified to draw a few comparative inferences about the conditions governing ethnicity in those early days through analogy with the usages of non-state communities of the more recent Sudan. On this, admittedly, rather speculative basis, one would conclude that probably all groups participated in a wide and complex web of interactions with their neighbours. Individuals and small groups often moved from one community to another, intermarriage was common, goods and ideas were exchanged, political relations of friendship or hostility conducted, and the merits of diverse technological or cultural practices continually examined and re-evaluated. Many people spoke two or more languages, or had kinsmen who did; specific ethnic identities, as perceived by the people who lived them, tended to be fluid, often multiple, and almost always open to renegotiation should the circumstances warrant. Within this setting of highly permeable cultural boundaries, an ethnic group constituted a community of people who shared a commitment to a specific strategy for obtaining a livelihood and the defence of that way of life against competitors; group members identified themselves through the adoption of sets of specific cultural symbols, notably linguistic, but also artistic, musical, and ideological or religious (Ambrose 1982; Mack 1982).

The development of food production and of pastoral strategies of livelihood account for much of the modern distribution of peoples in the Sudan and for this reason it has been convenient to introduce these themes first. In a broader sense, however, both developments may also be interpreted as expressions of a longer and more fundamental series of changes in Sudanese society, a sequence of increasingly rigorous amendments to the social constitution. Each in turn was generated in response to increasing levels of pressure by the growing human population upon the available natural resources. In each case the new constitution sought to strengthen the society as a whole by enlarging it in size and by subordinating some to others.

Communities of Palaeolithic hunter-gatherers were of necessity mobile and small; material culture was simple and the basic social unit was a band of about 30–50 members. Since it was advantageous to belong to as wide a network of potentially helpful kinsmen as possible, bilateral descent principles prevailed. Band members observed a sharing ethic of reciprocity by which all had a claim to what fortune provided; government was conducted informally through the social pressures implicit in small-group relationships and both men and women played a prominent role in it. The aged or infirm were expected to remove themselves from society 'when they became a burden to themselves and others,' as the Berta expressed it.

The Nilo-Saharan peoples of the Mesolithic aquatic culture also relied upon hunting and gathering, but in at least some favoured locations their mode of livelihood supported much larger sedentary populations than had hitherto been possible and single communities numbering 1000 people or more were known. The introduction of food production had a similar demographic effect upon those who practised it. In either case, it was necessary to amend the social constitution in ways that subdivided the larger community into small groups, upon whom the burden of maintaining social discipline and the responsibility of reciprocal sharing would devolve. In most cases this was accomplished by adopting unilineal systems of descent, which had the effect of dividing society into groups who identified themselves through reference to a real or putative common ancestor. The constitutions of food-producing societies also introduced new forms of social inequality. The aged were no longer an insupportable burden, but rather enjoyed a social eminence that increased with years; often, the leadership of a community would be conducted by a council of lineage elders, while the young were taught virtues of deference and sacrifice for the sake of the old. Further, the new social arrangements postulated that men held rights in women and that these rights could be exchanged among men in return for bridewealth (usually in the form of livestock or metal goods) or specified quantities of agricultural labour (brideservice). Some Sudanese communities, exemplified by the Bongo, preserved constitutional arrangements of this general type into recent times.

With the adoption of specialized pastoral modes of livelihood, cattle (or camels) were elevated to a central position in the value system of society and almost all social relations came to be understood in terms of this form of property. Pastoralism was very demanding of land and when the physical or social barriers to easy geographical expansion arose, society became intensely competitive. Disciplined cohesion within and unity against outsiders were attained through the development of new military and governing organizations that cut across the dividing lines of lineage; these age sets divided all the men of a community within specified age brackets into corps and assigned them responsibilities appropriate to their respective stations in the hierarchy of seniority.

By the turn of the present era most of the then available potentials of the pastoral mode of livelihood had been attained and one Sudanese community after another found itself no longer able to resolve its internal tensions through geographical expansion. Pastoralists were often obliged to accommodate a renewed subsidiary emphasis upon crop production and many communities, including some with a history of pastoral emphasis and some without, and root-crop cultivators as well as grain farmers, were obliged to seek new forms of social organization. The older institutions of the lineage and the age set did not necessarily disappear at once, but their functions were subordinated to a new group of leaders who defined their jurisdiction in terms of territorial districts and their privileges in terms of hereditary priestly or chiefly titles. These new leaders achieved their distinctions through victory over rivals of similar qualifications in an often intense competition to gather supporters through military prowess and the seizure and redistribution of wealth. Meanwhile, as new hereditary elites appeared at the apex of society, there also emerged an underclass of individuals with permanently restricted rights and lowly status. Examples of this process of transition toward territorially based, predominantly agrarian, socially stratified societies ruled by hereditary elites may be found among the Azande, Bari, Anuak, Shilluk, in the Nuba Mountains, the highlands of the far west and, most conspicuously, in the northern Nile Valley. The leaders of some of these societies ceased to redistribute their material gains, rather investing them in a permanent coercive apparatus to maintain themselves in power indefinitely. They had made themselves the monarchs of African kingdoms.

Agrarian states

The rise of kingdoms in several parts of the Sudan modified the organization of agriculture and transformed the political and social setting within which ethnicity would henceforth be defined. In the northern Nile Valley the transition to state forms of government is evident in the remains of the Kerma culture (sixteenth century BC); this indigenous tradition was absorbed for an age by the imperial expansion of Egypt (1500–1000 BC), but re-emerged as the autonomous viceroyalty of Kush (1000–500 BC) and the independent kingdoms of Meroe (500 BC–AD 300), medieval Nubia (AD 300–1300), and Sinnar (AD 1500–1821) (Adams 1977). A second tradition of state formation developed in the western highlands centred on Jebel Marra. In the absence of archaeological studies comparable to those of the Nile banks, comment must be limited to a recent period documented by literary and oral sources; these speak in passing of successive kingdoms of the Daju (before 1500) and the Tunjur (1500–1650), and in greater detail of the Keira dynasty (1650–1874) and the restored Darfur sultanate of Ali Dinar (1898–1916) (Kapteijns 1985, p. 13).[1] Comparable states also developed in the southern Sudan, notably among the Azande and Shilluk.

Despite their respective individual characteristics the rise of these precolonial states usually had very similar implications for ethnicity and agriculture. The new form of society was divided along class lines between a group of rulers, known as *hukkam* in Islamic kingdoms, for example, and a class of subject commoners, for which the corresponding term was *masakin* (literally, 'poor people'). The ruling establishment comprised a royal family, an interrelated landholding nobility, and privileged individuals of special status such as titled slaves, religious officials, and distinguished foreigners. While Sudanese kingdoms derived their basic ideological rationale for existence from the realities of Sudanese life, the culture of the elite often

[1] All dates are approximate.

incorporated elements adopted from the dominant Mediterranean tradition of the day; as ages passed this might bring to Sudanese courts the deities and written scriptures of Ancient Egypt, Christianity, or Islam, and some of the regalia and luxury goods of a Mediterranean court. Imported ideas and artefacts among Sudanese elites reinforced a broader set of sumptuary laws that maintained social distance between rulers and commoners. In several cases the hindsight of modern observers has perceived an erstwhile ruling elite as an 'ethnic group'; the corporate identities of modern groups such as the Funj and Musabba'at, Abdallab and Ghodiyat, may be understood in these terms.

Typically, each state was based initially upon commoners of one ethnic group—Dar Masalit upon the Masalit and the Shilluk kingdom upon the Shilluk, for example—but many extended their authority over a heterogeneous and polyethnic subject class. Each group of subjects derived from the state the right to occupy and support itself from the resources of a specified territory, subject to certain conditions. A fundamental obligation of each subject group was to contribute to the support of the ruling establishment through labour services and the delivery of taxes in kind, notably in the form of grain and meat. Further, tax demands were deliberately structured to guide the allocation of labour in the exploitation of agricultural resources according to the potential value of each territory to the state; thus, for example, riverbank cultivators were obliged to grow cotton through the exaction of taxes specifically in that form, while the inhabitants of areas where gold, honey, or ivory could be found were similarly obliged to commit effort to their acquisition. Local and regional exchange among commoners in basic resources was accepted, but no indigenous class of professional merchants was tolerated; the system of taxation was structured to gather precious goods of potential export value such as gold or ivory in the hands of the government, which administered a monopolistic foreign commerce. The coercive powers of the state served to uphold the social hierarchy through the confiscation of wealth from the excessively ambitious, the punishment of visible social climbing in defiance of the sumptuary laws, and the enslavement of unproductive or deviant individuals. Obedient cohesion among groups of subjects was fostered by holding each community collectively responsible for the behaviour of all its members. Most modern ethnic groups of the northern and central Sudan and some southern communities also, derive their sense of collective identity and many preferred cultural norms and values from their historical experience as subjects of a precolonial state (Kapteijns 1985, pp. 18–61; Spaulding 1985, pp. 3–119; Kapteijns and Spaulding 1982, pp. 29–62).

From the fourteenth century AD the elites of most kingdoms in northern and central Sudan adopted Islam as a state cult. Arabic, in addition to its religious pre-eminence, became the preferred regional lingua franca of trade and administration and medium for the conduct of diplomacy and intellectual discourse. Kings vied in the adoption of Islamically distinguished titles and embraced claims to eminent Arab ancestries. By the eighteenth century the commercial values and usages of the Mediterranean world began to subvert the royally-regulated pre-capitalist agrarian economies of the Sudan. The appearance and proliferation of new market towns testified to the rise of an insurgent middle class within the old body politic and the large Nile Valley kingdom of Sinnar disintegrated amidst the endless feuds of middle-class war-lords. As both middle class elements and communities of erstwhile subjects advanced claims to Arab identity, the collapse of Sinnar gave birth to many of the modern Arab tribes of the northern Sudan (Spaulding 1985, pp. 139–298).

The first colonial period

In 1820 the armies of Muhammad Ali, ruler of Egypt, overran Sinnar and, subsequently, as the century progressed, extended Egyptian authority to the approximate boundaries of the modern Sudan (Hill 1959). A major preoccupation of this colonial regime, known locally as the Turkiya, was the procurement of slaves and its most lasting legacy was the division of the Sudan into a juristically Islamic, directly-administered northern zone and an anarchic southern slave-catchment area. In the north the rigorous imposition of land taxes in coin and the establishment of orthodox bourgeois Islamic courts favoured the consolidation of landholding by monied middle class interests, who often preferred to foreclose on their indebted free cultivators and replace them with imported southern agricultural slaves. The logic of the Turkiya decreed that many displaced free northern cultivators would seek to restore their fortunes as slave traders, termed *jallaba*, in the south (Spaulding 1982). By 1900 prevailing ethnic stereotypes would assert that the sedentary Arab groups of the northern Sudan had an ingrained propensity toward the practice of commerce and that all agricultural labour in their homelands was traditionally performed by slaves.

Official Egyptian governmental involvement in the southern Sudan took diverse forms, pursued inconsistent objectives, and generated ambiguous consequences (Gray 1961). At unofficial levels, however, continuity was provided by the spread and consolidation of the *zariba* system, under which companies of northern *jallaba* gunmen and their local allies divided vast tracts of the southern Sudan into fiefs to be exploited commercially through the export of slaves and ivory and to be reorganized socially into predatory, slave-based Islamic principalities (Cordell 1985; Triulzi 1981).[1]

[1] These two excellent studies analyse the experiences of southern communities under nineteenth century lords at the western and eastern extremities of the area dominated by the *zariba* system.

Many smaller ethnic groups of the modern southern Sudan, exemplified by the Mandara, Feroge, Jabalawin, or Burun, derive some of their sense of corporate identity from their status within one or another of these polities (for case studies see Triulzi 1981; Santandrea 1964; James 1979). Some *zariba* lords deliberately recruited supporters from among northern refugees of their own ethnic backgrounds and in the course of often bitter rivalaries among *jallaba* the issue of Ja'ali, Dongolawi, or Rubatabi identity sometimes assumed an importance quite unfamiliar to the north (El-Zubeir 1970; Hallam 1977).

The Mahdiya

In 1881, the Sudanese Mahdi received his divine calling; by his passing in 1885 the Turkiya had been destroyed and his successor, the Khalifa Abdallahi, ruled most of the Sudan until the second colonial conquest of 1898 (Holt 1970). The Mahdist movement of spiritual cleansing drew upon a general revulsion against the Turks and specifically tapped state-building energies and skills forged among the *zaribas* of the south-western frontier. The men who rallied to the Mahdist cause across the length and breadth of the land were socially analogous to the colonial elites they supplanted; prince succeeded prince; brother replaced brother, slave officers changed masters, and merchants and technicians transferred loyalties. The temporal goal of the insurgency, symbolized by the great colonial metropolis of Khartoum, was seizure rather than eradication of the apparatus of government; no sooner was the colonial city overrun than it was socially reconstituted across the river as Omdurman, while key technical services of the old regime were maintained *in situ* and intact. In the eyes of many contemporaries, the accession of the Khalifa represented a dramatic reversal of ethnic fortunes; the Mahdist elites of northern Nile Valley origin were subordinated as key positions were given to Abdallahi's kinsmen from among the Ta'aysha Baqqara and important subordinate commands were given to officers from the west.

The specific events of this turbulent age lie beyond the scope of this essay, but they contributed to the modern ethnic composition of the Sudan through an extensive process of migration and mixing. Repeatedly, large armies raised in one part of the country would be marched to another, often accompanied by the soldiers' families and allowed to settle into permanent garrisons. The lords of outlying communities were systematically summoned to the capital through a policy of elite hostage-taking called *tahjir*, while those near the centre of authority who fell from grace were often dispatched to Rejaf to redeem themselves in exile through service in the remote south. Throughout the Mahdiya the sensibilities of a religiously aroused age challenged the latitudinarian complacency of traditional cultures in which allegiance to Islam had usually been understood in communal and corporate terms. Ancient, long-tolerated folkways such as the drinking of beer and the conduct of rain-making rites were now taken as proof of unbelief and, not infrequently, punished accordingly. Each individual, as an individual, was held strictly accountable to the standards of Islamic orthodoxy as promulgated by Mahdist authorities and as personified by the middle class of the Nile Valley represented locally, throughout the Mahdist state, by the ubiquitous *jallaba*. Among Muslims of the eastern, southern, and western periphery, an important legacy of the Mahdiya has been an aspiration to the attainment of middle class respectability through imitation of the *jallaba*, an effort entailing not only rigorous adherence to the formal requirements of the faith as inculcated by preachers, but also the adoption of religiously extraneous Nile Valley customs such as female circumcision.

The second colonial period and independence

In September 1898, with the definitive defeat of the Mahdist armies at Karari, the Sudan came under colonial rule. While Egypt and Britain were, in theory, co-rulers of the Sudan, in reality the Khedive's authority was negligible. British rule, building on trends that had begun during the Turkiya, revolutionized Sudanese agriculture and developed an orthodoxy about ethnic identity that still pervades the scholarly literature today.

It was a basic premise of the colonial official mind—exemplified by men like Rudolph von Slatin and Harold MacMichael—that the 'tribe', with its own territory (*dar*), law, and chief, was the most significant social unit of Sudanese society. This emphasis on the 'tribal' nature of the Sudanese rather than their long tradition of subordination to state authority was somewhat ironic; after all, the British filled the slot traditionally reserved for central governments in Sudanese society with relative ease. The colonial administration, however, was not blind to class differentiation and in so far as the 'better class of natives' on whom all hopes for progress were pinned consisted of members of the 'tribal' aristocracies, the concept of class and 'tribe' did not conflict.

However, there existed also a middle class of merchants who did not fit the colonial concept of 'tribal' society and the administration's attitude toward this cosmopolitan social group was ambiguous. On the one hand, merchants advanced the policies of the government; by spreading the use of coin currency and stimulating consumer demands for commodities such as sugar, tea, and cloth, they contributed significantly to government attempts to bring the Sudanese into the labour market. Thus in 1918 Harold MacMichael, noting the lack of minted money in the recently acquired western Sudan, advised the provincial governor that 'the need for a trader or two' was great (Kapteijns 1985, p. 233). On the other hand, merchants were an anomaly in 'tribal' society as perceived by the British. Merchants, almost by definition, did not restrict their movements to one *dar*. Their influence was, moreover, often large enough to threaten the authority of local community leaders and their example presented an alternative to those commoners who were not contented with their lot and aspired to social mobility. The administration's ambiguity towards the merchants led to a policy of vacillation. When economic development was foremost on the government's mind, the merchants were favoured and encouraged, but when the administration was mainly concerned with preserving what it perceived as the status quo, its attempts to freeze 'tribal' society led to the exclusion of the merchants. The 'closure' of the South in the late 1920s must be seen at least partly in this light (Collins 1983).

Slaves were another class of people toward which government attitudes and actions were ambiguous. In general, the administration was loathe to interfere with the existing social order and did all it could to keep slaves in their place. The Vagabonds Act of 1905 offers an example of this official determination to discourage runaway slaves (Daly 1986, p. 237). The reconstruction of the economy after the conquest and the government's development efforts, however, caused a perpetual shortage of wage-labour and thus created an economic niche for which slaves seemed suitable and willing candidates. Initially ex-slaves and runaways were recruited into the army and the administration commonly accepted the labour of slaves, whose wages were paid to their masters. Many slaves, however, left their owners and sought out wage-labour opportunities on their own initiative (Daly 1986, pp. 231–9; Warburg 1986, pp. 263–4).

For the first fifteen years of its existence British colonial rule in the Sudan was heavily subsidized by Egypt (Daly 1986, pp. 194–7). The increasingly obvious anomaly of this situation and the continuing official stinginess of Britain forced the Sudan Government to devise ways to make the Sudan pay for itself. Taxes had been introduced immediately after the conquest, but only a programme of economic development, synonymous in the official mind with the development of an export sector, seemed to offer a solution. Initially the government was wary of foreign capital and the political influence it might acquire, opposed to the alienation of land for commercial agriculture, and sceptical of the possibilities of attracting labour without creating a large, landless proletariat. Its fears were assuaged by the promising results of experimental schemes in which plantation labourers received a fixed share of the profits as small-holding tenant farmers. Already before the First World War several irrigated schemes were under way in the Tokar and Gash Deltas, and the Gezira Cotton Scheme, a cotton plantation which by 1954 covered nearly a million feddans (1 feddan = 1.038 acres), existed in blueprint (O'Brien 1984; Grabler 1986, pp. 311–36; Daly 1986, pp. 420–8).

The Gezira Scheme was a partnership between the Sudanese Government, foreign capital represented by a London-based company called the Sudan Plantations Syndicate (SPS), and the Sudanese tenant farmers. The latter, who contributed the labour, received forty per cent of the net profits on the cotton crops. The Government, acting as landowner and responsible for the irrigation works, also received forty per cent, while the SPS, charged with management, received twenty per cent. The Gezira Scheme became the prototype for many other agricultural projects in the Sudan, both irrigated and rain-fed, and both government sponsored and private.

Government expectations concerning the Gezira Scheme and its sister projects proved vain in two respects. The administration had hoped that the tenant system would give rise to a self-sufficient peasantry and prevent the emergence of a proletariat of wage-labourers and an elite of large landowners. In reality social differentiation of this type did occur and it was even—after the fact—officially sponsored and actively encouraged by the government. In the period 1925–55 there arose a class of agricultural capitalists. It consisted of two groups; men who controlled Gezira tenancies but did not perform agricultural labour and often acted as financiers and absentee 'landowners'; and owners of large private pump schemes. Among the latter the religious leaders of the Ansar and Khatmiya, political movements with strong religious ideologies, figured prominently.

From the beginning the Gezira Scheme and other projects in irrigated agriculture depended heavily on wage-labour, both seasonal and settled. Labourers came from many areas: the Blue Nile and Kassala Provinces, the parts of Kordofan that lay on the railway line and, particularly, western Sudan and beyond (Northern Nigeria and French Equatorial Africa). It was due to this labour migration that the ethnic map of the plantation areas of the White and Blue Nile Valleys became the ethnic mosaic of today. The numbers of western immigrants called Fellata—the non-pejorative term is Takari—were in 1928 already estimated at 80 000, with an annual increase of 3000 (O'Brien 1984, pp. 6–7). The government actively encouraged and regulated the immigration and settlement of West African wage-labourers, but the presence of this underclass of perceived or real foreigners caused resentment among the Gezira tenants, particularly in the 1940s. By then the flow of migrant labour no longer needed government encouragement and while the westerners continued to be essential to the success of irrigated agriculture, they were officially discriminated against in a variety of ways; the Sudanese Nationality Act of 1948 was constructed in such a way as to exclude from citizenship even westerners born in the Sudan (Duffield 1983, pp. 45–59; O'Brien 1984, pp. 15–18).

The development of irrigated cash crop agriculture thus led to an underclass of people who were ethnically labelled as Fellata and whose corporate identity derived to a large extent from their common experience and discrimination as immigrant wage-labourers. The Fellata, who speak several languages (including Hausa and Kanuri) and derive from many different areas, are but one example of the social construction of ethnicity in the colonial period (Duffield 1983, p. 51; Grabler 1986, pp. 152–3, 158; O'Brien 1986). Much of the stigma attached to being a southerner or westerner in northern Nile Valley Sudan derives from the near absence of capitalist development in all those parts of the Sudan whose participation in irrigated agriculture remained limited to the supply of unskilled wage-labour.

The new elite that came to the fore in the course of Condominium rule consisted of various groups: an educated elite, which consisted of students, teachers, and professionals and was concentrated in the northern towns; a military elite, which had been partly recruited from the Shayqiya, whom the British regarded as a 'natural warrior tribe', and which came to power in the first military coup of 1958; and the agrarian elite of cotton lords, already mentioned. This new elite had its roots in the area where capitalist transformation had occurred first and had proceeded furthest: the northern Nile Valley. Irrespective of an individual's family background, middle class or elite status was associated with northern Nile Valley origins; the claims to cultural and religious superiority that the middle class had used to set itself apart ever since its emergence in the eighteenth century increasingly separated the people of the northern Nile Valley from peoples in the west, east, and south.

The politics of the period preceding and following Sudanese independence in 1956 were largely dominated by two parties: the Umma, guided by Abd al-Rahman al-Mahdi, head of the religious family of the Mahdi and leader of the Ansar, and the National Unionist Party (NUP), influenced by Ali al-Mirghani, head of the Mirghani family and leader of the Khatmiya. Both groups appealed to religious and ethnic loyalties for mass support and maintained a system of political patronage that was to an extent ethnically based. Abd al-Rahman's Umma built wide support in the western Sudan and among those westerners who had flocked to the cotton plantations out of economic necessity and inspired by millenarian beliefs centring on Abd al-Rahman. The NUP, supported by the Khatmiya, extended its influence over the north and east, its original missionary field and the home of many Sudanese active in the export trade. While the two parties have been portrayed as religious movements and ethnic alliances, their political programmes must also be understood in terms of the class interests of their leaders, the cotton lords. The temporary coalition government that they formed early in 1956 and their merciless suppression of the strikes by the tenant farmers of the private pump schemes are a case in point (Ali and O'Brien 1984, pp. 215–18).

In the late 1960s the economic basis of the agricultural elite was drastically transformed. One section, represented by the Umma, came to dominate the large-scale, mechanized production of rain-fed grain for the internal market, while another, represented by the NUP, sought to make the import–export sector dominant in the economy (Ali and O'Brien 1984, pp. 218–20; Ali 1983). After a brief flirtation with the Soviet Union, in 1972, the Sudan turned toward the West. The vast amounts of Arab and American capital that began to flow into the country led to an overemphasis upon production for export; the concomitant decline in food production exacerbated the effects of the drought of the early 1980s (O'Brien 1985) and the ensuing famine, in its turn, intensified ethnic and regional tensions as thousands of refugees from famine descended upon the towns of the northern Nile Valley, in particular the capital. Meanwhile, ethnicity and ethnic conflict were brought to centre stage by a largely urban interest group that adopted Islamic fundamentalism as its political programme. Once this group, drawing on a section of the intelligentsia, professionals, and financial interests, captured the mind of the Sudan's autocratic ruler in the early 1980s, the Islamic northern and non-Islamic southern Sudan entered on a collision course.

The civil war between North and South has been characterized as a religious, ethnic/racial, cultural, political, and economic conflict. It is all of these things. The legacy of ethnic/racial, cultural, religious, political, and economic dominance of the northern Nile Valley developed in the context of the processes of capitalist transformation and class differentiation which began in the late eighteenth century and intensified under colonial rule. The history of ethnic groups and ethnicity in the Sudan from *c.*1800 to the present must be understood within this framework.

References

Adams, W. Y. (1977). *Nubia: corridor to Africa.* Princeton University Press, Princeton.

Adams, W. Y. (1982). The coming of Nubian speakers to the Nile Valley. In *The archaeological and linguistic reconstruction of African history* (ed. C. Ehret and M. Posnansky), pp. 11–38. University of California Press, Berkeley.

Ambrose, S. H. (1982). Archaeology and linguistic reconstructions of history in East Africa. In *The archaeological and linguistic reconstruction of African history* (ed. C. Ehret and M. Posnansky), pp. 104–57. University of California Press, Berkeley.

Ali, T. M. (1983). The road to Jouda. *Review of African Political Economy*, No. **26**, 4–14.

Ali, T. M. and O'Brien, J. (1984). Labor, community and protest in Sudanese agriculture. In *The politics of agriculture in tropical agriculture* (ed. J. Barker), pp. 205–38. Sage Publications, Beverly Hills.

Bender, M. L. (1983). *Nilo-Saharan language studies.* African Studies Center, Michigan State University, East Lansing.

Collins, R. O. (1983). *Shadows in the grass: Britain in the Southern Sudan, 1918–1956.* Yale University Press, New Haven.

Cordell, D. D. (1985). *Dar al-Kuti and the last years of the trans-Saharan slave trade*. Wisconsin University Press, Madison.

Daly, M. W. (1986). *Empire on the Nile: the Anglo-Egyptian Sudan, 1898–1934*. Cambridge University Press, Cambridge.

Duffield, M. (1983). Change among West African settlers in northern Sudan. *Review of African Political Economy*, No. **26**, 45–59.

Ehret, C. (1979). On the antiquity of agriculture in Ethiopia. *Journal of African History*, **20**, 161–78.

Ehret, C. (1982). Population movement and culture contacts in the southern Sudan, *c*.3000 BC to AD 1000: a preliminary linguistic overview. In *Culture history in the southern Sudan: archaeology, linguistics and ethnohistory* (ed. J. Mack and P. Robertshaw), pp. 19–48. British Institute of East Africa, Nairobi.

El-Zubeir Pasha (1970). *Black ivory, or the story of El-Zubeir Pasha, slaver and sultan, as told by himself* (trans. H. C. Jackson), Negro Universities Press, New York.

Grabler, S. (1986). From concessionaire to shaykh: the shaping of colonial economic policy in Sudan 1898–1930. Unpublished Ph.D. thesis. University of Wisconsin.

Gray, R. (1961). *A history of the southern Sudan*. Oxford University Press, London.

Hallam, W. K. R. (1977). *Life and times of Rabih Fadl Allah*. Stockwell, Ilfracombe.

Harvey, C. P. D. (1982). The archaeology of the southern Sudan: environmental context. In *Culture history in the southern Sudan: archaeology, linguistics and ethnohistory* (ed. J. Mack and P. Robertshaw), pp. 7–18. British Institute of East Africa, Nairobi.

Hill, R. (1959). *Egypt in the Sudan*. Oxford University Press, London.

Holt, P. M. (1970). *The Mahdist state in the Sudan, 1881–1898* (2nd edn). Oxford University Press, Oxford.

James, W. (1979). *Kwanim Pa: the making of the Uduk people*. Clarendon Press, Oxford.

Kapteijns, L. (1985). *Mahdist faith and Sudanic tradition*. Routledge & Kegan Paul, London.

Kapteijns, L. and Spaulding, J. (1982). Precolonial trade between states in the eastern Sudan, *c*.1700–*c*.1900. *African Economic History*, **9**, 29–62.

Kapteijns, L. and Spaulding, J. (1985). *Een kennismaking met de Afrikaanse geschiedenis*. Coutinho, Muiderberg.

Mack, J. (1982). Material culture and ethnic identity in south-eastern Sudan. In *Culture history in the southern Sudan: archaeology, linguistics and ethnohistory* (ed. J. Mack and P. Robertshaw), pp. 111–30. British Institute in East Africa, Nairobi.

Newcomer, P. (1972). The Nuer are Dinka: an essay on origins and environmental determinism. *Man*, n.s., **7**, 5–11.

O'Brien, J. (1984). The political economy of semi-proletarianisation under colonialism: Sudan, 1925–1950. In *Proletarianisation in the Third World*, pp. 121–47. Croom Helm, London.

O'Brien, J. (1985). Sowing the seeds of famine: the political economy of food deficits in Sudan. *Review of African Political Economy*, No. 33, 23–32.

O'Brien, J. (1986). Towards a reconstruction of ethnicity: capitalist expansion and cultural dynamics in Sudan. *American Anthropologist*, **88**, 898–907.

Santandrea, S. (1964). *A tribal history of the western Bahr el-Ghazal*. Editrice Nigrizia, Bologna.

Saxon, D. E. (1982). Linguistic evidence for the eastward spread of Ubangian peoples. In *The archaeological and linguistic reconstruction of African history* (ed. C. Ehret and M. Posnansky), pp. 66–77. University of California Press, Berkeley.

Spaulding, J. (1982). Slavery, land tenure and social class in the northern Turkish Sudan. *International Journal of African Historical Studies*, **15**, 1–20.

Spaulding, J. (1985). *The heroic age of Sinnar*. African Studies Center, Michigan State University, East Lansing.

Stemler, A. B. L., Harlan, J. R., and Dewet, J. M. J. (1975). Caudatum sorghums and speakers of Chari-Nile languages. *Journal of African History*, **16**, 161–83.

Sutton, J. E. G. (1974). The aquatic civilization of middle Africa. *Journal of African History*, **15**, 527–46.

Thelwall, R. (1982). Linguistic aspects of greater Nubian history. In *The archaeological and linguistic reconstruction of African history* (ed. C. Ehret and M. Posnansky), pp. 39–52. University of California Press, Berkeley.

Thelwall, R. and Schadeberg, T. C. (1983). The linguistic settlement of the Nuba Mountains. *Sprache und Geschichte in Afrika*, **5**, 219–32.

Triulzi, A. (1981). *Salt, gold and legitimacy*. Istituto Universitario Orientale, Naples.

Warburg, G. (1986). Ideological and practical considerations regarding slavery in the Mahdist state and the Anglo-Egyptian Sudan, 1881–1918. In *The ideology of slavery* (ed. P. E. Lovejoy), pp. 245–69. Sage Publications, Beverly Hills.

Further reading

Niblock, T. (1987). *Class and power in Sudan. The dynamics of Sudanese politics, 1898–1985*. State University of New York Press, Albany.

Oesterdiekhoff, P. and Wohlmuth, K. (1987). *The development perspectives of the Democratic Republic of the Sudan: the limits of the breadbasket strategy*. Weltforum Verlag, Munich.

6

Land tenure

I. G. SIMPSON

Background

The Sudan has a large land area but remains relatively thinly populated. Nevertheless, the continuing rapid population growth and the development of large-scale mechanized agriculture are resulting in perceived shortages of land in various parts of the country, especially in the central belt. There has always been a strong demand for irrigated land, use of which gives freedom from the constraints of an uncertain and often limited rainfall confined to only a few months of the year. Once the demand for agriculturally-usable land exceeds the available supply, systems of landholding and tenure become increasingly important determinants of production. In recent years land tenure systems have become major issues of agricultural policy, especially in relation to the large-scale irrigation schemes and rain-fed mechanized farming.

Traditionally, the tribe was the principal unit of social organization with responsibility for the disposal of land within the boundaries in which it held suzerainty. In parts of the Sudan especially in the south, tribes and their subunits still allocate land and control grazing rights. However, since the conquest of the Sudan in 1898 and the formation by Britain and Egypt of a National Condominium Government, land tenure has been determined increasingly by central legislation. Details of the history of land registration are given in Bolton (1948), El Mahdi (1971), and Simpson (1955). In 1899, the Title of Lands Ordinance established a commission to adjudicate on claims to land in Khartoum, Berber, and other provinces, and to prepare a register of ownership. This ordinance established the principle of compulsory registration. A 1903 Land Acquisition Ordinance gave the government powers to acquire land for irrigation schemes and other public purposes. The 1905 Land Settlement Ordinance made general provision for the settlement and registration of claims to lands which were, or were alleged to be, waste, forest, or unoccupied and added the important proviso that all such land should be deemed the property of the government unless claims to the contrary were proved. The above ordinances were consolidated with others in the 1925 Land Settlement and Registration Ordinance. Transactions in registered land are subject to government approval and sales to non-Sudanese have only occasionally been allowed. Registered land cannot be taken as collateral against debt.

In spite of this early legislation, much agricultural land remains unregistered and, even if *de facto* rights of ownership exist, they are not recognized in law. The 1970 Unregistered Land Act introduced an important modification to earlier legislation, stating that 'any land of any kind occupied or unoccupied which has not been registered before the commencement of the act shall be the property of the government and shall be deemed to have been registered as such.' Effectively, this removed the 1905 proviso that government had title to land unless claims to the contrary were proved and made it easier to allocate land for large agricultural schemes regardless of claims to ownership.

The development of large-scale irrigation schemes, commencing in 1925 with the Gezira Scheme, made new forms of tenure necessary. Individual holdings, relatively small in size, were distributed to tenants while the prime developer, normally the government, took ownership either on a freehold basis or a leasehold basis. A rather different form of tenancy was instituted with the development in the 1950s of the semi-mechanized rain-fed cultivation of sorghum (and subsidiary crops) in the central rainlands. This was on land which had been utilized largely for nomadic grazing with only sporadic cultivation. Being unregistered the government claimed ownership and distributed the land in holdings large enough to make the use of a tractor economic, initially of 400 hectares (952 feddans).

The issue of nationality in relation to land tenure (see Duffield 1983, p. 57) is of some importance in the Sudan. As already noted, alienation to non-Sudanese has been exceptional and a similar policy has applied to the allocation of government tenancies, although during the 1930s depression, West Africans were allocated Gezira tenancies abandoned by Sudanese. Sudanese nationality, under the 1948 Nationality Act, is only automatic for people whose ancestors were in the Sudan prior to 1897. Revocable nationality is available to others with ten years continuous residence. There are large numbers of people, many of West African origin, who are not nationals, particularly in the central areas of the country where they form much of the wage-labour on irrigated-cotton schemes, and in the riverain zones of the Blue Nile and its tributaries where they are independent farmers. Duffield (1983, p. 45) suggests that these people, most of whom were born in the Sudan, account for about ten per cent of the population of the northern Sudan. They are unable to acquire legal rights to land, but are frequently involved in unofficial arrangements.

A more detailed discussion of tenure arrangements is made in the context of specific agricultural areas with a broad distinction between irrigated and non-irrigated areas.

Land tenure in the irrigated areas

Non cotton-growing schemes

The area of land which can be irrigated from the main Nile, north of Khartoum is restricted by the narrowness of the valley, generally less than two kilometres and, even today, the area under cultivation in the Northern Region is only about 250 000 feddans. A considerably smaller area around Khartoum and, to a limited degree, the areas along the Blue Nile, receive irrigation for crops other than cotton. There are also five major irrigated sugar schemes which will be referred to later. The discussion here relates primarily to the situation in Northern and Khartoum Regions.

Before the introduction of mechanical pumps, irrigation was by means of *saqqiya* (animal-worked, water-lifting devices on the bucket and chain principle) or *shaduf* (manually-operated devices using a counter-weighted bucket). These could provide water only for land near the river. However, cultivation also took place on *seluka* land, exposed by the receding river as the flood subsided, and in a few naturally flooded basins. With insignificant rainfall, non-irrigated cropping is impossible and there is little scope for pastoral activity. The limited irrigable land has to provide the main source of subsistence for the agricultural population and demand for it is intense. Private rights of ownership came into existence at an early stage. Awad (1971) states that Arab incomers introduced land ownership between the seventh and mid-thirteenth centuries. However, the application of Islamic laws of inheritance, which allow a share of the land to be given to all the direct descendants, has led to areas of land having many owners. Awad cites Miskin (1950), who suggests an average of 300 owners per 10-feddan plot in the Northern Province, now the Northern Region. The majority of owners are absentees who usually let their land to tenants on a share-cropping basis.

The introduction of mechanical pumps extended considerably the area which could be irrigated and new methods of organization were needed, given the considerable capital required to finance the larger pumps and to construct the necessary canals. Firstly, at various times from 1917, the Government has developed large pump schemes in which the land is allocated to tenant farmers. Currently, there are fourteen of these schemes with a total area of less than 100 000 feddans of which less than one-third is cultivated annually. Originally, share-cropping was adopted, but in 1939 a change was made to a system of water rates related to the areas under different crops. The present rates are highly subsidized and tenants are each permitted to have one feddan of permanent crops such as dates, citrus, and mangoes.

Secondly, pump schemes were operated by companies or private individuals. One of the first schemes of this type was at Zeidab where a

10 000-feddan scheme was developed in 1906 by the Sudan Plantations Syndicate, a company which later managed the Gezira Scheme. Zeidab is now government-owned and unlike other Northern Region schemes has continued to produce cotton. Censuses by the Sudan Government (1965/6) reported 49 company-operated schemes in the Northern Province in 1963, each with an average of 1091 feddans, and 11 in Khartoum Province in 1966, each with an average of 283 feddans. In these schemes, land other than that in permanent fruit gardens is let out on a share-cropping basis. Ali (1986) reports that often one-eighth of the crop is allocated to land rent and half goes to the pump owner with the tenant cultivator taking the remaining share.

Thirdly, irrigation schemes are operated by co-operative societies. According to the census, there were 90 such schemes in 1963 in the Northern Province, each with an average area of 508 feddans, while in Khartoum Province in 1966 there were 11 schemes, each with an average of 1025 feddans. The co-operatives provide inputs and tractor services to the farmer members at cost price and are responsible for the irrigation and overall management of the schemes. The crop proceeds, less operating costs, are divided on a 50/50 basis between the co-operatives and the farmers. Any profit accruing to the co-operative after deduction of its own expenditure is divided among the members (Ali 1986, p. 167).

The pump irrigation schemes took in land which was not in registered ownership and which is now leased by the government for periods of up to 99 years at a nominal annual rent of £S0.20 per feddan (Ali 1986, p. 166). Tenancies on the pump schemes are normally small in area and the 1963 census reported an average of 4.5 feddans per tenant in the Northern Province. In a survey of 70 vegetable growers at Khartoum, Simpson (1966) found the average size of a tenancy to be about 3.23 feddans which, with some double cropping, represented 4.68 feddans cultivated annually. The operation of pumps is subject to licensing under the Nile Pump Control (General) Regulations of 1951. Perennial and flood licences are issued, but the great majority are perennial licences. The licence lasts 10 years for pumps of up to 10 inches diameter and 15 years for larger pumps.

Apart from the schemes mentioned above there is a small number of large irrigation projects in the Khartoum Region developed for the production of fodder crops either under the direct auspices of the government or with its support. These include the Arab–Sudanese Dairy Company with 5500 feddans, the Seleit Food Production Company, which was allocated 27 000 feddans in 1977 but recently utilized less than half of this, and the Kuku Dairy with 3500 feddans. Unlike the first two, which are under direct management, Kuku was established by the government in 1962 to be run on a co-operative basis. Milk producers who are members of co-operatives are allocated 10-feddan tenancies on which they are expected to maintain 12 cows, although some do not maintain stock but sell their fodder on the market. A rent charge for land and water is levied on a monthly basis.

Cotton-growing schemes

The Gezira Land tenure arrangements in the Gezira Scheme, the first of the large-scale irrigation schemes developed primarily for cotton production, provided the model for all subsequent schemes of this type. The Gezira Scheme began in 1926 following pilot developments from 1911 at Tayiba and later Barakat on the Blue Nile, these in turn being based on the Zeidab Scheme. At a very early stage it was decided that the Gezira should be operated on a tripartite basis with the colonial Sudanese Government making land available for the scheme, constructing a dam on the Blue Nile at Sennar, and undertaking the major canalization; the Sudan Plantations Syndicate, a London-based company, managing the scheme; and individual tenant farmers taking responsibility for the cultivation of crops under the direction of the company. This tripartite system continues both in the Gezira and in other cotton schemes, although the place of the company has been taken by Public Corporations. The Sudan Plantation Syndicate's concession expired in 1950 and its functions were taken over by the Sudan Gezira Board. Gaitskell (1959) provides an historical account of the Gezira Scheme.

Land settlement and registration were carried out in the Gezira between 1910 and 1922 before work on the scheme commenced. With higher rainfall than to the north of Khartoum, the land in the Gezira had been used for rain-fed cropping and livestock grazing and, consequently, ownership claims existed. Once these had been established and registered, the government under the 1921 Gezira Land Ordinance entered into an agreement with the owners to lease the land for 40 years at a rental of 10 piastres per year, representing the annual value of non-irrigated land. As an alternative the government was willing to purchase outright at 80 piastres per feddan, (raised in 1942 to £S1.00 and in 1958 to £S1.40). Land on lease could only be sold or mortgaged to the government and, as a result, most of the land has come, finally, into government ownership. Apart from direct compensation for loss of their land, owners were given the first right to tenancies in the scheme with a maximum of two standard tenancies (then each of 30 feddans gross, of which 20 were cropped), but with rights of nomination for the largest owners. For details see Taha (1966).

The tenancies consisted of several separate 10-feddan plots (*howashas*) in a 90-feddan field (*number*) which formed the basic unit for irrigation. Each *number* was planted to a single crop within a six- or an eight-year rotation, a system which continues to the present. Initially, tenants had their plots in contiguous *numbers* but this is no longer the situation. Tenancies were allocated on an annual basis subject to renewal.

Under the tenancy agreement, tenants have to adhere to a prescribed crop rotation, undertake the construction of field irrigation channels, and carry out all cotton-growing operations apart from the initial ploughing, according to the instructions of the scheme's officials.

Tenants did not make direct payments for land or water. Instead, the cotton crop was subject to a sharing arrangement whereby the tenants received 40 per cent of the net proceeds after collective expenses had been debited to a joint account. In the early years the main expenses debited to the joint account were for the ginning of cotton and its marketing. The government took 35 per cent of the net proceeds for operating the irrigation system, interest on capital, and as taxation, while the Sudan Plantations Syndicate received 25 per cent. The tenant's share was paid in three instalments during the year with further payments, called appreciations, made when all of the season's cotton was sold and the accounts finalized. Deductions were made for the cost of seed and ploughing and for loans provided to help meet the cost of hiring labour for cultivation and harvesting.

The cotton-sharing system operated until 1981 but modifications were made to the partner's shares and the items chargeable to the joint account system; the latter are described by Ahmed (1981). Under the 1960 Gezira Scheme Act, the net cotton proceeds were distributed as follows:

Partners	Per cent
Tenants	42
Tenant's Reserve Fund (a reserve for adverse years)	2
Sudan Government	42
Sudan Gezira Board	10
Social Development Fund	2
Gezira Local Councils (in lieu of taxes)	2

The tenant's share was raised to 44 per cent in 1964 and 48 per cent in 1966, and the government's share was reduced correspondingly. Items added to the joint account included cotton seeds (1950), ploughing (1963), and initial cotton ridging (1968). Fertilizer application and crop protection measures, neither of which were required in the early years of the scheme, were added to the joint account at a later date and became an increasingly heavy burden. The advances given to tenants for the hiring of cotton-picking labour were also added. With yields static, these additions to the account meant a marked fall in the surplus available for distribution.

The cotton-sharing system had the advantage of sharing risks between the three partners and, with strictly defined allocations, there was no opportunity for the government to raise 'implicit taxation' by retaining a larger proportion of the gross proceeds, a frequent occurrence with crop marketing authorities elsewhere in Africa. However, the system was to the disadvantage of tenants with high cotton yields for there was no direct link between the amount of cotton a tenant delivered to the Board and the payments he received. In the 1960s, new unshared crops such as groundnuts and wheat were introduced to the rotation. The continuance of sharing for cotton alone was thought likely to discourage tenants from giving adequate time and attention to this crop.

In 1966, two reports (IBRD 1966; Simpson *et al.* 1966) advocated replacement of the cotton-sharing system by land and water charges levied at a flat rate per feddan on all irrigated crops. It was noted that if these charges were applied at a uniform rate over the whole scheme, tenants in the less productive parts would be penalized, but a basis for differential charging was never established. These proposals were not taken up, but the need for change was recognized. In 1972, a committee proposed retention of cotton-sharing but with tenants receiving an increasing share eventually rising to 100 per cent when cotton yields exceeded predetermined levels. The Tenants' Union backed this proposal and, in 1979, it was announced that the system would be introduced along with charges for water and services for the non-cotton crops. However, the World Bank, which was to provide part of the funding for a Gezira rehabilitation project, pressed for the introduction of land and water charges at varying rates per feddan according to crop and water usage. These charges were implemented by Presidential Decree from the 1981/2 season introducing what is known as the individual account system, whereby tenants are charged individually for the services provided by the Board. They receive full payment for their cotton at the end of each season on the basis of a pre-announced price. Then deductions are made from the payments to cover the costs of services and the land and water charges. It was intended that the land and water charges should be set at levels which would enable the government and Board to recover the full costs of operating the scheme. In practice this has not been achieved although the charges, at least up to 1987, have been raised annually in line with inflation and the depreciation of the Sudanese pound.

The charges are uniform throughout the scheme and, as foreseen in 1966, this has created problems. Tenants achieving low cotton yields do not obtain a sufficient payment to cover the charges and to repay advances from the Board. Consequently, they become increasingly indebted. In 1984 the managing director of the Gezira Board reported that 35 000 tenants, about one-third of the total, failed to cover their debts. Legislation passed in 1983 permitted dispossession of tenants more than three years in debt, but mass evictions are not a political option. There is an incentive for indebted tenants to sell their cotton at a discount to those who are not indebted and thus able to obtain a full return from any extra cotton sold to the Board. In spite of these problems, the new system has provided increased incentives for efficient tenants located in favourable growing areas. Cotton yields in the 1983/4 and 1984/5 seasons showed a significant recovery from the low levels of the 1970s, but were disappointingly low in 1985/6. However, it is impossible to isolate the impact of tenure changes from environmental and other factors. Unlike the joint account system, the individual account system imposes the entire risk of production on the tenant. Risk aversion may thus offset the incentive factor.

As already noted, the Gezira standard tenancy was originally of 30 feddans with 20 feddans cropped. In 1933 the size was raised to 40 feddans with the adoption of a wider rotation with more fallow, but the cropped area remained at 20 feddans. The 40-feddan tenancy remains the basic unit in the older part of the Gezira (the main Gezira) but to accommodate more tenants, half-tenancies of 20 feddans, with 15 under crops, have been allocated. These are now the most frequent tenancy. In the Managil Extension, which was developed between 1957 and 1962, 15 feddans, currently all cropped, is the standard unit. Further sub-division is not permitted. In spite of the small size of the majority of tenancies, on average well over half the agricultural work is done by hired labour.

Although in the past the tenancy was subject to annual renewal, now this is automatic and the rights of heirs to take over a tenancy are recognized. As a result, around 12 per cent of tenancies are held by women, mainly widows and a few by minors. Their tenancies are usually operated by nominated agents (*wakeels*). Transfers of tenancies, often unauthorized, are not uncommon on the basis of capital or annual payments by the transferee. These payments vary according to the cropping potential and location of the tenancy. It is probable that transfers have increased since the introduction of the individual account system.

Subleasing, usually on a share-cropping basis, is widespread. For a long time, it has been common for those allocated plots for vegetable gardens to sublease to a specialist gardener (*kholi*) and to split the proceeds on a 50/50 basis. More recently share-cropping of groundnuts and sorghum (*dura*) has become widespread. A 1983 survey revealed that 38 per cent of tenants were share-cropping groundnuts (Ahmed 1981). Most of the sub-tenants are westerners living in separate villages or camps with no opportunity to acquire a tenancy.

The Gezira system of land tenure seems unstable and it is doubtful whether the prescription of standard crop rotations for the whole scheme remains appropriate. In some parts of the area, the returns from cotton are too low to cover input costs. Continuing to grow cotton here produces no economic gain yet, without cotton, the Gezira Board has no automatic means of recovering the land and water charges. The combination of the officially recognized allocation of tenancies with unauthorized transfers and subletting provides little security and siphons off revenue as rent. It is possible that recognition of the existing land market and a relaxation of control in cropping would enhance the overall productivity of the scheme.

The White Nile and Blue Nile Pump Schemes The majority of these schemes were developed in the early 1950s by private licensees who installed pumps and constructed canal systems often borrowing much of the capital from the banks. In addition to the private schemes, eight government schemes were established on the White Nile to compensate families losing land to flooding after completion of the Jebel Aulia Dam. A Gezira-type partnership system was adopted whereby land was allocated to tenants usually in 15-feddan tenancies, of which 5 feddans was in cotton and 2.5 feddans in sorghum (*dura*), although inadequate water often restricted the area of the sorghum crop. Relationships between scheme owners and tenants were governed by the 1951 Nile Pump (General) Regulations and the 1960 Nile Pumps Control (Tenancies) Regulations. A joint account system was adopted with the owner receiving 56 per cent of the net cotton proceeds, the tenant 44 per cent. Owners paid a land tax equivalent to 5 per cent of cotton proceeds. While there were numerous small schemes, those with over 250 feddans of cotton accounted for over 80 per cent of the area in 1963 (Sudan Government 1965/6).

The profitability of the schemes fell after 1957 and new private investment ceased. In 1968, the larger schemes were nationalized with their management transferred to the Agrarian Reform Corporation. Since 1978 management has been in the hands of separate corporations for the White Nile and Blue Nile schemes. Both corporations were dissolved in 1986 and interim arrangements currently operate.

There has been a lack of investment in the schemes and they have deteriorated. In the mid-1980s the cotton area was about 120 000 feddans, barely half of the 211 000 feddans reached in 1963/4. Low cotton yields have rendered schemes unprofitable, and Duffield (1981, p. 54) reported that in the Blue Nile area most tenants were involved in rainland cultivation, in addition to their pump scheme crops. An individual account system, with land and water charges set at higher levels than in the Gezira to offset pumping costs, was introduced in 1981.

The New Halfa (Khashm el Girba) Scheme This gravity-irrigated scheme of 330 000 feddans gross area on the Atbara River was developed in the early 1960s following the decision to construct the new dam at Aswan and the 1959 revision of the Nile Waters Agreement permitting the Sudan to extract more water from the Nile and its tributaries (Heinritz 1978; Hoyle 1977). One of the purposes of the scheme was the resettlement of people from Wadi Halfa whose lands were lost by flooding after the completion of the Aswan Dam. However, of the 22 000 tenancies, 16 000 were made available to the nomadic tribesmen of the area (Shukriya and Beja). Part of the area was reserved for a government sugar plantation. The tenure arrangements were similar to those for a 15-feddan Gezira tenancy and Halfans also received a 1-feddan freehold plot. Full cropping with cotton, wheat, and groundnuts was intended, but there has been a reluctance to grow groundnuts. A joint cotton account system was adopted and it was intended that water rates should be charged for groundnuts and wheat. This, however, proved impossible to implement. Absenteeism is reported to be very high among the Halfan tenants. Many who found life on the scheme uncongenial have left, leaving their tenancy under a *wakeel* and worked by paid labourers mainly from the West. Nomadic tenants are said to be reluctant to live permanently on the scheme and many continue a nomadic existence with their animals, setting up their tents on the *howashas* in the cultivation and harvesting seasons (Hoyle 1977, p. 128).

An individual account system with land and water charges was adopted for New Halfa in 1981 as in the Gezira.

The Rahad Scheme This 300 000-feddan scheme which started operation in 1977 is the most recent of the major cotton irrigation schemes. Like the New Halfa Scheme, the area developed was largely unregistered grazing land and tenancies were allocated to local inhabitants. The basic tenancy is of 22 feddans with a two-course rotation of cotton and groundnuts or sorghum. The joint cotton account system has never been used in the Rahad Scheme. Instead, water rates were levied on the tenants and deducted from their cotton proceeds. Since 1981, land and water charges have been harmonized with those in the other schemes.

A feature of the Rahad Scheme has been the allocation of specialist tenancies for horticulture and dairying. 2000 feddans set aside for horticulture are divided into 5-feddan tenancies and 10 000 feddans to be used for fodder production were to be allocated as 10-feddan milk-producing tenancies. By early 1986, no rentals had been charged for these. The Es Suki Pump Irrigation Scheme of 64 000 feddans on the Blue Nile, south of Sennar is operated on a similar basis to the Rahad Scheme.

The Gash and Tokar Schemes These long established schemes are based on flush irrigation from the flood flows of the Gash and Baraka Rivers. The areas receiving water vary greatly from year to year according to the size of the floods. Cotton and sorghum have been the main crops but castor has been grown in the Gash for many years. The government owns the land in both schemes and plots are allocated annually to registered tenants with a degree of preference to members of the traditional tribes of the area. Some of the plots have a higher probability of receiving water and the allocation system is intended to ensure that these are not in the same hands each year.

Many of the plots are reported as sublet on a share-cropping basis. A joint account system was operated in both schemes with shares to tenants, government, the corporations responsible for the schemes, and the local councils.

Sugar schemes

The first scheme developed was at Guneid. This was opened in 1963, utilizing 37 000 feddans which although pump irrigated and on the east bank of the Blue Nile had become a part of the Gezira Scheme. The sugar scheme thus inherited a tenancy system and, for a time, payments to tenants for cane were related to those previously made for cotton. Currently, payment is based on the weight of cane harvested with deductions for the cost of mechancial operations, fertilizers, and pesticides. The later sugar schemes at New Halfa, Sennar, and Assalya are under direct management by public corporations and wage-labour is used. The very large Kenana Scheme is also directly managed.

Land tenure in rain-fed areas

Traditional farming

With only limited exceptions, land in the areas of rain-fed agriculture is unregistered and, consequently, in terms of the 1970 Unregistered Land Act deemed to be government owned. However, traditional forms of tenure continue to exist. In areas of very low population density, those wishing to clear land for cultivation can do so without hindrance. This probably remains the situation in some parts of the southern Sudan. Ellman (1972) referring to the Nuer tribe on the border with Ethiopia writes 'newcomers get land by requesting it from a kinsman or age mate in the village; in the same way young men get land from their fathers or from other members of the village whose lands adjoin unoccupied land. From the Nuer view point there is no shortage of most types of land.' Often rights of usufruct are granted by the traditional leadership of the village or tribe. Haaland (1980, p. 13), referring to the savanna areas of the west, states that cultivators' rights to their plots lapse two years after they were last cultivated. The Condominium Government recognized the right of tribal leaders, the *nazirs, omdas*, and *sheikhs* to adjudicate land disputes but these powers were removed formally by the 1971 Local Government Act. In areas of relatively high population density, private ownership is recognized although not registered. In the Nuba Mountains area, most of the population lived originally in the hill or *jebel* villages. Land adjacent to these villages was scarce while the plains were left unoccupied. Bolton (1948) notes that 'every tract of land that is or has been under cultivation in the Nuba Hills is privately owned; everywhere house-farms are bought and sold'. Haaland (1980, p. 78) in Southern Darfur comments, 'land which can be cultivated continuously for longer periods, especially land on the alluvial *wadi* beds, tends in fact to be considered as private holdings, implying the right to exchange it for money.' Reeves and Frankenberger (1981) refer to land which is regarded as individually owned and heritable in a village close to El Obeid.

Land not currently in cultivation in areas of communal tenure is accessible to anyone with free rights of pasturage for livestock. However, stock owners should prevent trespass on cultivated land, an issue leading to frequent disputes. Free access to pasture is essential for the nomadic systems which play an important role in the rainlands. Nomads are tribally organized with every tribe having its traditional grazing area (*dar*). Incursions by other tribes with their stock are resisted and open conflict is not uncommon. In order to avoid such conflict the Condominium Government arranged grazing settlements identifying the *dars* over which particular tribes had rights. These settlements lapsed when recognition of tribal rights ceased and grazing is now available to anyone with livestock. However, intervention by local officials to resolve disputes still occurs if problems of security are involved.

The ownership of *Acacia senegal* trees, which produce gum arabic, is an important matter in savanna regions. It used to be common for trees to be owned by a tribe and allocated to individuals for tapping with rents paid to the tribal authority (Abd al Ghaffar 1974). Now it appears that private ownership is more common. Some gum gardens are owned by the Forestry Department and rented out.

In areas where traditional systems prevail, land tenure systems hinder agricultural development. Sedentary cultivators are deterred from long-term improvements in the absence of officially recognized ownership, and free access to pasture has led to severe overgrazing and loss of carrying capacity. Nevertheless, the institution of individual rights of ownership and occupation can result in landlessness and increasing poverty as the population increases. In areas of marginal rainfall which are affected by extended drought, scope for population movement to areas with higher rainfall is essential if famine is to be avoided. Rapid population growth plus the growth of mechanized farming is putting considerable pressure on existing tenure systems and there is an urgent need for acceptable alternatives.

Large-scale farming

The extensive clay plains of the central Sudan are well suited to mechanized cultivation and initial attempts to grow sorghum there by this method were made by the Department of Agriculture in 1943 at a location near Gedaref (Simpson and Simpson 1978). Cultivation was with tractors but the weeding and harvesting of sorghum was done manually requiring a large labour force. Recruiting labour was difficult and in 1948 a participating cultivator system was introduced in the hope of solving the problem. The Gezira was taken as the model with plots of 28 feddans (20 feddans in higher rainfall areas) plus 4 or 5 feddans for private cultivation allocated as tenancies. Government tractors cultivated the land, but tenants were responsible for all subsequent operations and the crop was shared on a 50/50 basis. The system was not a success, tenants were blamed for neglecting their plots and in 1953 the system was abandoned.

In 1954, a departmental working party noted the heavy losses on the government schemes and recommended their closure, leaving future development to private owners (Sudan Government 1954). The size of the basic private farming unit was determined by the area which could be operated with a single tractor. It was set as a two kilometre square block of 952 feddans (400 hectares), subsequently often referred to as a '1000-feddan scheme'.

Large blocks of land were demarcated and allocated by Provincial Land Boards mainly to individuals, but also to co-operatives and companies. Individuals could possess up to two schemes and co-operatives and companies up to eight. These limits were often circumvented through the use of nominees or unofficial transfer and some individuals have operated twenty or more schemes. Initially, a nominal rent of one piastre per feddan was charged; by 1981, this had reached £S2.00, but was still low in relation to output values. Initially, leases were annual, but were soon extended to eight and eventually twenty years. The lessees who came mainly from the merchant class had to clear their land of trees and bushes before putting it in production. They rarely lived on their schemes but appointed a foreman (*wakeel*) for day-to-day management. The *wakeel* and a couple of guards (*ghaffirs*) are usually the only resident employees. Two tractor drivers each with an assistant are needed for two months and about eighty seasonal labourers per scheme are used for weeding and harvesting.

The demarcated blocks which were all on unregistered land, lacked sources of year-round drinking water and were without permanent settlement. They were regarded as empty lands but were, in fact, used by nomadic pastoralists and for some shifting cultivation. The intention was to have a rotation of four crop years followed by four fallow years. This would mean that those allocated schemes would need a further scheme after four years, but either demarcated land was unavailable or too far distant. Some 1500-feddan schemes were allocated with the intention that one-third should remain fallow, but operators have been reluctant to adopt rotational cropping.

The Mechanized Farming Corporation was established in 1967, and in 1970 it took over scheme allocation from the Land Boards. It had a particular responsibility for directed private schemes financed through

World Bank loans. Four large state farms with areas of 30 000 to 210 000 feddans were operated by the Corporation but these have now been transferred to companies.

Invariably there has been an excess of applications for new schemes since initial profits can be high and possession gives access to loans from the Agricultural Bank of Sudan (ABS). A considerable number of schemes have been developed without authorization, but the difficulty of movement on the clay plains during the wet season makes this hard to prevent. In Southern Kordofan in 1979, the writer was told that the opening of schemes outside the demarcated blocks was permitted with the approval of the Province authorities to whom a rental was payable but often not made. It is believed that the situation in Southern Darfur is similar.

A significant number of schemes were allocated to co-operative societies including some societies from Northern Region villages. Prior to 1970, the minimum membership of a registered co-operative was only twelve, which meant that societies were more often a means of spreading the cost of the investment needed to start a scheme rather than composed of working members. The co-operatives have not been generally successful and the majority have wound up.

Company involvement in mechanized rainland farming became more important in the 1970s, being stimulated by concessions under the 1976 Promotion of Agricultural Investment Act. In the Blue Nile Province, in particular, some extremely large land concessions (up to one million feddans), were granted to companies financed partially by foreign interests. The leases were for twenty-five years and were renewable and there appeared to be little regard for the interests of existing land users. Not all of the concessions have been taken up or fully utilized and some have lapsed or are being reviewed.

The area under mechanized cultivation reached six million feddans by 1983/4. The large-scale farms provided most of the marketed supply of sorghum for urban consumption and, in some years, also for export, together with a large tonnage of sesame and some short-staple cotton. Unfortunately, however, the system of tenure has not led to the development of stable farming systems. Relatively high production is obtained in the early years after clearance but yields sink rapidly to low levels and abandonment is common when the scheme occupants move on to clear new land. The spread of mechanical cultivation has disrupted traditional systems, especially nomadic pastoralism. A more stable form of tenure is needed to give scheme owners an incentive to adopt farming systems which can be sustained over time.

References

Abd-al Ghaffar, M. A. (1974). *Shaykhs and followers*. Khartoum University Press, Khartoum.

Ahmed, M. E. J. (1981). The joint account in the context of the Gezira Schemes production relations. Unpublished M.Sc. thesis. Department of Economics, University of Khartoum.

Ali, O. H. A. (1986). The irrigated subsector of agriculture in the Northern Region. In *The agricultural sector of Sudan: policy and systems studies* (ed. A. B. Zahlan and W. Y. Magar), pp. 161–78. Ithaca Press, London.

Awad, M. H. (1971). The evolution of land ownership in the Sudan. *Middle East Journal*, **XXV**, 212–28.

Bolton, A. R. C. (1948). Land tenure in agricultural land in the Sudan. In *Agriculture in the Sudan* (ed. J. D. Tothill), pp. 187–97. Oxford University Press, London.

Duffield, M. (1981). *Maiurno: capitalism and rural life in the Sudan*. Ithaca Press, London.

Duffield, M. (1983). Change among West African settlers in northern Sudan. *Review of African Political Economy*, **26**, 45–59.

El Mahdi, S. M. A. (1971). *A guide to land settlement and registration*. Khartoum University Press, Khartoum.

Ellman, A. O. (1972). *An agricultural and socio-economic survey of south Sudan refugee settlements and surrounding areas, Awraja, Ethiopia*. Institute of Agricultural Research, Addis Ababa.

Gaitskell, A. (1959). *Gezira*. Faber and Faber, London.

Haaland, G. (1980). *Problems of savannah development: the Sudan case*. Department of Social Anthropology, Occasional Paper, No. 19. University of Bergen, Bergen.

Heinritz, G. (1978). Social geographic problems in the Khasm El Girba project: Sudan. *Land Reform, Land Settlement and Co-operatives*, **2**, 25–35.

Hoyle S. (1977). The Khasm el Girba agricultural scheme. In *Land use and development* (ed. P. O'Keefe and B. Wisner), African Environment Special Report, No. **5**, pp. 116–31. International Africa Institute, London.

IBRD (1966). *The Gezira study mission*. Unpublished report, World Bank, Washington D.C.

Miskin, A. B. (1950). Land registration. *Sudan Notes and Records*, **31**, 274–86.

Reeves, E. B. and Frankenberger, T. (1981). *Socio-economic constraints to the production, distribution, and consumption of millet, sorghum, and cash crops in North Kordofan, Sudan*. College of Agriculture Report, No. 1. University of Kentucky, Lexington.

Republic of the Sudan, Department of Statistics (1965/6). *A report on the census of pump schemes (June–August 1963). Vol. II Part 1 Blue Nile Province, Vol. II Part 2 Northern Province, Vol. II Part 3 Khartoum Province*. Department of Statistics, Khartoum.

Simpson I. G. and Simpson, M. C. (1978). *Alternative strategies for agricultural development in the central rainlands of the Sudan*. Rural Development Study, No. 3. University of Leeds, Leeds.

Simpson, I. G., Yassin, M. M., and Dishoni, S. (1966). *The development of agriculture and organizational planning in the main Gezira area*. Final Report by the Working Party, Department of Agriculture (Ministry of Agriculture), Khartoum.

Simpson, M. C. (1966). *Khartoum's food supplies*. Department of Rural Economy, University of Khartoum, Khartoum.

Simpson, S. R. (1955). Land law and registration in Sudan. *Journal of African Administration*, **7**, 11–17.

Sudan Government, Ministry of Agriculture (1954). *Working party's report on the mechanized crop production scheme*. Sudan Survey Department, Khartoum.

Taha, T. E. J. (1966). *The land tenure system in the Gezira Scheme*, Sudan Gezira Board Working Paper, No. 11E. Sudan Gezira Board, Barakat.

7

Agricultural credit

D. HANSOHM

The formal financial system

In the Sudan the formal financial system consists of the Bank of Sudan (the central bank), 21 commercial banks (national, foreign, and joint venture), with 214 branches (1987 data), three savings institutions (the Sudanese Savings Bank (SSB), the Post Office Savings Bank, and the Premium Saving Bonds Project), three specialized banks (the Agricultural Bank of Sudan (ABS), the Industrial Bank of Sudan (IBS), and the Estates Bank of Sudan), and some developmental institutions, for example, the Sudan Rural Development Company Ltd. (SRDC). Most of these agencies are engaged in financing the agricultural sector, either directly or indirectly.

Credit required by public sector agricultural projects is financed by loans from the Bank of Sudan through seasonal budgetary allocations approved by the Ministry of Finance and Economic Planning (MFEP). This finance is available for seasonal production loans and for administrative expenses.

The Bank of Sudan estimates the costs and returns for projects requiring finance and then determines payment rates for the loan. Repayments are deposited directly with the Bank of Sudan after crop sale.

The activities of the commercial banks have increased in recent years. Between 1981 and 1987 the total value of bank deposits increased by 760 per cent, from £S943m. to £S7168m. During the same period, the advances to private borrowers increased by 497 per cent, from £S2827m. to £S14061m. (Bank of Sudan 1983–8) and the number of branches increased from 148 to 195. However, the advances continued to be concentrated on short-term loans (12–18 months): these were 79.2 per cent of the total advances in 1987. Furthermore, about two-thirds of bank deposits were on current accounts. About half the short-term loans finance import/export trade and about 20–30 per cent industry, but it is estimated that not more than one per cent goes to agriculture (Bank of Sudan 1983–8; Ahmed Humeida Ahmed Ali 1986; Sudan, Republic of 1986*a*). This pattern has not changed over the years. Commercial banks have neglected agriculture because of the risky nature of its production systems, farmers' lack of liquid assets, and especially because of the Sudanese Government's policy on agriculture. Commercial banks are not allowed to charge more than 12 per cent interest on credits to the agriculture sector and in cases of crop failure or other delays they have to borrow from the Bank of Sudan at a rate of 14.5 per cent. Furthermore, in general terms, government policy is biased against agriculture and rural areas, which results in low returns on capital in agriculture compared with high profits and high returns on capital in the tertiary sector.

Since 1984 and with the implementation of *sharia* law, which prohibits the charging of interest, the number of Islamic banks has increased. These offer different alternatives to interest-bearing loans, based on the principal of profit sharing. However, these banks follow the same pattern of loans, concentrating on trade activities and short-term investments.

The main source of institutionalized credit for the agricultural sector is the ABS. It was established by the Sudanese Government and it is the only specialized agricultural credit institution. Its objectives, as stated by the Agricultural Bank Act (1957), are to support agriculture and its incidental, accessory, ancillary or subsidiary activities by providing assistance in cash, kind, goods or services to persons who are primarily engaged in agriculture or its allied and subsidiary industries. These functions are:

(1) To grant loans for agriculture and its subsidiary services;
(2) To purchase and sell agricultural inputs;
(3) To purchase and sell crops;
(4) To undertake commercial banking activities such as handling short- and long-term accounts;
(5) To borrow from foreign and local institutions and to invite the public to buy shares in the ABS or in other companies.

Concerning the ABS's lending policies, the Act stresses the following points:

(1) Preference is given to small and medium cultivators and co-operatives;
(2) The ABS helps to increase agricultural productivity by supplying improved seeds, insecticides, and fertilizers, and by providing extension services;
(3) The ABS will assist producers with the handling, storage, and sale of their crops and will provide expert market information.

Through its 32 branches and sub-branches, the ABS provides short-term (seasonal) loans as well as medium- and long-term loans, for which the maximum repayment periods are 15 months, 5 years, and 10 years, respectively. The share of long-term loans (for developing new schemes and for conducting major improvements to old ones) was always insignificant and since 1982 there have been no long-term loans; the ABS has concentrated on short-term lending in the production and marketing sectors. The share of short-term loans has varied between 66 and 90 per cent and was 70.3 per cent in 1984 (ABS 1984, p. 7). Some medium-term loans have been granted for the purchase of agricultural implements and livestock, the improvement of irrigation systems and farm buildings, and the establishment of enterprises allied to agriculture.

The ABS started its operations in 1959 with paid-up capital of £S5m. By the end of 1985 this had risen to £S44.1m. In addition to this, the ABS is authorized to draw short-term loans from the Bank of Sudan, secured by ABS assets. However, as the interest rate on these loans was increased from 6.5 per cent per annum to 14.5 per cent by 1983, this source of liquid assets was no longer advantageous. Other sources of capital are external loans and grants secured from international lending institutions such as the World Bank, the African Development Bank, the International Fund for Agricultural Development, the Organization of Petroleum Exporting Countries (OPEC), and also from bilateral agreements with different countries. Furthermore, the ABS raises capital from its trading activities, by providing crop storage facilities, and from the interest on agricultural loans and commercial commissions (the ABS has five commercial lending branches). Deposits are not an important source of capital. At the end of 1984 they amounted to £S6.69m., 60 per cent of which was in current accounts (ABS 1984, p. 15).

Although the ABS was established to serve the needs of small- and medium-sized farmers, in the early years the major portion of its business was concerned with large, private cotton schemes in the White Nile and Blue Nile Provinces. However, in the late 1960s, the indebtedness of these schemes increased and the ABS divested itself of them. The Government took over their organization and set up the Agrarian Reform Corporation. The ABS's emphasis then shifted to financing imports of farm machinery and other farm inputs and to short- and medium-term loans, mainly for mechanized farming schemes producing sorghum and sesame in rain-fed areas.

About three-quarters of the total funds of the ABS are directed to farmers cultivating farms of 1000–1500 feddans in the areas demarcated under the Mechanised Farming Corporation. In 1984, ABS finance covered 35 per cent of the 3000 farmers in these areas (Sudan, Republic of 1986*b*, p. 60). Loan values are intended to cover 70 per cent of a farmers' requirements but, according to farmers, they meet only 25 per cent of their needs.

Most of the remaining ABS finance goes to irrigated agriculture and only a small part (6–7 per cent) to 'traditional' agriculture. The ABS assumes that its credits cover only 3 per cent of the traditional farmers' financial needs, compared with 20 and 50 per cent for the irrigated and mechanized farming sectors, respectively (personal communication). The main reason for this is the traditional farmers' lack of collateral—the ABS will accept land, buildings, crops, implements, shares and bonds, and guarantees by the Government, public institutions, autonomous agencies or co-operatives. However, since land is traditionally not registered, traditional farmers can rarely provide any sort of collateral. Therefore, the ABS endeavours to reach these farmers by various other arrangements, through co-operatives, farmers' unions, local councils, village councils, farmers' groups, and project authorities. The experience of credit disbursement and repayment by the ABS has been very good in the Jebel Marra and Nuba Mountains regions, where a close relationship with the ongoing rural development projects was established. In these cases the extension services were used to select borrowers, distribute the ABS loans and supervise their use and repayment. The credits for a camel/plough package distributed by the Jebel Marra project were repaid in 1983 and 1984 at a rate of 99–100 per cent, but in 1985 the rate was 85–100 per cent due to drought. The picture was similar for the Nuba Mountains project.

It has been recognized that close contact between lender and borrower increases the success of a credit programme. However, Ahmed (1980) calculated that costs incurred by the ABS in lending to an average-sized borrower ranged between 28 and 30 per cent of the loan, without taking inflation into account. At the same time, as with the commercial banks, the ABS is not allowed to charge more than 12 per cent interest per annum for agricultural loans. While the objective of the Government's interest policy was to provide cheap credit in order to modernize agriculture, the banks' (including the ABS) reaction has been to direct funds away from farmers to more profitable activities such as trading, to shorten the loan term structure, to shift more of the lending cost to the borrower, and to concentrate loans on a few borrowers who can offer security. It is recognized increasingly that successful agricultural credit programmes must be based on cost-covering interest rates. Abundant cheap credit cannot compensate for low incomes or low returns on capital investment in agriculture, for cheap credit does not make an unprofitable investment profitable. Cheap credit often has an adverse effect on income distribution for it is captured largely by the well-to-do. For example, the credits for mechanized farming went largely to traders. For recent general discussions on agricultural credit see Donald (1976), Von Pischke *et al.* (1983), Adams *et al.* (1984), and Adams and Vogel (1986). In the Sudanese context, a successful agricultural development policy has to remove negative biases in order to make agriculture more profitable. This would enable realistic interest rates to be paid rather than agriculture relying on cheap, subsidized credit.

Like many rural finance institutions, the ABS has faced serious repayment problems with many of its customers, especially during years of drought, e.g. 1985. Up to 1981, the general repayment ratio was 80 per cent for all types of loans. Later it declined to less than 30 per cent for some types of loan in some branches (Ahmed Humeida Ahmed Ali 1986, p. 337). This was attributed to productivity factors, poor marketing facilities, land ownership arrangements, but also to inadequacy and late delivery of ABS credits.

In 1974 the Government set up the Sudanese Savings Bank (SSB) with its headquarters in Wad Medani. The objectives of the SSB were to serve the low-income groups by encouraging local savings and by using these savings for local investments in areas neglected by commercial banks. The SSB's capital (£S0.5m. at its inauguration) was obtained interest-free from the Bank of Sudan. In 1984 the capital was £S3m. The deposits amounted to £S56.7m., of which 58.4 per cent were current accounts. This put the SSB in a far better position than commercial banks to promote investments. Total loans and credit facilities amounted to £S38.3m. in 1984 (SSB 1985) and £S10.0m. (25.8 per cent) went to agriculture.

In 1982 the SRDC group was formed to promote and develop small- to medium-sized agricultural and industrial projects in the rural areas. It is financed by the Government, the SRDC, the Bank of Sudan, four Sudanese commercial banks, and three European development banks. In 1987 the SRDC was involved in thirteen agricultural, thirteen industrial projects, and two service sector projects (SRDC 1988, pp. 28–30).

The informal financial system

The formal financial system provides only a negligible part of the credit used by farmers. All groups of farmers are to some degree dependent on informal sources of finance, but this is especially true of traditional farmers; it is estimated that not more than three per cent receive institutional credit (Sudan, Republic of 1986*b*, p. 62).

Informal credits are given by relatives, friends, traders and shopkeepers, landlords, and professional money lenders. The most important of these are the *sheil* credits provided by traders and shopkeepers, and these form an integral part of the marketing system. Several types of *sheil* are practised, but usually the local merchant advances a loan to a grower (in cash or kind) which is repaid at harvest with a specific quantity of the crop yield equivalent to the loan. Livestock (especially in the West), *hashab* trees, and labour may also be included in the *sheil* practice.

However, in this system the interest rates are much higher than those of institutionalized credits. Money lenders sometimes realize profits of several hundred per cent and for many farmers this constitutes a debt trap resulting in chronic indebtedness. When a farmer enters a *sheil* contract because he lacks grain or basic foodstuffs, he is unlikely to have sufficient food the next year because the return on his harvest will be lower under the *sheil* system. Furthermore, the terms of *sheil* contracts are less favourable when the farmers' borrowing capacity is poor.

It has been argued that the high interest rates are justified by the high risks involved (Wilmington 1955), but empirical data seem to disprove this. The high level of interest rates seems to be determined by the oligopolistic/monopolistic (as well as the oligopsonistic/monopsonistic) position of the traders/moneylenders.

Using data from the Gezira Scheme, Saad el Medani Ahmed (1983) confirmed the hypothesis that charges for *sheil* credits are excessive. He calculated actual *sheil* charges of 726 per cent per annum, far in excess of the cost components he considered, namely: opportunity cost, administrative charges, premium for risk of default, and premium for inflation. Adding these, he arrived at a 'justified' interest rate of 122 per cent per annum. He argued that it was not these cost components, but the minimum subsistence needs of the tenants that should determine the level of interest rates: if the interest rate was so high as to endanger the borrower's subsistence, this would erode the base of merchants' lending activities.

The *sheil* system is blamed for the slow growth rate of Sudan's agriculture, because it reduces production incentives. Nevertheless, in view of the marginal role of institutionalized credits, it does play an important role in removing the financial constraints faced by small farmers. Without the system, large numbers of traditional farmers would have given up farming.

In the past, measures taken by the Sudanese Government to control or reduce *sheil* have not been successful and they will remain unsuccessful until sufficient institutionalized credit is available. Much can be learnt about the efficient implementation of agricultural credit distribution from the traditional informal systems. Several factors make them highly viable and efficient. There is a fairly close social relation between the lender and the borrower making it easier for the lender to assess the borrower's creditworthiness, indebtedness to others, previous yield records, the reputation of the tenant, and other factors. All these factors, in addition to the social nearness to the borrower, determine the interest rate and administrative costs are very low. The repayment rate is much higher than in institutionalized credit systems because it is socially undesirable not to repay informal loans. This does not apply so much to institutional loans.

In order to match the continuity, reliability, flexibility, and local adaptability of the informal systems, agricultural credit institutions not only need to increase funds, but also need to be innovative with regard to mechanisms of lending.

Administrative measures alone (an obligation for commercial banks to earmark a specific percentage of their resources for the agricultural sector, an obligation to open branches in rural areas, and an obligation to afford soft loans to farmers (Sudan, Republic of 1986*b*, pp. 64–5)), will not be successful in increasing agricultural lending.

New lending mechanisms such as those operated successfully by the ABS and the SSB, and described earlier, offer the greatest promise for the future.

References

Adams, D. W., Graham, D. H., Von Pischke, J. D. (ed.) (1984). *Undermining rural development with cheap credit*. Westview Press, Boulder.

Adams, D. W. and Vogel, R. C. (1986). Rural financial markets in low-income countries: recent controversies and lessons. *World Development*, **14**, 477–87.

Agricultural Bank of Sudan (1983–1985). *Annual Reports 1982–1984*. Agricultural Bank of Sudan, Khartoum.

Ahmed Humeida Ahmed Ali (1986). Finance and credit. In *The agricultural sector of Sudan* (ed. A. B. Zahlan), pp. 332–57. Ithaca Press, London.

Ahmed, A. H. (1980). Lender behavior and the recent performance of rural financial markets in the Sudan. Unpublished Ph.D. thesis. Ohio State University.

Bank of Sudan (1983–1988). *Annual Reports 1982–1987*. Bank of Sudan, Khartoum.

Donald, G. (1976). *Credit for small farmers in developing countries*. Westview Press, Boulder.

Saad el Medani Ahmed (1983). *Interest rates in the informal credit markets of underdeveloped rural areas. The case of the 'sheil' credit in the Sudan Gezira Scheme*, Rural Development Series Paper, No. 1. Department of Rural Economy, University of Khartoum.

SRDC (Sudan Rural Development Company Ltd.) (1988). *Annual Reports and Accounts 1987*. Sudan Rural Development Company Ltd., Khartoum.

SSB (Sudanese Savings Bank) (1985). *Annual Report 1984*. SSB, Khartoum.

Sudan, Republic of (1986*a*). *Strategy for development of rain-fed agriculture*. MFEP, Khartoum.

Sudan, Republic of (1986*b*). *Strategy for development of rain-fed agriculture*, Annex 3, Institutions and Finance Task Force Main Report. MFEP, Khartoum.

Von Pischke, J. D., Adams, D. W., Donald, G. (ed.) (1983). *Rural financial markets in developing countries: their use and abuse*. Johns Hopkins University Press, Baltimore.

Wilmington, M. W. (1955). Aspects of moneylending in Northern Sudan. *Middle East Journal*, **9**, 139–46.

8

Food preferences and nutrition of the people

M. Z. ALI

Outlook for food production and the food industry

Sudan, with about 200 million feddans of fertile arable land and 55 million heads of camels, cattle, goats, and sheep, is, potentially, a prosperous country in terms of national supplies of food raw materials. The varying climatological zones, ranging from the humid tropics to the barren desert, also provide different niches for the production of various kinds of animals, crops, and foods.

From a nutritional point of view, however, the challenge facing the country is to secure a year-round supply of food for all Sudanese people, a total of 23 million, living in all parts of the country, an expanse of land covering one million square miles. This means that the country has to supply the daily requirements of essential nutrients, such as carbohydrates, proteins, fats, minerals, and vitamins, in addition to potable drinking water.

Since the 1960s, there has been a widely publicized notion that Sudan could, one day, become a 'world bread basket'. However, this dream has all but vanished in the wake of the recent drought which resulted in drastic food shortages and led to mass starvation, under-nourishment, and wholesale death of humans and animals. The country that was supposed to become a 'world bread basket' had, in a few years, become a major recipient of national and international relief supplies and food aid.

Nevertheless, the national consumption of raw materials of finished products, containing the basic nutrients (carbohydrates, fats, and proteins) could be calculated and then compared with the national production of cereals, sugar, oils, and other protein sources. By comparing food production with food consumption in the Sudan, one might conclude whether the Sudan could become a 'bread basket' and to what section of the world population.

The daily intake of carbohydrates (400 g per person) is provided by bread and bread-like preparations from sorghum, millet, wheat, cassava, yams, minor quantities of rice, and sugar.

Only 10 per cent (equivalent to 20 million feddans) of Sudan's arable land is utilized for crop and animal husbandry by traditional practices or irrigated agriculture. Of these 20 million feddans, about 13 million feddans are allotted for the cultivation of sorghum varieties, which yield an average of 270 kg per feddan (MANR 1986*a*). Sudan manages to harvest 3.6 million tonnes of sorghum grain, equivalent to 37 per cent of the total African production of sorghum. The world estimated production of sorghum and millet, in 1986, is in the region of 800 million tonnes.

Estimates of the area and total production of millet (*dukhn*) in the Sudan have reached 4.11 million feddans and 429 000 tonnes, respectively. Millet, second only to sorghum in nutritional preference in the Sudan, dominates the rain-fed tenancies of traditional agriculture.

The question of storage and strategic stocks of sorghum and millet grain remains to be seen. At present in Sudan there are only two modern silos for mass storage of grain, one at Gedaref, the centre of mechanized rain-fed production, the other at Port Sudan, mainly for grain destined for export. In the traditional sector, ground pit storage is still practised.

Mass rural migration to urban centres has dramatically increased demand for bread made from wheat flour. This change is partly because of the large amount of time and fuel required for the traditional home-made preparations of bread, but also because it has become fashionable to eat bread made from wheat flour.

According to the MANR (1986*a*), home production of wheat during the 1985/6 season amounted to 200 000 tonnes, and the country had to import 400 000 tonnes of wheat flour to satisfy home consumption. Home production of cassava and yams was estimated to be around 40 000 tonnes during the same season, but their consumption is restricted mainly to the ethnic communities of southern and south-western Sudan, where they are used in local food preparations.

Sudan's production of 200 000 tonnes of wheat is extremely low when compared with world production of 500 million tonnes. The Arab world, of which the Sudan is a part, now produces about 25 million tonnes, a figure far below its actual need.

In addition to the problems of irrigation and pest control affecting wheat cultivated in the Gezira Scheme (the main area of wheat production in the Sudan), the author also observed other difficulties in some large-capacity milling and processing plants. For example, two wheat flour mills at Wad Medani and two other mills in Khartoum were found to be aiming for an extraction rate of 40–45 per cent during late 1985, compared with the global figure of 72–80 per cent.

Annual home production of refined cane sugar is estimated at about 400 000 tonnes, though the actual need is in the region of 600 000 tonnes.

There are five sugar refineries, at Assalya, Guneid, Kenana, New Halfa, and Sennar, but only at Kenana, is the term refined sugar an accurate description of the product. Problems of quality are possibly due to poor management and technical difficulties and, according to government plans, some $US180m. are currently being invested in a rehabilitation programme in the hope of re-establishing the industry.

The major sources of edible fats in the Sudan are sesame, groundnuts, and cotton seed, and the basic daily nutritional requirement for an adult man is 100 g fat, sufficient to release 900 calories.

The estimated area of oilseed crops in the 1985/6 season was 4.24 million feddans, and these yielded 688 000 tonnes (MANR 1986*b*). However, this area is probably inflated due to the inclusion of cotton, which is grown not only as a source of edible fat, but also for the domestic textile industry and for export. Sesame is grown on 2.5 million feddans, while groundnuts are grown on 951 000 feddans. Annual production of sesame and groundnuts is in the region of 131 000 and 274 000 tonnes, respectively.

The potential annual capacity of the local oil mills is thought to be one million tonnes, but actual production is only about 160 000 tonnes of oil, with an efficiency of 8 per cent of the designed capacity (Salih, personal communication). Moreover, the prices offered to producers may be lower than the floor prices set by the Sudanese Government, a pricing policy which might have a negative influence on producers' decision making and expectations (MANR 1986*b*).

Daily requirements of protein (100 g per person) are normally secured from both animal and plant sources. The census of the national herd (Sudanese Meat and Livestock Marketing Corporation 1981), shows a distribution of 18 million cattle, 21 million sheep, 13 million goats, and 2.6 million camels, but about 90 per cent of this animal wealth is reared under nomadic and semi-nomadic patterns of land use, a situation which hinders productivity. Cattle, goats, sheep, and camels are estimated to produce 365 000 tonnes of meat annually.

Improvements to the existing system of marketing should take into consideration the social and economic aspects of herd owners. Owning cattle is still considered to have prestige value and it is also an indicator of the true strength of the tribe. In some western and southern parts of the Sudan it is also the currency in marriages.

Salih (personal communication) has estimated that the annual potential for fish production in the Sudan is in the region of 110 000 tonnes, but only 30 000 tonnes are utilized. This is because most of the fish harvest (some 45 000 tonnes) is found in rough, unreachable water courses, such as those of the vast swampy Sudd area of the Upper White Nile.

Milk production and its industry are still very undeveloped, and are characterized by seasonality and scarcity, being restricted mainly to the rainy season (*kharif*). However, annual production is estimated to be 1.5 million tonnes and domestic consumption is thought to be about 68.8 kg per person per year (MANR 1977). Reyad (1977) has estimated that during 1975/6 cheese production was 8 million kg and the production of ghee was 5 million kg.

The weak infrastructure of the Sudanese economy reflects badly on the performance of the agro-based industries. The dairy industry, in particular, suffers from lack of paved roads, electricity supply, and adequate fresh water. The locally-produced cheese, which is a white brined variety, is far from being standardized and also suffers from the absence of strict enforcement of hygienic regulations.

Faba bean (*foul masri*) is a major source of plant protein and is widely consumed by urban communities during breakfast and supper meals because of its ease of preparation. The MANR had estimated the area under faba bean to be around 41 000 feddans and the total production 33 000 tonnes.

Fruits and vegetables, rich in minerals and vitamins, are known to be an important part of the human diet, but recent estimates of their production in the Sudan are scarcely available. The MANR had provided some estimates but they excluded southern Sudan. About 654 000 tonnes of vegetables were thought to be produced on an area of 154 000 feddans, averaging 4.2 tonnes per feddan; slightly less, about 511 000 tonnes fruit, with an average yield of 4.6 tonnes per feddan, were grown on a total area of 112 000 feddans. Domestic canning, dehydration, and extraction industries are constantly facing severe shortages due to the low production of fruit and vegetables. State-owned factories, as at Karima in the Northern Region, and at Wau in the Southern Region, were designed initially in the 1960s to process this vanishing harvest, canning and dehydration being the main activities. The Kassala plant, for onion dehydration, was erected during the same period.

The private sector is even more reluctant to enter the fruit and vegetable processing business. The Saeed and Widaa plants, are rare examples of privately-owned processing concerns, but most of the production lines in these factories are, at present, inoperative, due to lack of raw materials and technical difficulties.

The food-processing industry in the Sudan is doomed to fail if crop-related constraints, technical aspects, and marketing systems are not carefully revised in an attempt to secure a continuous flow of raw materials to the manufacturing plants. Low capital investment and managerial difficulties are some other features of the local agro-based industries. The performance of the food industry in the Sudan must surely affect the patterns of food consumption and it also reflects on the eating habits and preferences of the Sudanese people.

Preferences and habits of food consumption

Socio-cultural considerations

Sudan is inhabited by various ethnic groups, ranging from purely African and Nilotic tribes to groups of Arabic origin. This diversified cultural set-up manifests itself in the traditional practices and taboos of food preparation. Jeffille (1968) described these taboos as either harmful, harmless, or good, where beneficial food habits need to be promoted.

In the Sudan, Modawi (1982) has stated that the indigenous and traditional culture was essentially African, especially in the southern part of the country. Through the ages, this culture has undergone successive moulding and transformation by the introduction of other cultures and traditions, which have included Egyptian, Turkish, Indo-Pakistani, and colonial influences and which have interacted with the then present culture to leave their mark on the methods of food preparation and the selection of cooking ingredients.

Islam is the dominant religion in northern Sudan, while Christian and pagan rituals prevail in the south. However, the consumption of alcoholic beverages and pork is restricted, according to Islamic teachings. Other phenomena, related to supernatural powers have tabooed the consumption of some traditional foods.

A broad classification of food habits and attitudes among various Sudanese cultures is given by Modawi (1982). In addition to ordinary foods, he recognized other categories such as (i) prestige food, taken by rich people and described as being composed of fatty ingredients, (ii) celebrative foods for special occasions and (iii) other religiously defined foods, such as those consumed by Muslims during Ramadan.

It is very important to consider the cultural acceptance of food, since it comprises the items actually eaten by various communities in the Sudan.

Common food dishes

Cereal-based food constitutes the major portion of the Sudanese diet, amounting to more than half of the daily intake. Different sorghum varieties, known locally as *dura*, are milled either by mechanical, electrical or manual means. The flour is then fermented in water to prepare *kissra* and *assida*, both bread-like varieties. Millet (*dukhn*) forms the second most popular cereal, especially in western Sudan, while cassava, yams, and maize are the basic foodstuffs in southern Sudan. Bread from wheat flour is widely used by urban communities. Another wheat flour preparation known as *gurrassa*, which is a pancake-like preparation, is very popular in northern Sudan.

A typical traditional Sudanese main course is composed of either *kissra, assida*, or *gurrassa* (depending on the region) and meat/vegetable stew. The sauce or stew is of two main kinds. One is known locally as *tabikh*, and consists of meat, oil, onion, tomato paste, together with a vegetable such as okra, potato, sweet potato, pumpkin, egg-plant, purslane, or jewsmallow, and various other kinds of peas and beans. The second stew known as *mufruka*, is prepared from meat and onion, usually with dried or fresh okra. Alternatively, leafy vegetables, such as jewsmallow, cowpea leaves, chard, or other various vegetables are used. This would appear somewhat slimy to a European.

In areas where meat supplies are scarce, sun-dried fish (*kajaik*), is cooked with onion and dried minced okra; the stew is consumed by peasants in rain-fed areas in Sudan. The faba bean dish, *foul masri*, is a very popular breakfast and supper dish among urban communities. To well-boiled beans, are added condiments, such as fennel, together with oil and sometimes boiled eggs. The dish is easily prepared, and forms a convenient snack food to take in working areas as a late breakfast; it can also be bought ready prepared from local shops. Among rural communities, spiced fresh or sour milk is boiled with dried minced okra to make other types of stew.

Salads are very important in the Sudanese diet and a salad containing tomatoes, cucumber, and onion is served with the main dish. Other salad vegetables include rocket, lettuce, carrots, beets, and to a lesser extent, egg-plant, used in yoghurt-based salads.

Soups of meat and vegetables are regarded as a delicacy, being consumed only by a very narrow sector of urban communities.

Descriptive survey of important traditional foods and beverages

One extensive study of traditional foods and beverages has been reported recently by Dirar (1987), but a thorough scientific study of the production of various traditional foods is still lacking. In the following sections an attempt is made to define some of the traditional foods, describe their methods of preparation and of cooking and mention their specific occasion of use. All of these are home-made products and food preparation is the domain of women.

Kissra This product consists of an unleavened fermented batter made from sorghum flour and water, cooked into thin sheets (Badi *et al*. 1987). The batter which has been fermented naturally overnight is spread on to a hot, oiled plate with a wooden spatula and sheets of *kissra* are removed within 30–40 seconds. *Kissra* and *assida* are eaten with meat/vegetable stew.

There is a great demand for *kissra* in towns and the production process has been modernized in an attempt to make more economical use of fuel and to obtain more controlled fermentations.

Assida This is a porridge-like preparation made from fermented or fresh batters of sorghum or millet flour. The dough is stirred continuously in a container of hot water until the consistency of the product becomes too difficult to work. This is a popular food and is sometimes referred to as *kissra*, especially in rural communities in the north.

Gurrassa Gurrassa is made from wholemeal or decorticated wheat flour. A fermented or fresh batter is spread thickly over hot plates and the *gurrassa* is cooked in the same way as *kissra*. The product is very popular in northern Sudan, along the banks of the Nile.

Stew and sauce preparations Stew is the second important component of a traditional dish. Beef, poultry or fish, could be the basic ingredient, together with vegetables and oil. Typically, the stew is prepared by frying the onion in oil and adding the selected vegetable. Usually, a dish takes the name of the vegetable used. An array of spices is used to flavour the stew, including red and black pepper, coriander, garlic, and fennel. Okra, especially in the dry powder form (known locally as *weika*), is very popular in traditional dishes, its viscous nature being a very acceptable characteristic of Sudanese traditional foods. The use of *weika* is also very common among west African communities.

Sun-dried strips of meat (*shermute*), from beef or game animals, is very common in both rural and urban centres and, again, okra is an important ingredient of the dish.

Some traditionally fermented Nilotic fish products are important ingredients of the Sudanese diet, especially for communities living near the Nile. *Faseikh* and *terkeen* or *meloha* are fish products preserved in brine. Kosti, a town some 300 km south of Khartoum and situated on the bank of the White Nile, is a major processing centre for *faseikh*.

Sour milk stew, with or without meat, is prepared from fermented milk (*robe*), to which okra, spices, and onion are added. The product is usually eaten with *assida*. Fresh milk stew is very popular among nomadic and semi-nomadic groups. A stew of onion, okra, and spices is a dish common among the poor who cannot afford to buy meat.

Other vegetable substitutes are very common in traditional stews. These include various peas and beans such as lentils and cowpeas.

Dirar (1984) described an unexplored meat substitute of a fermented leafy plant, *kawal* (*Cassia obtusifolia*). *Kawal* is used extensively by the Fur tribe of Darfur Region, but the product is also used as a condiment in urban cooking.

Milk and dairy products Milk production and the processing of dairy products are still largely a cottage-based industry, practised by nomadic and semi-nomadic communities.

Sour milk (*robe*), is produced traditionally by an overnight spontaneous fermentation of milk. The product is churned in a sheepskin to extract butter for ghee processing and a sour buttermilk is left.

A special fermented camel's milk (*gariss*) is very common among camel herders of Northern Kordofan and eastern Sudan. The product is obtained by a continuous fermentation of stirred camel's milk and the product is somewhat similar to *koumiss* and *keffir* milks. It is claimed to have medicinal value in curing visceral leishmaniasis and the drinker of *gariss* is thought to tolerate water shortages.

Cheesemaking in the Sudan has been surveyed recently by Ali and Robinson (1982). The main type of cheese is a white cheese, produced by enzymatic curdling of naturally soured milk. The cheese is kept in brine prior to use and is known colloquially as *gibna baida*. This brine white cheese is a very popular snack food and is also used in dressings with faba beans. There are some other types of cheese such as cream cheese and *madaffara* but they are only produced in negligible quantities. The latter is of Syrian origin and textured by stretching the heated coagulum into flakes.

Non-alcoholic beverages Various types of beverages are made traditionally from wheat, sorghum and milk. Most of these products are prepared during ceremonies and for special festivities or occasions. During the postnatal period, mothers are given special foods including *nasha*, a thin porridge of sorghum or wheat, sometimes fortified with proteinaceous fenugreek seeds, or flavoured with spices such as ginger and cinnamon. The product is claimed to have fattening qualities and to increase the amount of milk secreted. It is also very popular during Ramadan for the evening breakfast, as a warm soup equivalent.

Hilu mur is a brownish spicy extract of fermented, malted, baked sorghum batter. The solid *hilu mur* is prepared from equal amounts of milled and malted sorghum lots. Some fruits and spices such as dates, *aradeib* (*Tamarindus indica*), *karkadé* (*Hibiscus sabdariffa*), cinnamon, and black pepper are used as additives (Agab 1987). The drink is prepared by soaking thin sheets of *hilu mur* in water to produce an extract which is usually sweetened with sugar. Some improvements to the quality and means of processing are suggested by Bureng (1980).

Another beverage food is *abré*. This is a sour whitish broth of fermented, dehulled sorghum flour. This product and *hilu mur* are served during Ramadan.

A reddish filtrate of *karkadé* is another popular drink which is now competing with bottled or canned soft drinks. Various other squash and juice products are made from domestic fruits, such as that of the tamarind. Diluted sour milk is used to quench thirst. In western Sudan, the drink is sometimes mixed with coarsely milled middlings of millet.

TABLE 8.1. Average nutrient intake per capita in different villages

	Food energy (cal)	Proteins (g)	Ca (mg)	Fe (mg)	Vitamin A (I.U.)	Thiamine (mg)	Riboflavin (mg)	Niacin (mg)	Vitamin C (mg)
Omara									
Intake per head per day	1539	39	388	10.6	3340	0.79	0.65	9.6	44
Requirement per head per day	2038	60	553	12.7	3572	0.86	1.28	12.6	27
Intake inadequacy	76%	65%	70%	83%	94%	92%	51%	77%	162%
Gotia									
Intake per head per day	1789	44.3	368	11.4	3790	3.68	0.69	8.2	26
Requirement per head per day	2030	58.0	491	13.7	3607	1.03	1.21	12.0	27
Intake inadequacy	67%	76%	74%	83%	132%	357%	57%	68%	96%
Sangaat									
Intake per head per day	1666	44.4	313	13.0	3364	0.83	0.47	10.0	64
Requirement per head per day	1981	57.0	519	12.7	3448	0.92	1.26	11.7	24
Intake inadequacy	84%	78%	65%	102%	98%	90%	37%	85%	267%
El Dighenat									
Intake per head per day	1554	47.8	388	11.9	1961	0.82	0.39	9.8	47
Requirement per head per day	1915	56.0	557	15.0	3471	0.80	1.93	11.6	25
Intake inadequacy	81%	85%	68%	73%	57%	103%	20%	84%	188%
El Gobba									
Intake per head per day	1631	36.1	334	11.0	2029	0.87	0.48	9.6	53
Requirement per head per day	1924	56.8	531	12.9	3461	0.93	1.26	11.7	25
Intake inadequacy	85%	64%	72%	85%	59%	84%	38%	82%	212%

Suliman Ali									
Intake per head per day	1449	34.8	305	11.1	1226	0.97	0.41	8.2	49
Requirement per head per day	1900	57.0	534	13.2	2464	0.92	1.25	11.2	25
Intake inadequacy	76%	61%	57%	84%	50%	105%	33%	69%	196%
Hasir Extension									
Intake per head per day	3229	123.9	556	24.6	3088	1.94	1.92	19.3	67
Requirement per head per day	2026	57.8	532	15.9	3603	0.95	1.24	15.2	26
Intake inadequacy	159%	214%	105%	155%	86%	204%	155%	127%	258%
El Fitihab									
Intake per head per day	2603	50.0	494	39.5	2891	2.26	1.82	14.0	62
Requirement per head per day	1975	57.0	533	13.0	3534	0.91	1.30	13.6	26
Intake inadequacy	132%	88%	93%	304%	82%	248%	140%	102%	238%
Soba West									
Intake per head per day	3614	100.1	712	28.0	1525	3.07	2.20	20.3	92
Requirement per head per day	2045	56.1	547	14.4	3592	0.99	1.27	12.8	26
Intake inadequacy	177%	178%	130%	194%	42%	310%	159%	159%	354%
Idd Hussain Village									
Intake per head per day	3199	88.0	557	28.9	2353	2.07	1.41	22.0	110
Requirement per head per day	1979	56.4	515	13.0	3481	1.05	1.25	12.3	26
Intake inadequacy	161%	156%	108%	218%	67%	197%	113%	171%	423%

Source: Ministry of Agriculture and Natural Resources (1986*a*).

Alcoholic beverages Despite recent tendencies to enact Islamic laws which prohibit alcohol consumption, domestic brewing is still practised in some parts of the Sudan. These include areas of the Southern Region, the Nuba Mountains, and Southern Blue Nile.

Merissa (a weak alcoholic drink) is a popular food beverage and is even becoming a generic name for other alcoholic beverages made locally. The product is described by Dirar (1978) as a beverage containing a high level of suspended starch and its preparation is technically complicated in comparison with other African beers. Other beverages of cereal and honey syrup, for example *assaleya* (from millet and sorghum), are brewed locally.

Sudanese date wines were investigated by Ali and Dirar (1984). They recognized three different wines made from home fermentations of dates. Two wines, *sherbote* and *nabit*, prepared from date syrup, are produced by relatively quick fermentations taking about two days. A third wine, *dakkai*, is prepared from whole date fruits and takes four days. These immature young wines are consumed normally at weddings and other ceremonies.

A distillate of fermented date and sorghum beverages, known locally as *araki*, is an illegal, (under the counter) alcoholic drink prepared in the Sudan. Many people have resorted to drinking *araki* since alcohol consumption was made illegal in September 1983.

Consumption patterns: deficiencies and costing considerations

The lack of comprehensive nutritional surveys has made it difficult to give an elaborate account of dietary consumption per capita for Sudanese people. However, individual nutritional surveys are being carried out continuously by the Nutrition Division of the Ministry of Health, in an attempt to calculate the average intakes of various food items.

Table 8.1 shows the daily intake of nutrients in different villages, in or around the capital, Khartoum.

The work of Culwick (1951) is an important contribution on the dietary habits and consumption patterns of the people on the irrigated Gezira Scheme. She found that the range of caloric intake was 1500–2750 calories, while protein intake was 58–71 g per person per day. She observed some calcium deficiency and deficiencies of vitamins A, C, and riboflavin.

Malnutrition is a common problem especially among pre-school children in the Sudan (Sukkar 1971; Shazali 1974). The non-availability of food at the family level is not the major factor of malnutrition, since some reports have shown that caloric and protein intakes were adequate in the Sudanese diet. Shazali (1982) indicated that infections, diarrhoeal diseases, ignorance, taboos, and traditional practices of feeding are the real factors behind clinically diagnosed malnutrition, especially among children.

Dietary practices among Sudanese women groups during the periods of

TABLE 8.2. Proportion of household expenditure for food and drink spent on various items, for different income groups in urban Khartoum, 1978/9

Item	Income group (£S per annum)				
	500 %	500+ %	1000+ %	2000+ %	3000+ %
Cereals	7.7	7.9	6.1	4.8	4.8
Cereal products	12.6	11.6	11.9	11.8	10.8
Starches and roots	2.9	2.5	2.1	1.6	1.4
Potatoes	1.8	1.7	1.4	1.2	1.1
Pulses	2.7	3.0	3.6	3.2	2.2
Vegetables	15.8	15.3	14.8	12.4	12.4
Fruits	2.2	2.9	3.4	4.2	4.8
Meat	22.6	22.9	23.7	24.3	25.6
Fish	0.1	0.4	0.5	0.7	0.8
Dairy products	8.1	9.3	9.8	9.7	10.8
Eggs	0.0	0.2	0.4	0.8	1.0
Oils and fats	5.5	5.9	6.0	5.7	4.5
Sugar and sweets	10.2	10.1	8.7	10.4	9.4
Miscellaneous foods	1.7	1.6	1.6	1.4	1.4
Tea, coffee, etc.	7.1	4.1	3.9	3.4	3.6
Non-alcoholic drinks	0.0	0.3	0.4	0.6	0.6
Alcoholic drinks	0.2	0.3	0.5	0.6	0.6
Cigarettes and tobacco	0.5	1.5	2.4	4.2	5.7
Groundnuts and seeds	0.1	0.2	0.2	0.2	0.1
Total	100.0	100.0	100.0	100.0	100.0

Source: Ministry of Finance and Economic Planning (1982).

post-partum and lactation were investigated by Osman (1982), who has emphasized the wisdom behind certain good practices which increased remarkably the consumption of high-protein foods.

Dietary habits and preferences are by no means the only decisive factors in the daily purchase of food items. Factors, such as the food prices might also influence purchasing decisions. Table 8.2 gives an idea of the daily expenditure on various food stuffs by the citizens of Khartoum.

It is worth mentioning that expenditure on meat, vegetables, cereals, sugar, and dairy products consumes a sizeable proportion of the income of many groups. The expenditure on vegetables has been observed to decline with the increase of group income.

It has been impossible within the scope of this chapter to include all Sudanese traditional foods and their place in the daily life of different cultural and ethnic groups. There is also very little information on technical aspects and the real costs of traditional foods, processed at cottage industry level, or prepared in households. The present work concentrated only on common food items, consumed by a wide range of Sudanese groups.

It is worth mentioning, however, that some artful traditions and unique methods of food preparation and food preservation, especially those with meat and cereals, have developed through the ages in the Darfur Region of western Sudan. Such food has received the admiration of all who have had the opportunity to sample it. Such unique cuisine reflects a deep knowledge of recipes and of various novel ingredients, all intended to extend the shelf-life of some perishable products, such as those required by travellers engaged in long journeys.

Acknowledgement

Thanks are due to Professor O. Giha of Gezira University for his constructive criticism during the preparation of the chapter.

References

Agab, M. A. (1987). *Hilu mur* drink from *Sorghum bicolor*. Unpublished paper presented to the UNESCO Regional Training Course on Fermented Foods of the Arab World, 1–15 February, 1987, Shambat, Sudan.

Ali, M. Z. and Dirar H. A. (1984). A microbiological study of Sudanese date wines. *Journal of Food Science*, **49**, 459.

Ali, M. Z. and Robinson, R. K. (1982). Survey on cheesemaking in the Sudan. *Sudan Notes and Records*, **LXIII**, 154–8.

Badi, S. E., Bureng, P. L., and Monawar, L. Y. (1987). *Kissra upgraded technology*. [Pamphlet, obtainable from: Department of Agricultural Extension, Khartoum North, Sudan.]

Bureng, P. L. (1979). A Study on the malting characteristics of sorghum grain and the preparation of the fermented food hullu-murr. Unpublished Ph.D. Thesis. University of Reading.

Culwick, G. M. (1951). *Diet in the Gezira irrigated area, Sudan*. Government Press, Khartoum.

Dirar, H. A. (1978). A microbiological study of Sudanese '*Merissa*' brewing. *Journal of Food Science*, **43**, 1683.

Dirar, H. A. (1984). *Kawal*, meat substitute from fermented *Cassia. Economic Botany*, **38**, 342.

Dirar, H. A. (1987). Fermented foods and beverages of the Sudan. Unpublished paper presented to the UNESCO Regional Training Course on Fermented Foods of the Arab World, 1–15 February 1987, Shambat, Sudan.

Jeffille, D. B. (1968). *Infant nutrition in the subtropics and tropics* (2nd edn). World Health Organization, Geneva.

MANR (Ministry of Agriculture and Natural Resources) (1977). *Statistical year book*. MANR, Khartoum.

MANR (1986*a*). *Agricultural situation and outlook*. Vol. II, No. 2. MANR, Khartoum.

MANR (1986*b*). *Agricultural situation and outlook*. Annual report 1985/86. MANR, Khartoum.

MFEP (Ministry of Finance and Economic Planning) (1982). *Household income and expenditure survey 1978–80*. Department of Statistics, Khartoum.

Modawi, O. (1982). Traditional practices in childbirth in Sudan. In *WHO/EMRO Technical Publication No. 2, Volume 2*. WHO/EMRO, Alexandria.

Osman, A. K. (1982). Dietary practices and aversions during pregnancy and lactation among Sudanese women. In *WHO/EMRO Technical Publication No. 2, Volume 2*. WHO/EMRO, Alexandria.

Reyad, F. (1977). In Proceedings of the 8th Veterinary Conference, Sudan Veterinary Association, Khartoum (In Arabic).

Shazali, H. E. (1974). Unpublished M.D. Thesis. University of Khartoum.

Shazali, H. E. (1982). Nutritional taboos and traditional practices in pregnancy and lactation, including breast feeding. In *WHO/EMRO Technical Publication No. 2, Volume 2*. WHO/EMRO, Alexandria.

Sudanese Meat and Livestock Marketing Corporation (1981). *National herd distribution*. Sudanese Meat and Livestock Marketing Corporation, Khartoum.

Sukkar, M. Y. *et al.* (1971). The nutritional status of children, rural Khartoum. *Sudanese Medical Journal*, **9**, 23.

9

Government and infrastructure

P. R. WOODWARD

System of government

Sudan has had very unstable government since Independence in 1956, having been beset by repeated changes from liberal democracy to military rule and back again; two periods of civil war (the second of which is still not concluded at the time of writing, in December 1989); and a number of major changes to the structure of regional and local government and administration. It was proposed after the overthrow of former President Numeiri in April 1985 that, shortly, there should be a constitutional conference to review the whole future of Sudan's system of government, but until that took place there should be a parliamentary constitution similar to that which was in operation before Numeiri took power in 1969.

Sudan began its third period of parliamentary government in 1986 with fresh hopes (the others were from Independence in 1956 to 1958 and from 1964 to 1969), but they were not to last. The headship of state was held by a 5-man State Council, which was intended to have a non-political and non-executive character (though this non-political formula was not observed in practice in the 1960s). Parliamentary government was broadly on Westminster lines though with a unicameral Constituent Assembly of 301 seats rather than Britain's bicameral parliament. The Assembly itself was composed of representatives of single-member constituencies elected by a 'first past the post' method. In the election of April 1986, 264 seats were contested, while civil war prevented the election from taking place in 37 of the 68 seats in the South. Twenty-eight seats were reserved for the graduate constituencies and these were elected by postal ballot by graduates of Sudan's universities. At the time of the election, however, many of Sudan's graduate voters were working outside Sudan, especially in the Gulf, though many of them still voted. The Constituent Assembly then chose a prime minister who had the support of the majority of the members and he in turn selected government ministers to serve in his cabinet. The Government of the Sudan as a whole was accountable to the Constituent Assembly. In the election of 1986 the Umma Party won the most seats and its leader,

Sadiq al-Mahdi, was elected by the Assembly as prime minister. Since the Umma lacked an overall majority he formed a coalition government with the Democratic Unionist Party and other smaller parties, while the National Islamic Front became the official opposition.

While attractive on paper, this form of liberal democracy, has always had two main problems in the Sudanese context, both of which soon emerged once more. The first is the character of the political parties, for their representation is sectional rather than national and some elements, such as Southerners, feel excluded by the system. The second problem is the balance between the parties, for almost invariably no party wins a majority and weak coalition government results.

Sudan's major parties, the Umma Party and the Democratic Unionists, have deep historical roots and have re-emerged with continuing strength whenever liberal democracy has been restored. The Umma Party is a complex organization which grew from the movement which supported the Mahdi in his nineteenth century revolt against the Turco-Egyptian rulers. It was re-built before Independence and was turned into a political party by Sayed Abd al-Rahman al-Mahdi. The Umma Party has been financed largely from the private irrigation schemes of the Mahdi family and its closest allies along the White Nile but it draws electorally on many spiritual adherents, especially in western Sudan. Through the Ansar sect, these supporters are capable of mounting violent demonstrations in the capital and in 1976 were responsible for an attempted coup.

Traditionally, the Unionists have been associated with a rival Islamic sect, the Khatmiyya, which has most support in the north and east of the country. In addition, the Unionists have much support among the commercial class who control Sudan's rapacious and deep-rooted trading networks.

The main challenge to the two old parties has come from the National Islamic Front, backed by the Muslim Brothers. It is highly ideological, receiving consequent student support and it is well supplied with funds to help small businessmen. Most of its support is in urban areas, especially Khartoum.

However, none of these essentially sectarian-based parties have made any headway in the Christian and animist southern Sudan, but this region has not produced strong parties of its own either. In the South there is not the breadth of support generated by sectarianism to transcend the deep ethnic divisions and parties that have started have lacked the resources in what is the poorest area of the country. No adequate regional representation has been developed and, in consequence, the Southern Region has felt excluded from, and often exploited by, the political system of the country as a whole. This resulted in a long civil war in the 1960s which ended by agreement at Addis Ababa in 1972. In spite of the overthrow of the former military regime in 1985—the main target of the Sudan Peoples Liberation Army (SPLA) which is supported in the South—there has been no agreed end to the second civil war which started in 1983. The war re-opened for three main reasons, all directly related to Numeiri: his policies towards the building of the Jonglei Canal on the White Nile and towards the oil fields discovered at Bentiu, led to fears of economic exploitation; his interference in re-dividing the Southern Region was seen as politically damaging; and the introduction of *sharia*, Islamic law, was seen as discriminating against the non-Muslim south. Many hoped that Numeiri's departure would soon lead to peace, but the SPLA was suspicious of the interim arrangements when, after the elections, the new Government failed to withdraw the *sharia*. (The SPLA wanted a return to a secular constitution for all, while the northern parties sought a reformed version of *sharia* for the Muslim majority.)

The other long-standing weakness of liberal democracy in Sudan has been the balance of the parties in the Assembly. At every election since Independence (1958, 1965, 1968, and 1986) there has been no clear majority and as a result coalition governments have been formed. These have been unstable, with disagreements between the parties and a great deal of intra-party factionalism as well. In consequence, successive governments did little to contribute to effective development policies. The failure to solve the problems led to continued conflict in the south, failed to halt the economic decay, and resulted in an eventual military coup on 30 June 1989, under President Beshir.

Military rule, however, is unlikely to solve Sudan's problems, even though it may last for substantial periods, as did Abboud's regime from 1958 to 1964 and, particularly, Numeiri's from 1969 to 1985. Sudan is too big and too heterogeneous to be ruled by coercion and, besides, the army itself is often unstable under military rule. Military rule, especially Numeiri's rested on an increasingly personalized series of alliances with different groups in society: different in terms of ideology, class, and regional background. Yet, because these alliances were necessarily selective relations in a heterogeneous and competitive society, they were often divisive. The main preoccupation of the Sudanese Government under the military was to manoeuvre for survival rather than to pursue effective policies, especially with regard to the poor who, by definition, lack any constituency in the non-accountable ruling circle. As a result, when one group or another failed to deliver development, as the soldiers purport to wish, it was replaced by another, thus cultivating more enemies. Numeiri ran the whole gamut of recruiting and discarding every significant political grouping until, finally, in the midst of a collapsed economy, they all demonstrated their opposition collectively on the streets and brought his downfall, just as they had brought the downfall of Abboud in the 'October Revolution' of 1964. Since President Beshir seized power, he has sought the co-operation of sections of the Muslim Brotherhood and has exacerbated relations with the SPLA (in spite of attempted negotiations), whilst achieving little on the economic front.

Administration

Just as central government has changed radically several times since Independence, so the structure of administration has altered. The imperial legacy was that of a lightly staffed structure based on central control of provincial administration and responsibility for law and order but only for very limited services. In the twenty years following Independence there was a ten-fold increase, to 120 000, in the number employed in the bureaucracy and modernization required a sweeping extension of social services. The needs of successive governments to try and establish a firm grip on power resulted in the administration becoming highly centralized. Later, in an effort to reduce centralization, regional government was fostered, first for Southern Sudan, in 1972, and then for the whole country in 1980. The Southern Regional Government was part of the peace settlement of the long civil war in the south, but after 1980, when regional government was introduced throughout the north, the Southern Region was divided into three. The Regional Governments are thus Northern, Eastern, Central, Darfur, Kordofan, Bahr el Ghazal, Upper Nile, and Equatoria, although the last three, all in the south, share an umbrella tier in an effort to compromise between those for and those against the re-division of the old Southern Region. Under Numeiri, regional governors were partly elected and partly selected. Following his overthrow, regional councils were disbanded and central government appointed regional governors pending consideration of the whole future of regional government as part of the planned review of the constitutional and governmental system.

Any future reform will need to tackle some of the obvious weaknesses of the regional government experiments hitherto. In practice, there was insufficient effective devolution of power from Khartoum to the regions, especially on financial matters. Lack of an adequate local tax base left regional government heavily financially dependent on the central government and, as Sudan ran ever deeper into economic crisis in the 1980s so the shortage of money became a growing impediment. Regional governors and senior officials had to spend an increasing amount of their time in the capital trying to acquire finance to pay salaries, let alone for development. In any case development planning at the regional level was handicapped by the inadequate character of national planning, which failed to provide a framework for regional planning and, at the same time, obstructed regional autonomy in planning. Frustrated at these shortcomings, some regional governors, notably those of Darfur and Kordofan, pursued their own attempts at attracting resources from the oil-rich Gulf States. Planning at both the regional and national levels was also constrained by the failure to integrate departments. At the national level, planning powers were shuffled around continually within the Khartoum bureaucracy, sometimes as a separate ministry or department and, increasingly, as economic crisis loomed, coming under the control of the Ministry of Finance. At a regional level too it was difficult to undertake integrated planning. In the 1970s, although more powers were, in theory, devolved to the provinces, the departments involved remained hierarchical and centralized and, in consequence, there was little effective interdepartmental co-operation at the regional level.

Local administration has also varied from time to time. The British-run colonial state had sought to utilize tribal structures for native administration, but with the approach of independence there were moves to encourage more democratic local government. In many rural areas, however, native administration survived effectively. Under Numeiri there was a great effort to wipe out what had become essentially local autocracy and to replace it with a pyramidal structure of councils from local councils, of which there were 4000, through rural or town to district councils which, in turn, were supervised by the lynch pin of the province council (roughly half each of the present regions). Here, government-appointed provincial commissioners were to mesh with the people's representatives to promote social change and development and, indeed, the number of employees of local government and the volume of services did expand significantly. However, the character of local administration was not changed in the direction intended. The intended meshing of more dynamic and progressive administrators with local representatives freed of the constraints of local autocracy was not a great success. The administrators still tended to dominate in relations with the local representatives rather than regarding them as partners in a new experiment; thus in so far as there was a 'May Revolution' (as proclaimed by Numeiri's regime) it remained largely a revolution from above. At the bottom, although the remnants of native administration were supposed to have been cleared out, their influence in the rural area was still strong and there were often back doors, such as the involvement of relatives and friends in the newly-formed, single party, the Sudan Socialist Union, which ensured that they could retain an involvement in local affairs.

It was largely because of these shortcomings of the reform of local government in the 1970s that, even before Numeiri fell, the poorer regions of northern Sudan, Darfur, Kordofan, and eastern Sudan, all began to consider some form of reversion to native administration, especially for the nomadic and semi-nomadic peoples. This action was encouraged during the famine of 1984–5 by the realization that government at all levels had very limited knowledge of, let alone control over, much of what was happening in the countryside. Thus, following the introduction of regional government in 1980, the Area Councils, which replaced the former province councils, were permitted to set up nomadic chiefs (*muawin idari*) under the direct supervision of the Area Council offices. These were charged with a number of responsibilities with regard to law and order, both within tribes and in relation to other tribes, and to the environment. Some critics regarded this as a retrograde step, but for others it was a necessary recognition of the realities in large areas of the countryside and the failure of the reforms of the 1970s.

At all levels of government there have been common problems of bureaucracy arising not only from the multiplicity of councils mentioned, but also from the more general shortcomings of bureaucracy in Sudan, which are similar to those elsewhere in the Third World. Seniority has been more important than ability to attain promotion; training for junior staff has been poor; remuneration for middle-grades has been inadequate to compete with the private sector (and in the 1970s the lure of the oil-rich Gulf); and favouritism and corruption became ever more rife under Numeiri. Following his downfall there were initially few major attempts at a shakeup, though after President Beshir's coup in 1989 there were reports of numerous dismissals.

In addition to the bureaucracy, Sudan developed a large number of parastatal organizations. The giant Gezira Corporation, the largest agricultural scheme under single management in the world, or so Sudan liked to boast, was followed by many other schemes, especially from 1969 to 1971 when there was a wave of nationalization. By the late 1970s, such schemes supplied work for a further 130 000 people on the public payroll, but they suffered from the same kinds of shortcomings as the bureaucracy. In the last years of Numeiri there were attempts to sell off some schemes, if necessary to foreign buyers, and to incorporate expatriate managers. This was rarely an attractive proposition and, in consequence, there is still a substantial parastatal sector.

Transport infrastructure

One of the largest of the parastatals has been Sudan Railways, for many years the backbone of the country's transport system. Initially it was designed primarily to reinforce security and it consisted of a single track railway linking the Central Region with Egypt and Port Sudan, with a loop to Kassala in the east and a branch to El Obeid in Kordofan. After Independence, it was extended to Nyala in Darfur in the far west and to Wau in the south. Sudan Railways were also responsible for the main river network which went round the great bend of the Nile in the north, from Abu Hamed to Dongola and up the Nile from Kosti to reach Malakal and on through the Sudd to Juba. This network of rail and boat was at best exiguous and since Independence has proved inadequate. As the volume of trade and travel has grown the single track railway has proved to be of inadequate capacity, while the rails themselves have poor foundations and the system regularly suffers from washouts in the rainy season. The rail network suffered its final humiliation in 1985 when relief supplies to the starving in Kordofan and Darfur were delayed by rail failures west of Kosti. The river services have been hampered by inefficiencies similar to those of the railways and, since 1983, the civil war has usually prevented boats reaching Malakal and especially Juba.

The national airline, Sudan Airways, is based in Khartoum and operates domestic and international passenger and freight services. It maintains its own small fleet of Boeing 707s and 727s as well as some smaller aircraft mostly Fokker. Internationally it flies mainly to the Middle East (Egypt and the Arab Gulf) and to London and East Africa. Domestically its services are erratic due not only to problems of fuelling and servicing aircraft, but also to the problems of maintaining air strips and security in southern Sudan—a Sudan Airways flight was shot down by guerrillas in August 1986.

During Numeiri's period in power there were great extensions to the road network. Previously there had been only one asphalt road of 200 km, but by 1982 this had been extended to nearly 2200 km, linking the centre of the country to Port Sudan. In addition, there are about 16 000 km of graded tracks in the north, but these are very rough, are often difficult in the rainy season, and take a heavy toll on vehicles. In the south there are 2–3000 km of all-weather gravel tracks but the renewed civil war will have damaged many of these. The growth of road transport is a world phenomenon but in Sudan such growth also owes much to the limitations of the rail network and the cost and erratic nature of the main domestic airline. In consequence, people and goods took to the roads and the petty bourgeoisie soon realized that it was a very profitable investment to buy trucks and buses; for many it was a better prospect than mechanized agriculture. The power of the lorry owners' cartel was evident during the famine of 1985 when Bandaid, forced by the voracious demands of the existing owners, had to buy its own fleet of trucks. Although this large truck and bus business suffers periodically from shortages of spare parts and fuel, it was essentially trucks, and not the railway or later the planes, which carried the bulk of famine relief in both east and west.

Services

The semi-collapse of Sudan Railways in particular, and the rise of road transport, are indicative of the way in which the state and parastatal organizations have been unable to meet peoples' needs and deliver the services for which they hope. In spite of the growth of expenditure and personnel by the state, the provision of services is very uneven with the lion's share going to the towns, especially those in the centre of the country. Thus, opportunities for education and some basic health services are much greater in the capital and in the Central Region than they are elsewhere. The towns in more remote areas, such as the Southern Regional capitals of Wau, Juba, and Malakal, and also Nyala and El Fasher in Darfur have suffered severely as a result of their failure to provide services which people came to the urban areas to seek out. Such remote centres were often inundated with the Sudan's own citizens seeking services and, in the south in the 1980s, with citizens seeking sanctuary from civil war. In addition, hundreds of thousands of refugees from conflicts in Ethiopia, Uganda, and Chad in particular, flocked towards towns in Sudan.

In the countryside the situation was worse, especially in remote areas. Services were limited to a few poorly staffed and equipped schools and clinics and, apart from this, little was seen of government other than intermittent attempts to collect animal and market taxes; in this area government's ability declined notably after the ending of native administration. By the 1980s it was becoming clear that even law and order was in sharp decline and not just in the south where civil war was re-opening. The repeal of native administration limited the government's ability to control the countryside, while rising population and animal numbers, as well as growing desertification, led to increased pressure on and growing competition for natural resources. In consequence, there was growing anarchy and violence in the remote rural areas, fuelled by the increased availability of automatic and semi-automatic weapons, partly as a result of conflicts which spilled over into Sudan from neighbouring states.

This picture of declining actual services and growing chaos in the countryside was an expression of the political and economic weakness of the Government of the Sudan as a whole by the 1980s. This weakness was manifestly demonstrated not only by the famine of 1984–5, but by the reluctance and then inability of the Government to respond to it. Just as Numeiri's Government was forced, reluctantly, to turn to the outside world in its hour of need so the outside world will also have a major part to play in the country's present re-building. For many years, foreign relations have been a vital dimension of government, influencing internal affairs significantly.

External relations

After Independence, Sudan had flexed her international wings on occasions, especially under the premiership of Mohammed Ahmed Mahgoub, who hosted the post Six-Days' War gathering of Arab leaders in 1967. For the most part, however, the structure of politics under the party-based liberal–democratic system, and the lack of economic reform which maintained pre-independence economic ties with the West, meant that Sudan did not need a very active foreign policy.

Under Numeiri this situation changed dramatically, largely because his government became increasingly narrowly based and he needed outside friends to prop up his regime. In particular, he cultivated the West and especially the USA, with the aid of a vitriolic anti-communist policy, criticizing the Soviet Union and allowing relations with Marxist Ethiopia to deteriorate following the 1974 revolution. Good relations with the USA helped Numeiri in his dealings with the International Monetary Fund and the World Bank on which he relied increasingly as his economic problems grew. Regionally, Sudan developed closer relations with the oil-rich Arab Gulf States and in the 1970s there were high hopes that Arab investment could transform Sudan into the 'bread basket of the Arab world'. With the reality of this economic dream Sudanese society would be transformed and the Arab states, a number of which imported food, would be in little danger of having to pay for raising oil prices by constraints on food. When it became clear that this strategy for Sudan had largely failed, relations with the Arab states weakened, especially when Numeiri supported Egypt on the Camp David Agreement with Israel. During the 1970s, Egypt became Sudan's closest ally and from 1974 there were even plans for the long-term integration of the two countries. Egypt sought as always for the security and the development of the Nile, while Numeiri's regime sought political support. Egyptian assistance for the defence of the regime grew, though she declined to become involved when faced with renewed civil war in southern Sudan in the 1980s. By cultivating the West and Egypt, Numeiri became the enemy of Ethiopia and Libya; he opposed Libyan activity bitterly not only in Sudan but Chad, which became a haven for Qadaffi's opponents.

Following Numeiri's downfall the Transitional Military Council and the newly elected prime minister, Sadiq al-Mahdi, embarked on a more balanced policy. Relations with Libya re-opened, there were attempts at talks with Ethiopia, union with Egypt was suspended, and new relations were sought with both super-powers. In reality, policy was not just about balance, it was about crisis management with civil war and a collapsed economy. The improvement in relations with Libya was designed, successfully, to remove Libyan support for the SPLA fighting in the South; contact with Ethiopia did not evoke a similar response, and she was blamed increasingly for the continuation of the conflict. The Arab states were wooed for more urgently needed financial support and that was one factor in suspending the union with Egypt, the partner in Camp David. Although Sudan's contact with the USSR served notice on the USA that, unlike the days of Numeiri, Sudan was no longer a lap-dog (though in reality the tail had often wagged the dog), the realization that the USA was the key to much greater economic aid than the USSR meant that relations had to be maintained with the West first for famine relief and then for economic reconstruction. For their part, Western governments are reluctant to give much while the war in the south remains unresolved.

If Sudan is to develop, there is a need to re-build her system of government to overcome the past shortcomings of both liberal democracy and military rule. There is also a need to develop Sudan's governmental capacity by establishing a more stable form of administration. The system could re-emphasize regionalism and there are those in the South and the West who would welcome some form of federalism to break down past dominance by the centre. This would involve sweeping change in the constitution and bureaucracy, change which many feel is necessary in order to build a stable future for Sudan, with improved services for its people. Sudan will have to achieve much of this for herself, but in the present economic situation she cannot succeed without international support. If that fails and if the Sudanese state continues to decay and perhaps collapse, the consequences will affect not only Sudan as a country, but also a strategically important area of international politics.

Further reading

Nelson H. D. (ed.) (1982). *Sudan: a country study*. Area Landbook Series. US Government, Washington.

Muddathir Abd Al-Rahim, Badal, R., Hardallo, A., and Woodward, P. R. (ed.) (1987). *Sudan since independence: studies of the political development since 1956*. Gower, Aldershot.

Woodward, P. R. (ed.) (1990). *Sudan since Nimeiri*. Routledge, London.

Woodward, P. R. (1990). *Sudan 1898–1989: the unstable state*. Lynn Rienner Boulder and Lester Crook Academic, London.

10

Agriculture in the national economy

C. G. GURDON

Introduction

Agriculture plays a pivotal role in the economy, providing a livelihood for over 80 per cent of the population. Its share of the labour force is still over 60 per cent and its share of total exports is over 95 per cent. The country's efforts to achieve economic progress and transformation are dependent on the performance of the agricultural sector and Sudanese industry is also almost entirely dependent on agriculture as a source of raw materials.

The position of agriculture in the national economy is considered in terms of its share of gross domestic product (GDP), foreign and domestic trade and employment, and its contribution to the industrial sector. Agriculture's use of scarce resources such as investment capital, foreign aid, power and water are also discussed, and the latest available statistics are presented.

Share of gross domestic product

The predominance of the agricultural sector in the Sudanese national economy is illustrated by its share of GDP. Until 1962 it contributed over half of GDP, since when its share has fallen gradually to about 33 per cent. This fall, however, is related directly to the growth of the service sector, from 28.5 to 59 per cent during the same period (UNIDO 1983, p. 5). The service sector includes government spending and its increase in size demonstrates Sudan's 'pseudo development'—having a large service sector at a low level of development. Perhaps a more accurate reflection of the importance of the agricultural sector in Sudan's current GDP is a comparison with other non-service sectors. In 1987/8 agriculture at factor cost accounted for £S1935m. or 32.5 per cent of GDP which totalled £S5958m. at constant 1981/2 prices. This compared with only £S465m. for manufacturing and mining, £S637m. for transport and communications, and £S700m. for all government services (Bank of Sudan 1989). The contribution made

TABLE 10.1. Gross domestic product at factor cost at 1981/2 prices (£Sm.)

Sector	1983/4	1984/5	1985/6	1986/7	1987/8
Agriculture	2159	1651	2153	2196	1935
Commerce	887	805	843	915	869
Manufacturing and mining	402	405	443	461	465
Transport and communication	664	615	617	666	637
Construction	386	339	312	315	332
Electricity and water	102	117	126	127	129
Government services	558	565	692	651	700
Other services[a]	927	809	807	867	891
GDP	6085	5306	5993	6198	5958
Annual change in GDP	—	−12.8	12.9	3.4	−3.9
Price index (1981/2 = 100)	171.2	243.1	345.2	483.0	627.9

[a] GDP at factor cost, constant prices.
Source: Bank of Sudan (1989, p. 7).

by the agriculture sector to GDP stands head and shoulders above all other parts of the economy (Table 10.1). It can be seen from Table 10.1 that the fall in agricultural production, which was caused primarily by the prolonged 1984/5 drought, affected the whole economy and reduced national GDP.

When analysing agriculture's share of GDP, it should be remembered that this sector makes a major contribution to other sectors of the economy. Most of Sudanese industry is involved in processing agricultural primary products and agricultural products also make up the majority of goods that are transported around the country. Agriculture's share of government services, water, power, and other elements of GDP also need to be taken into consideration in any analysis of the sector's importance.

Foreign and domestic trade

For Sudan, the importance of agricultural exports (Tables 10.2 and 10.3) cannot be over-estimated. Although the exact proportion varies from year to year, depending on production levels and world market prices, agriculture has consistently contributed over 80 per cent of Sudan's exports during the 1980s.

The most striking point in Table 10.2 is the major fluctuations in the level of exports from year to year. Cotton exports, which accounted for about half of total exports throughout the 1970s, collapsed in the early 1980s but have since largely recovered. The effect of the 1984/5 drought is obvious from the decline in the level of *dura* (sorghum) exports. Similarly, the increase in livestock exports can be attributed to the drought. The lack

TABLE 10.2. Sudan exports by commodity

Commodity	Value (£Sm.)						Percentage of total exports					
	81–2	82–3	83–4	84–5	85–6	86–7	81–2	82–3	83–4	84–5	85–6	86–7
Cotton	60.1	201.7	447.5	417.8	337.1	473.4	17.4	33.2	50.1	49.3	42.8	42.1
Gum arabic	37.0	54.7	83.9	77.8	67.1	250.0	10.7	9.0	9.4	9.2	8.5	22.2
Groundnuts	42.1	18.5	21.5	21.7	12.2	6.1	12.1	3.1	2.4	2.6	1.6	0.5
Sesame	36.9	59.7	87.9	80.8	90.4	76.4	10.7	9.8	9.8	9.5	11.5	6.8
Livestock	43.5	82.6	91.2	124.3	124.2	19.8	12.6	13.6	10.2	14.7	15.8	1.8
Dura	57.0	98.5	41.1	—	1.4	100.6	16.5	16.3	4.6	—	0.2	8.9
Hides and skins	6.4	12.0	16.8	29.2	32.5	33.3	1.8	2.0	1.9	3.4	4.1	3.0
Castor seeds	1.0	—	0.7	—	—	—	0.3	—	0.1	—	—	—
Karkadé	2.5	3.7	4.6	4.3	7.6	23.8	0.7	0.6	0.5	0.5	1.0	2.1
Cake and meal	30.0	19.0	22.4	17.7	8.5	24.7	8.7	3.1	2.5	2.1	1.1	2.2
Groundnut oil	8.7	6.2	19.3	5.4	—	—	2.5	1.0	2.2	0.6	—	—
Others	20.8	50.2	56.6	68.3	105.7	117.2	6.0	8.3	6.3	8.1	13.4	10.4
Total	346.0	606.8	893.5	847.3	786.7	1125.3	100.0	100.0	100.0	100.0	100.0	100.0

Source: Bank of Sudan (1988, p. 32).

TABLE 10.3. Sudan exports, 1986–8

Commodity	1986		1987		1988	
	Quantity	Value (£Sm.)	Quantity	Value (£Sm.)	Quantity	Value (£Sm.)
Cotton (480–lb bales)	936907	366.7	738788	455.2	816197	978.4
Groundnuts (t)	1090	2.5	7283	10.1	69080	86.5
Sesame (t)	29051	58.9	60587	134.8	98874	269.0
Gum arabic (t)	19141	141.7	18587	267.1	18563	281.6
Dura (t)	30490	13.9	534249	248.8	237137	106.7
Livestock[a] (head)	244315	71.5	71171	42.9	205299	128.0
Hides and skins (t)	7743	33.7	5305	39.0	7776	69.5
Cake and meal (t)	41521	14.2	102176	50.5	155165	103.0
Others (value)	—	130.1	—	248.7	—	268.2
Total	—	833.2	—	1497.1	—	2290.9

[a] Excludes exports of camels to the Arab republic of Egypt.
Source: Bank of Sudan (1989).

of fodder and water forced herdsmen to sell their livestock on both the domestic market and to exporters. Following the drought, the export of livestock was prohibited in order to replenish Sudan's herd and stabilize the price of livestock and meat on the domestic market (Economist Intelligence Unit 1986*a*). Because of these major fluctuations in the level of exports, it is important to look at trade statistics for a number of years in order to gain an accurate impression of Sudan's agricultural trade.

Sudan also has significant agricultural imports (Table 10.4). These fall into two major categories—food, of which the major imports are wheat and wheat flour, tea, coffee, and sugar, and agricultural inputs such as farm machinery and pesticides.

Unfortunately, there has been a steady growth in the level of wheat and wheat flour imports. This is almost entirely the result of increasing urban demand for white bread, as opposed to Sudan's traditional brown bread made from *dura*. The import of vast quantities of food aid in the form of wheat and wheat flour during the 1984/5 drought has increased demand and so the Sudanese Government contemplated increasing the price of white bread to finance the continuing price subsidization of brown bread (Economist Intelligence Unit 1986*b*, p. 16). The Government has proved very cautious in implementing this plan because it was food price rises in the urban areas that sparked the revolution which overthrew ex-President Numeiri.

Another noticeable feature of Sudan's food imports was the continuing decline in the level of sugar imports. In 1980, when world sugar prices were at their highest, almost 14.8 per cent of Sudan's total import bill went on

TABLE 10.4. Sudan: imports, 1986–8

Commodity	Value (£Sm.)		
	1986	1987	1988
Petroleum products	292.3	497.9	1093.1
Manufactured goods	481.4	501.0	930.1
Machinery and equipment	405.7	484.9	776.6
Transport equipment	434.1	368.9	509.3
Chemicals	341.2	248.1	475.8
Wheat and flour	120.5	199.6	649.1
Tea	71.9	39.8	68.6
Coffee	12.2	4.0	24.7
Sugar	—	52.7	10.3
Other food	155.5	117.5	172.4
Drinks and tobacco	14.3	13.6	56.9
Textiles	71.1	84.9	125.9
Total	2400.2	2612.9	4892.8

Source: Bank of Sudan (1989).

sugar (Gurdon 1986, p. 22) and the country was importing almost 301 234 tonnes at a cost of almost £S117.6m. This fell dramatically to a point where only 6118 tonnes of sugar were imported at a cost of £S10 280 in 1988 (Bank of Sudan 1989). There are two reasons for this fall—increasing domestic production and very low world sugar prices. The controversial giant Kenana Sugar Scheme is now producing almost 330 000 tonnes, making Sudan almost self-sufficient in sugar. At the same time the four government-owned sugar plants are being rehabilitated. This should soon enable Sudan to export substantial quantities of sugar, if and when world sugar prices recover from their current historic low (Economist Intelligence Unit 1986*b*, pp. 19–20).

The civil war in southern Sudan has halted domestic production of tea and coffee which are grown in the tropical highlands of the Nile–Congo watershed in Western Equatoria Province. Although the area was not initially affected by the war, transport of tea and coffee to northern Sudan is exceptionally difficult. Consequently, it has been necessary to increase the level of imports in order to satisfy urban demand in northern Sudan.

Agricultural machinery and other inputs constitute the second major category of agricultural imports. However, in recent years because of Sudan's perilous economic situation, agricultural imports have been financed largely by foreign donors. In 1988, for example, Sudan imported 7465 tonnes of insecticide (£S176m.), 95 939 tonnes of fertilizer (£S100m.), 48.4 million jute sacks (£S97m.), and 914 tractors (£S62m.). In addition, a significant proportion of the £S171m. of imported machinery spare parts was for the agricultural sector (Bank of Sudan 1989). The sector also took a major part of the imported petroleum products, trucks, railway wagons, and other goods.

While the role of agriculture in foreign trade is very significant, its role in domestic trade should not be forgotten. In this area, statistics are very scarce but some indication of the scale of domestic trade can be gauged from figures for 1979/80, when a total of 790 666 sheep, cattle, goats, and camels were sold for a total value of £S46.9m. In the same year 451 296 tonnes of *dura*, groundnuts, and other crops were sold for a total value of £S27m. (MFEP 1983, pp. 173–4). As the majority of Sudan's population are subsistence farmers, these statistics represent only the excess crops and livestock sold in the urban markets throughout the country. It is clear, however, that the domestic trade in agricultural products is very substantial.

Employment

The agricultural sector remains overwhelmingly dominant in the Sudanese labour force, but its proportion is gradually decreasing. Table 10.5 illustrates that in 1979/80 agriculture employed 3 432 600 people or 65.8 per cent of the economically active population (UNIDO 1983, p. 7). This was more than five times as many as the service sector and almost nineteen times as many as the industrial sector. Agriculture's share of the labour force will undoubtedly decrease as government bureaucracy and the service sector expand, but there is no doubt that agriculture will remain the dominant sector for employment in the coming decades. In 1976/7, the total agricultural labour force of 3 435 275 was divided between 2 538 017 men (73.9 per cent), and 897 263 women (26.1 per cent). This meant that agriculture employed 63.04 per cent of the men and 87.92 per cent of the women in the labour force (MNP 1977).

Once again it should be noted that a large proportion of the personnel employed in other sectors are indirectly employed in the agricultural sector. For example, many of the 183 000 Sudanese who work in the industrial sector are engaged in processing agricultural crops. These include those in sugar production, fruit and vegetable canning, textiles, tanning, grain-milling, and the oil and soap industries. It is obvious, therefore, that the agricultural sector is by far the largest employer in the country.

Agricultural industries

As already noted, a significant proportion of Sudan's industrial production involves the processing of its agricultural crops. As seen in Table 10.1, manufacturing and the very small mining sector accounted for almost ten per cent of GDP, compared with only about one per cent of GDP at

TABLE 10.5. Distribution of economically active persons by sector 1976–7 to 1979–80

Sector	1976–7		1979–80		Percentage change 1976–7—1979–80
	Persons ('000)	Percentage share	Persons ('000)	Percentage share[a]	
Agriculture	3435.3	68.5	3432.6	65.8	−2.7
Industry and mining	185.1	4.46	183.3	3.5	−1.8
Utilities	45.6	0.91	59.2	1.1	13.6
Construction	92.3	1.84	107.6	2.1	15.3
Commerce and finance	245.7	4.9	220.8	4.2	−24.9
Transport and communication	169.0	3.37	198.8	3.8	29.8
Services	521.6	10.4	679.8	13.0	58.2
Unallocated	320.5	5.12	340.5	6.5	20.0
Total	5015.0	99.5	5222.6	100.0	

[a] Estimate.
Source: UNIDO (1983, p. 31).

TABLE 10.6. Percentage structural change in manufacturing value added, 1967/8 and 1977/8

Sub-sector of manufacturing	1967/8	1977/8
Food, beverages, and tobacco	32.9	40.1
Textiles and apparel	20.4	11.8
Wood products	4.3	5.0
Paper and paper products	2.2	4.4
Chemicals	11.9	20.1
Non-metallic minerals	4.4	5.0
Basic metal industries	7.6	4.6
Machinery equipment	0.1	8.1
Other	16.2	0.9

Source: UNIDO (1983, p. 14).

Independence in 1956 (UNIDO 1983, p. 7). Until the nationalization of key manufacturing units in 1970 the industrial sector was dominated by private enterprise. Today the public sector plays the leading role although private and foreign investments are encouraged.

As Table 10.6 illustrates, the largest subsectors of industry are food, beverages, and tobacco followed by textiles and clothing (UNIDO 1983, p. 15). Between them they accounted for over half the industrial production in 1977/8, a proportion which has probably increased since that date with the growth of the sugar industry. This has enabled Sudan to become almost self-sufficient in sugar and, when world prices recover, the country could become a sizeable sugar exporter.

Besides sugar factories there are four old Soviet-built canning factories which were built in the wrong locations and are operating at less than half their capacity, although they may be rehabilitated by the Soviet Union. Around the capital there are seven plants producing fodder and poultry feed and seven government-owned grain-milling companies with a total annual production capacity of 477 000 tonnes. In addition, there are twelve private mills, the majority in central and eastern Sudan which is the granary of the country. There are 75 vegetable oil mills with a design capacity of 911 700 tonnes, but they operate at little more than half their full design capacity (UNIDO 1983, pp. 50–3).

Besides food processing there is the soap industry, which uses vegetable oil as its basic ingredient and has 14 modern and 50 traditional factories. There are also three public sector and nineteen private sector tanneries, but few are operating at full capacity (UNIDO 1983, pp. 54–6). The largest non-food agricultural industry is the important textile industry. In 1982/3 it consisted of eight government-owned textile mills and eight privately-owned mills which together should have been able to produce 110 per cent of Sudan's domestic demand for cloth and yarn (Economist Intelligence

Unit 1986*c*). Like the vast majority of Sudan's industrial sector, however, the textile plants operate at only 25–50 per cent of capacity.

There are many reasons for the failure of the Sudanese industrial sector. They include the severe shortages of trained manpower, raw materials, and the foreign exchange needed for importing essential intermediate inputs. Production costs have increased due to a series of devaluations of the Sudanese pound and these, coupled with infrastructural bottlenecks, have crippled the industrial sector. As a result Sudan has been forced to use scarce foreign exchange to import goods that the country has the capacity to produce. One of the Sudanese Government's priorities in recent years has been the rehabilitation of the industrial sector. Despite its failures, agriculture still plays a crucial role in providing the raw materials for Sudan's manufacturing industries. As in the case of most developing countries, it can be expected that food processing will remain the basis of Sudanese industry in future years.

The economic effects of the 'bread basket' strategy

One of the most important aspects of the role of agriculture in the national economy is the legacy of the so-called 'bread basket of the Arab world' strategy. The plan, which called for the rapid expansion of agricultural production in order to create a food surplus which could be exported to the rest of the Arab world, proved much more costly than initially envisaged. The three pillars of the 'bread basket' theory were Sudanese natural and demographic resources, Western technology, and Arab investment. The latter failed to meet expectations for a variety of economic and political reasons, while the cost of Arab oil increased substantially. When Sudan resumed diplomatic relations with Egypt in the light of the Camp David Accord, aid and investment from Arab allies, primarily Saudi Arabia and the Gulf States, fell by almost 90 per cent. Of the 100 projects which the Arab Authority for Agricultural Investment and Development (AAAID) selected for investment as part of a massive £S2287m. programme, only seven were ever started (Gurdon 1986, p. 23).

The result of the fall in Arab investment, as well as the indigenous reasons for the failure of the 'bread basket' strategy, was that the Sudanese Government got into a spiral of foreign debt from which it could not extract itself. The external debt rose substantially in the early 1980s because of a combination of factors. These included the Government having to finance the completion of the 'bread basket' projects, the second oil price explosion, and falling cotton exports which led to an escalating trade deficit. In 1978 the International Monetary Fund (IMF), introduced a rolling three-year programme to rehabilitate the economy but, despite the austerity measures introduced, the foreign debt total is now estimated to be around $US13 billion.

The 'bread basket' strategy remains very controversial. In view of the economic and environmental problems that it created, and the fact that it is linked to disgraced ex-President Numeiri, current opinion is generally hostile to the strategy. Any balanced analysis, however, must realize that exogenous factors were as responsible for the failure of the strategy as were the indigenous problems, mismanagement, and corruption. It should be recognized that Sudan's economic future will remain inextricably linked to the agricultural sector and that this has not been negated by the apparent failure of the 'bread basket' strategy.

Investment and aid for the agricultural sector

Even since the apparent failure of the 'bread basket' strategy which sought to invest very substantial sums of capital in the agricultural sector, this sector has remained the focus for both domestic and foreign investment and aid. The IMF's programme focused on the rehabilitation of existing plant and agricultural schemes, while placing a moratorium on new projects, consolidating and expanding infrastructure, and correcting price distortions that had constrained cotton production and exports. Since its introduction, investment has focused on the agriculture and transport sectors which accounted for over half the total investment between 1983/4 and 1985/6 (Economist Intelligence Unit 1986*c*, p. 15). One of the principal reasons for the investment in transport and infrastructure has been to improve the transportation of agricultural crops. This requirement was emphasized by the inability to transport food aid from Port Sudan to western Sudan during the 1984/5 drought.

As the economic crisis has worsened in recent years, Sudan has become almost entirely dependent on foreign aid and investment. Once again it is agriculture and infrastructure which are receiving the majority of funds. Currently there is a general recognition that the success of the Sudanese economy is largely dependent upon the ability of the agricultural sector to produce a surplus which can be exported. In recent years the emphasis has shifted from the large-scale 'bread basket' type projects to the traditional small-scale farming sector. It is hoped now that production will be boosted by increasing price incentives for small-scale farmers. Consequently, foreign aid donors are turning away from capital intensive projects, such as the £42m. (sterling) British aid-financed thermal power station outside Khartoum, towards small-scale rural agricultural projects. Although they are less tangible symbols of assistance and there are smaller benefits to the donor country, it can be expected that this trend will continue in the coming years.

Despite the substantially increased level of aid to Sudan in the wake of the 1984/5 drought, it must be recognized that, increasing agricultural production and exports alone will not solve Sudan's economic crisis. Like all primary product producers, Sudan will continue to remain a victim of the wildly fluctuating world agricultural prices. For example, while the fall in sugar prices in recent years helped reduce Sudan's sugar import bill, the economics of exporting its future surplus sugar are far less attractive than they were when the Kenana Scheme was first conceived. Like sugar, Sudan's other export crops including cotton, *dura*, groundnuts, gum arabic, and sesame as well as livestock are subject to price fluctuations. At the same time the cost of Sudan's imported manufactured goods tend to increase far more rapidly than its exports thereby weakening its terms of trade.

The crippling debt burden will, however, remain Sudan's overwhelming economic problem in the coming years. It has reached totally impossible proportions whereby the country's annual debt service obligations are more than three times as great as its average total export revenue. Therefore, even if agricultural production and exports doubled it would not cover its debt servicing obligations. In the 1985/6 budget, which was the last comprehensive one, it was planned that over half of total government expenditure would be used for debt servicing, although this target was not met (Economist Intelligence Unit 1986*c*, p. 41).

In the long-term the country's economic crisis can be solved only if Sudan's creditors are prepared to write off some of its debts. While governments and individuals may congratulate themselves for giving more aid to Sudan and other African countries, the fact is that developing countries are paying more debt interest and principal to the West than they are receiving in aid. Sudan's economy will remain crippled by its debt servicing obligations unless the burden is reduced. Only then, and when the civil war ends can its considerable agriculture potential and its newly found oil reserves, contribute successfully to the development of Sudan's economy.

Agriculture's use of scarce resources

Having examined agriculture's direct role in the national economy, it is important to study briefly the sector's use of scarce resources. These resources include water, fuel, electric power, and skilled personnel. Unfortunately, there are very few statistics which provide a breakdown of the use of scarce resources in agriculture, so it is possible only to indicate such use in rather general terms.

Water is obviously one of the most important resources for agriculture. At the time of writing about 1.7×10^6 hectares are under irrigation and it is planned that a further 0.68×10^6 hectares will be irrigated eventually (Whittington and Haynes 1985, p. 132). Currently, Sudan is using between 15 and 16 milliards of water compared with its allocation of 18.5 milliards under the 1959 Nile Waters Agreement with Egypt (Whittington and Haynes 1985, p. 139). Almost all of this is used for irrigation and it is highly unlikely that it will use up all of its allocation before the turn of the century. Since Sudan's principal cities and industries are located on the Nile, there should not be any major water shortages in the near future. Problems of water shortage do occur in the non-irrigated rain-fed areas, but urban and industrial use of water remains very small outside the Gezira heartland.

The agricultural sector's energy consumption has remained very small. In 1980 it used 94 500 tons of oil equivalent (TOE, 1 TOE = 41.8×10^9 joules), almost all of which was in the form of gasoil. This represented only 1.35 per cent of the estimated national total energy consumption. The latter total excludes both sugar processing, which is included in the industrial statistics and the transportation of crops and agricultural labour. Of its 90 900 t of gasoil consumption in 1980, 57 900 t were used in agricultural operations, 29 300 t for irrigation pumping, and 3700 t for canal maintenance (MEM 1983, p. 35). In 1980 the agricultural sector used 78.9 Gwh (gigawatt hours) of electricity which was 11.39 per cent of the total national electricity sales (MEM 1983, p. 32).

If the use of wood and charcoal for domestic use are taken is consideration, total energy consumption looks very different. Together with other biomass, wood and charcoal accounted for 5.533m. TOE or 79 per cent of total energy consumption (MEM 1983, p. 30). This demonstrates the importance of Sudan's reserves of wood, and provides a partial explanation for the country's desertification problems.

For the major irrigated crops, fuel consumption for irrigation varies from the equivalent of two gallons per feddan for *dura* at Kosti to 34 gallons per feddan for sugar at Sennar. The requirement is dependent upon the water requirement of the crop and the necessary pumping head. In general, irrigation fuel requirements per feddan are three to four times greater along the Blue Nile compared with the White Nile, due to the much greater range in the pumping head along the Blue Nile. Fuel consumption for agricultural operations, which includes everything from land preparation through to harvesting and threshing, varies far less. It ranges between one and three gallons per feddan for all crops except mechanically-picked cotton, for which picking alone requires four gallons per feddan (MEM 1983, pp. 37–8).

Besides water and power, one of the most important resources used by the agricultural sector is its skilled personnel. Like any developing country, Sudan requires agricultural specialists to help improve its methods of production. In 1979/80, Sudan had three agricultural technical schools with three Sudanese and 34 non-Sudanese teachers. In the same year the University of Khartoum had 1150 students in the Faculty of Agriculture and 171 graduates. The small University of Gezira, which specializes in agriculture and rural development, had 158 students studying agricultural sciences while Khartoum Polytechnic had 411 (MFEP 1983, pp. 47–60). Besides the students, the Sudanese Government employs hundreds of agricultural specialists and extension officers. In addition, the Agricultural Research Council (ARC), which was founded in 1970, is involved in encouraging, promoting, and developing research in all government agricultural departments.

Agriculture is, therefore, a major user of Sudan's scarce resources. Irrigated agriculture is the major user of water and will continue to be so in the coming decades. While its direct use of energy is comparatively small, it is increasing annually. Far more important, however, is the use of wood and charcoal for domestic cooking. Sudan's forest resources supply 79 per cent of the country's total energy requirement—a fact which should not be overlooked. Other scarce resources including skilled technicians and specialists are vital for the agricultural sector. This use of scarce resources as well as other factors, including agriculture's use of transport, fertilizers, insecticides, plant and machinery should be taken into consideration in any analysis of the agricultural sector.

Conclusions

It has been seen that agriculture is the principal component of Sudan's GDP and accounts for over 90 per cent of the country's export earnings. Agriculture still provides a livelihood for over two-thirds of the population and it will continue to do so in future decades. There are a few major towns in the Gezira heartland but over most of the country, agriculture dominates the rural economy. Even Sudan's industrial sector is dominated by the processing of food and other crops.

Development planning, particularly since the early 1970s, has focused on the agricultural sector. Unfortunately, the 'bread basket' strategy suffered from a number of problems and this contributed to Sudan's escalating debt crisis. Despite this, both the IMF and Sudan's other creditors and aid donors have continued to stress the importance of agricultural production and exports.

Irrigated agriculture takes the vast majority of the country's available water resources and while the sector's use of fuel is comparatively limited, it is increasing gradually. Its supply of fuel, in the form of wood and other biomass, is of vital importance.

Agriculture is currently the most important sector of the Sudanese national economy and although the secondary and tertiary sectors will undoubtedly expand, it will remain totally predominant until at least the turn of the century and probably far longer.

References

Bank of Sudan (1984). *Foreign trade statistical digest annual 1983*, Vol. 16, No. 4. Bank of Sudan, Khartoum.

Bank of Sudan (1988). *Economic and financial statistical review—January–March 1988*, Volume 30, No. 1. Bank of Sudan, Khartoum.

Bank of Sudan (1989). *Twenty ninth annual report 1988*. Bank of Sudan, Khartoum.

Economist Intelligence Unit (1986*a*). *Sudan country report*, No. 2. Economist Intelligence Unit, London.

Economist Intelligence Unit (1986*b*). *Sudan country report*, No. 4. Economist Intelligence Unit, London.

Economist Intelligence Unit (1986*c*). *Sudan country profile 1986–87*. Economist Intelligence Unit, London.

Gurdon, C. G. (1986). *Sudan in transition: a political risk analysis*. Economist Intelligence Unit, Special Report No. 226. Economist Intelligence Unit, London.

MEM (Ministry of Energy and Mining) (1983). *Sudan national energy assessment executive summary*. MEM, Khartoum.

MFEP (Ministry of Finance and Economic Planning) (1983). *Statistical abstract*. Department of Statistics, Khartoum.

MNP (Ministry of National Planning) (1977). *The six year plan of economic and social development 1977/78–1982/83*, Vol. 1. MNP, Khartoum.

UNIDO (United Nations Industrial Development Organization) (1983). *The Democratic Republic of the Sudan*, Industrial Development Review Series. UNIDO, Khartoum.

Whittington, D. and Haynes, K. E. (1985). Nile water for whom? Emerging conflicts in water allocation for agricultural expansion in Egypt and Sudan. In *Agricultural development in the Middle East* (ed. P. Beaumont and K. S. McLachlan), pp. 125–49. John Wiley & Sons, Chichester.

11

Rural small industries and crafts

D. HANSOHM

In discussions of socio-economic development, especially in the African context, the actual and potential role of small industries (including crafts) is recognized increasingly. Many governments as well as supra-national organizations give high priority to the promotion of small industries—at least in the plans.

The Organization of African Unity's Lagos Plan of Action (OAU 1982), for example, urges the creation of a 'network of small and medium-scale industries as well as actively promoting and encouraging the informal sector'. In the Sudan, it has been proposed to include the promotion of small industries, especially in rural areas, in a basic programme of economic reform which is believed to be necessary in order to halt and reverse the trend of economic deterioration (ILO 1987; Hansohm and Wohlmuth 1989; Wohlmuth and Hansohm 1986). The new Sudanese Government, elected in 1986, declared the promotion of small industries as one of its priorities. To this end, a special department is being set up in the Ministry of Industry.

Defining rural small industries and crafts

Many different definitions of this sector are in use. They differ according to the kind of activities as well as in the size of enterprises. Besides 'small-scale industry' and 'rural small-scale industry', the concepts of 'small-scale enterprise', 'rural non-farm activities' and 'informal sector' are relevant.

Small-scale enterprises, in the broadest sense, include all non-agricultural enterprises and most of these are service activities, predominantly trade. In the narrower sense, focusing on economic development, the term is restricted to productive activities, such as industry and repair services (Page and Steel 1984, p. 12). The World Bank, which has done most research on small-scale enterprises in Africa, defined these recently as 'enterprises engaged in activities involving barriers to entry in the form of human or physical capital that do not have ready access to institutionalized credit and incentives without special assistance' (Page and Steel 1984, p. 13).

Informal enterprises are defined by seven characteristics in a basic discussion document, the International Labour Office (ILO) Kenya Report (ILO 1972, p. 6): ease of entry, reliance on indigenous resources, family ownership of enterprises, small scale of operation, labour-intensive and adapted technology, skills acquired outside the formal school system, unregulated and competitive markets. These characteristics hold true for many of the small industry enterprises but, while the informal sector includes service activities, the small industry sector includes enterprises which are formally registered.

The concept of 'rural non-farm activities' implies that many of the activities which are of interest are below the level of enterprises: they are part-time activities and are a source of income (principal or secondary) additional to agriculture. This is especially true for traditional industries and crafts.

Concerning the upper limit of the small industry sector, this is usually defined by the number of employees, the capital, or a combination of both. While the first criterion may be misleading (it might include capital-intensive modern enterprises with a high level of technology and few employees; this would not be advantageous for socio-economic development), it has the advantage of being easy to determine. In Sudan, the most common definition of small industries (Department of Statistics, MFEP 1974) includes enterprises with less than 25 employees.

The term 'rural areas' is taken to mean all areas except for the 'Three Towns' (Khartoum, Omdurman, and Khartoum North), so regional and provincial capitals and other towns may be regarded as rural centres. Even the Three Towns still possess many rural characteristics, so that for many types of small industry there are no significant differences from those in the rural areas.

The rural small industries cover a wide range of different types. They can be broadly divided into traditional and modern industries. Traditional industries have a long history and relied totally on local resources (blacksmithing, wood work, tannery and leatherwork, plaiting of dom palm leaves and straw, artistic handicrafts, carpet-making, pottery). At least until the beginning of the twentieth century they covered almost totally the demand for manufactured goods in Sudan. Some trades were regionally specialized, for example, leatherwork in Darfur (Oesterdiekhoff 1984, p. 12). Modern small industries have come into existence more recently (modern carpentry, apparel industry, car and machinery repair, metal workshops, electricians, oil mills, electric bakeries). Some of these are modernized agro-processing industries, others were established by modern demands.

Developmental advantages of the rural small industries sector

The case for promotion of rural small industries rests on several, interrelated arguments (Hansohm and Wohlmuth 1987 p. 173ff.; Page and Steel 1984, pp. 5–9):

1. Efficiency. These enterprises appear to be more efficient (or can be made more efficient) in their use of available resources, when compared with large enterprises.
2. Employment. Small industries provide the bulk of non-agricultural productive employment. They are also more labour-intensive than large enterprises.
3. Domestic savings. Small industries can provide an outlet for personal savings at low income levels which would otherwise not be realized.
4. Provision of commodities. Rural small industries provide commodities especially directed to the needs of the low-income population.
5. Equitable income distribution. Rural small industries contribute to improve the income distribution both by providing income to the lower-income population and by supplying their needs.
6. Promotion of indigenous entrepreneurs. Because of their number, small industry enterprises provide a far larger training ground for entrepreneurs.
7. Conserving scarce managerial abilities. Small industries offer this possibility through a less complex organization of production than large-scale firms.
8. Adapted technology. Small industries use technologies which are more appropriate and adapted to local resources and conditions.
9. Skill diffusion. Because of their technologies and labour-intensity, small industries are better suited for this purpose than large-scale enterprises.
10. Regional development. Because their activities are more geographically dispersed, rural small industries can be an agent for regional development.
11. Minimizing rural–urban migration. They help to minimize rural–urban migration by creating income in rural areas.
12. Domestic inputs. These are used to a greater extent by small industries, especially in rural areas.
13. Linkage effects. Small industries, especially in rural areas, have comparatively stronger linkages with the agricultural sector, via the processing of agricultural products, the supply of agricultural inputs and the supply of goods for consumption by the agricultural population.

These points are highly interrelated and reinforcing. Their relevance will depend on the specific circumstances of a region and the type of industry.

Size and role of rural small industries and crafts

There are only very few records for small industries in Sudan. Existing statistics cover only a fraction of the enterprises, especially in rural areas, and they rarely cover part-time non-farm activities. This is partly because informal enterprises, which constitute a large part of small industries, are not registered by definition. As the formal small industries are concentrated in the Three Towns, the statistics are biased towards that region.

The limited statistics on rural industries is well illustrated by their regional distribution (Table 11.1) The lack of statistics on small industries is demonstrated most clearly in the Southern Region, but it is also true for other regions.

TABLE 11.1. Regional distribution of small industries in 1981–82[a]

Region	No. of units	% Share
Khartoum	878	45.7
Central	529	27.5
Kordofan	152	7.9
Darfur	63	3.3
Eastern	185	9.6
Northern	110	5.7
Southern	5	0.2
Total	1922	100

[a] Enterprises with less than 25 employees, excluding those engaged in bakery and grain milling.
Source: Sen (1985, p. 31).

The total number of 6412 small enterprises can be subdivided into a number of different activities (Table 11.2).

In terms of number of enterprises, gross output, and employment the 'Food, tobacco, and beverages' category is the most important. Its activities are concentrated on bakeries, grain mills, oil milling, fruit juice, local sweets, noodles, and macaroni. The category, 'Fabricated metal products and machinery', second in gross output, concentrates on auto-servicing and repair workshops and sheet metal workshops (furniture, doors, windows etc.). In the third ranking category, 'Chemicals and chemical products', indigenous soap and perfumes are prominent.

Small industries as a whole play a major and positive role in the industry

TABLE 11.2. Distribution of small industry enterprises according to categories of activity (1981–2)[a]

Category	Number	%	Gross output (£S'000)	%	Employment	%
Food, tobacco, and beverages	686	35.0	190 133	66.4	6105	31.9
Textile, weaving, apparel, and leather	77	3.9	11 864	4.1	984	5.1
Wood and wood products, furniture	173	8.9	8369	2.9	1467	7.7
Paper, paper products, printing	77	3.9	3244	1.1	769	4.2
Chemicals and chemical products	69	3.4	30 376	10.6	1232	6.4
Other non-metallic products	68	3.4	7909	2.8	2059	10.7
Basic metal industries	4	0.02	159	0.05	64	0.3
Fabricated metal products and machinery	820	41.5	34 223	12.0	6451	33.7
Total	1974	100	286 227	100	19 185	100

[a] Different from the number of enterprises in Table 11.1 due to 'errors and omissions'.
Source: Sen (1985, p. 30).

sector. They account for 95 per cent of all enterprises, 27.4 per cent of employees and 34.3 per cent of gross output. The United Nations Industrial Development Organization's industrial survey (UNIDO 1985) estimates that 30 per cent of small-scale industries are located in Khartoum Province (in contrast to 60 per cent of large industries). Compared with large industries, small industries produce 10 times more per unit of investment, need half of the investment to create a job and their value added per worker is about three times more (Sen 1985, p. 28). Large industries appear to be heavily capital-oriented and grossly unproductive.

In spite of being outdated and confined to the northern regions, the Handicraft Industries Survey of the Department of Statistics is a valuable source of data on a specific sector of the rural small industries. The survey defined a handicraft unit as a unit engaged in the production of goods and services without using modern equipment. It identified a total of 19 022 handicraft units and analysed a sample of 3966 (Table 11.3).

TABLE 11.3. Regional distribution of handicraft units (1970–1)

Region	No. of units	% of total	No. of workers	% of total	Value of production	% of total
Khartoum	2736	14.33	9084	24.93	8998	41.45
Blue Nile	4928	25.90	8948	24.55	5290	24.37
Kassala	2347	12.34	4126	11.32	2204	10.15
Kordofan	5980	31.44	8388	23.02	2524	11.63
Northern	1206	6.35	2954	8.11	1703	7.84
Darfur	1825	9.64	2941	8.07	990	4.56
All northern regions	19022	100	36441	100	21709	100

Source: Department of Statistics, MFEP (1974).

By comparing the number of handicraft enterprises in Table 11.3 with the number of small industries in Table 11.1, one can see that the handicraft survey provides a better assessment of the rural industries. Here 85.7 per cent of the handicraft enterprises, 75.07 per cent of the total number of workers, and 58.55 per cent of the value of production are outside Khartoum Province. However, the survey seems to be confined to full-time activities and, as the part-time craft and industry activities are more common in rural areas, it must be assumed that the handicraft survey is also biased towards the urban area. It can also be assumed that the survey was confined to major centres in the rural areas, so for that reason, it seems likely that the activities which are more common in small villages are not represented adequately.

The handicraft survey divides the crafts into 10 categories (Table 11.4). In terms of the numbers of enterprises and workers, 'Clothes' is the most

TABLE 11.4. Distribution of handicraft units by category of activity (1970–1)

Category	No.	% of total	No. in Khartoum	% of total	No. of workers	% of total	Value of production	% of total
Food products	1189	6.25	184	6.72	6243	17.13	8965	41.30
Drinks	2792	14.68	164	5.99	3737	10.26	897	4.13
Clothes	9308	48.93	1532	55.99	13096	35.94	4053	18.67
Furniture	941	4.95	88	3.22	1946	5.34	1337	6.16
Leather, plastic, and tannery	1232	6.48	56	2.05	2220	6.09	915	4.21
Ornaments and blacksmith products	1733	9.11	168	6.14	3466	8.51	2457	11.32
Palm leaf products	670	3.52	8	0.29	716	1.96	53	0.25
Pottery	465	2.44	180	6.58	3669	10.07	626	2.88
Tobacco	520	2.73	320	11.70	716	1.96	431	1.99
Unclassified	172	0.91	36	1.32	632	1.74	1975	9.10
Total	19022	100	2736	100	36441	100	21709	100

Source: Department of Statistics, MFEP (1974).

important category, both for Khartoum and the rural areas. The food production categories ('Food products' and 'Drinks') rank second in terms of the numbers of enterprises and workers, but first in terms of the value of production. Other important categories are 'Ornaments and blacksmith products', 'Furniture' and 'Leather, plastic and tannery products'.

So far, no survey has been done to elucidate the full range of rural industries and crafts as well as their distribution by region and by size of settlement (activities by nomads are not included).

The impact of government policies on rural small industries

Like other African States, Sudan has given priority to the promotion of small-scale industries, especially in rural areas, in its different plans. In the *Five Year Plan (1970/1–1974/5)* the need to modernize and to promote the traditional industries and the handicrafts and to widen the productive employment opportunities for the local people was emphasized (MNP 1977, p. 150). The objectives of the *Six Year Plan of Economic and Social Development (1977/8–1982/3)* were formulated more explicitly: 'The distribution of investment opportunities over the various parts of the country lay special emphasis on the backward areas in order to achieve the highest possible level of social justice and balanced growth . . . Taking the necessary steps to reduce migration from rural areas and pastural lands to the town and urban centres by introduction of rural and environmental industries which depend on the utilization of local raw materials . . .' (MNP 1977, pp. 150–1).

Unfortunately, these statements contrasted with neglect in terms of policy measures and allocations (Hansohm and Wohlmuth 1987, p. 183ff., 1989, p. 21ff.) Furthermore, there was no clear responsibility for the promotion of small industries within the administration. More important than the lack of positive promotion for rural small industries was the indirect impact of government policy in several other fields. As in other African countries, there are certain policy biases and market distortions curtailing the development of rural small industries in Sudan. The Sudanese Government has a massive influence on the industrial policy sector through its own strong involvement in industry and through a comprehensive set of regulations and incentives for private investments. Large, capital-intensive and highly import-dependent industries, concentrated in the Central Region, were promoted. These industries remained sectorally fragmented and had weak linkages with each other as well as with other outside sectors, especially agriculture (Oesterdiekhoff 1983, pp. 168–80). The industrial policy was neither sectorally integrated nor co-ordinated with the agricultural policy. This contributed to the weak character of linkages and resulted in paradoxes like material shortages for agro-processing factories.

During the industrialization process the emphasis was shifted from processing raw materials to production based on import substitution. This was especially true for the private sector. The overall capital intensity of production increased because of a rise in capital input in existing production processes as well as the establishment of new industries in luxury goods and semi-finished products which are characterized by higher capital intensity. The process of capital intensification and weak employment growth was accompanied by and inter-dependent with a process of deteriorating income distribution.

The capital intensity was favoured by a steady and increasing overvaluation of the currency and by the provision of subsidized credits. Because of high inflation rates, these often had negative interest rates. Small industries, however, have hardly any access to these credits. In contrast, they have to rely on self-finance or to deal with traditional money-lenders which implies extremely high interest rates.

In order to facilitate the emergence of a domestic industry, the tariff rates for capital goods and raw materials were held low while the domestic production, mostly of luxury goods, was highly protected. This policy was a disincentive for an overall reduction of import dependence. If we assume that rural small industries are less dependent on imported capital goods and raw materials, this policy grants an artificial advantage to urban, large-scale industry.

The second instrument for promoting industrial growth is the allocation of scarce foreign exchange. This allocation system discriminates heavily against small producers who have hardly any access to foreign exchange. Moreover, most of the government procedures are so complicated that the formal procedure in itself constitutes a barrier for small producers.

Many other subsidies to encourage industrial investment discriminate directly against small producers who are not able to make use of them, or for whom they are irrelevant. Moreover, they are part of a development model which is directed at foreign markets and at higher income domestic markets, while restricting the market for small industry products which is believed to be concentrated on the low-income population. Furthermore, the development of rural small industries was restricted by the urban bias of development policies which impoverished the rural areas. The characteristics of agricultural policy are similar to those of the industrial policy. It concentrates heavily on the modern sectors of irrigation, mechanized farming and cattle-ranching and neglects the traditional agriculture, which still provides a living to more than two-thirds of Sudan's population.

At the same time, the industrial crisis in particular, and the economic crisis in general, have resulted in a boom of informal activities. Different factors have contributed to this: on the one hand, rising unemployment has forced people to join the informal sector (especially in the urban area), while on the other hand, products and services rendered by the informal sector substitute for those of the formal sector because they are cheaper and because of the eroding supply capacity of the formal sector.

Financing rural small industries and crafts

The access of rural small industries to institutionalized credit is very limited. The commercial banks concentrate their activities on short-term lending (consistently more than 70 per cent of activities) and the finance of import/export trade (almost 35 per cent). Less than one-third of the short-term loans go to industry; in 1987, the share declined to 17.8 per cent (Bank of Sudan 1983–8). However, this sum goes, almost exclusively, to the major urban area, the Three Towns. The stagnant share of loans directed to industry is primarily determined by the unfavourable environment for industrial investments which contrasts with potential high profits in the tertiary sector. This is even more pronounced for small and rural industries.

Since 1980, however, the Faisal Islamic Bank of Sudan has begun to take interest in medium-term financing of industries in the small business and artisans sectors. A special branch for this purpose was opened in Omdurman and is operating successfully. However, it does not cover rural areas and it has not altered the pattern of the Faisal Bank's activities significantly, which are still concentrated on trade.

In 1961, the Government established the Industrial Bank of Sudan (IBS) in order to fill this finance gap and relieve the capital constraint for industrial development. The IBS started operations with a capital of £S3m. and by 1986 the paid up capital amounted to £S18.7m. During the period 1962–86, 265 loans and 85 supplementary loans were approved. The total IBS contributions came to £S24.8m. which amounted to 55.6 per cent of the projects' total investments (IBS 1967–87). Ranked by total investment, the most important categories are Food processing and ice factories, Oil mills, Textiles, Knitwear and ready-made clothes, Building materials, and Flour mills. It is remarkable that traditional industries and very small industries are not covered. Furthermore, the activities of the IBS are highly concentrated in the urban area (Table 11.5).

In order to service the rural areas, the Sudanese Government formed the Sudan Rural Development Company Ltd. (SRDC) group in 1982. Its

TABLE 11.5. Regional distribution of IBS loans, 1962–86

Region	% Share
Khartoum	54.1
Eastern	15.1
Darfur and Kordofan	4.0
Central	2.2
Northern	2.9
Southern	1.8

Source: IBS (1987).

operational area is the whole country except for Khartoum Province. The task is to promote and develop small to medium-sized agricultural and industrial projects. In 1987, the SRDC was involved in thirteen industrial, thirteen agricultural, and two service sector projects with a sum of £S6.51m. (disbursed investment). Seventy-seven per cent of this credit sum went to the industrial projects (SRDC 1988) and these, as in the case of the IBS, were also comparatively large projects in the modern categories of industrial activity. The smallest loan approved is £S31 000.

Existing programmes and institutions for promoting small industries and crafts

The rural small industrialists and craftsmen lack adequate institutional representation. Neither in the Trade Unions nor in the Federation of Industrialists and Businessmen is there a strong group supporting their interests. Their sole institution is the Sudanese Craftsmen and Small Enterprises Union (SCSEU) which is not, at present, an effective means of representation. Firstly, most of the members are in the Three Towns and the few branches outside are in some of the major regional centres. Secondly, only a fraction of the craftsmen and small entrepreneurs are members. The institution is very weak. Surveys show that these people use other channels to make their problems known to public authorities (Curtis 1980, p. 22ff.). The main reason to join the SCSEU seems to be the option to get access to real estate at favourable conditions (far below the market prices). This option is, however, possible only for a fraction of those interested, and they are those in possession of capital, not necessarily craftsmen by qualification or small industrialists (Hansohm and Wohlmuth 1987, p. 184ff.).

There is a number of training institutes (Vocational Training Centres, Youth Training Centres, the National Co-operative Development and Training Centre) and projects operating at the micro-level intended to promote traditional rural industries and crafts by co-operative organization and production, training etc. They are financed by the Sudanese Government multilateral organizations, bilateral donors and non-governmental organizations. No systematic evaluation has yet been done, but the available evidence allows the following four conclusions (Hansohm and Wohlmuth 1987, pp. 183–6):

1. Ideas from outside are often imposed on local communities.
2. Socio-economic factors are not usually considered adequately, especially where they concern the role of women.
3. Most of the projects have had only marginal, if any, success.
4. The programmes strongly reflect the view that the constraints to rural

small industrial developments are within the sector itself, and ignore the external constraints, especially the impacts of the Government policy in different fields

Characteristics of rural small industries and crafts

As yet few studies have been done on small-scale industrial activities, either of a formal or informal character. Available evidence will be summarized in this chapter in order to throw some light on characteristics which are of special importance in view of the presumed developmental advantages of the sector (see above). Material for this chapter is drawn from field studies by the author in western Sudan (Hansohm 1986, pp. 241–65, 1988) and from reports on Darfur (Bakhit *et al.* 1986), on two areas in eastern Sudan (Babiker 1982), on Wad Medani (Anand and Nur 1985; Bilal 1985), and on Juba (Jenkins 1981). For comparison, two studies on the Three Towns—Omdurman (Hansohm, unpublished) and on Khartoum North (Curtis 1980)—are included.

Most of the production is destined for the final consumption demand. Few industrial activities supply production inputs, the exceptions being tannery (producing for the leather industry) and blacksmiths/metal workshops as far as they produce agricultural implements and spare parts. Thus the inter-industrial linkages appear to be low. This is in contrast to the experiences of South Asian economies, especially India, where subcontracting (of large firms to small firms) provided a stimulus to small industry growth.

However, it is assumed that the potential for productive linkages between agriculture and industry would be great if an appropriate policy were implemented. Local blacksmiths have provided a wide variety of agricultural tools, albeit on a primitive level. The main constraint to increasing the technological level (intermediate technologies) seems to be the limited purchasing power of farmers (Silsoe College 1986). The Jebel Marra rural development project is experimenting with the introduction of camel-drawn ploughs on small to medium farms, assisted by a credit programme. The ploughs are produced locally and the experience has been encouraging.

Most of the production is directed at the local market, if not for home consumption. Only few items seem to be traded inter-regionally, for example, shoes from Darfur. Nothing seems to go to export. Most products, traditional and modern, are produced to order and marketed by the producers at the place of production.

Traditionally, rural small industries produced goods predominantly for lower-income groups of the population. This is no longer the case for many of the products or services seem to be used by all income groups and some are even directed at higher-income groups, for example, modern carpentry and car repair. This would appear to contradict the assumption that small industries are directed at low-income groups and that their development is, therefore, constrained by the deteriorating income distribution. However, this is not necessarily the case, for the studies are limited to entrepreneurs and do not give a representative view of the sector's development. At the same time, if there is a shift away from inferior goods, this might be interpreted as a sign for high development possibilities.

For many of the examined enterprises/activities, the demand perspectives are evaluated positively. One reason for this is that there is little competition with large-scale domestic enterprises, partly because of the low capacity utilization of large industries. This points to possible stimulating effects of the economic crisis on the development of rural small industries. Similarly, the limited infrastructure in the country acts as a barrier and gives comparative advantages to those industries which are based on local resources. However, the overall declining purchasing power of the population is, in many instances, more than outweighing those comparative advantages.

The incomes of rural small industries differ widely. Incomes from many activities appear to be high compared with those from agriculture and other activities in the formal sector. In contrast to the formal sector, incomes have increased in recent years partly due to declining wage levels. This corresponds with the fact that a surprisingly high number of entrepreneurs (29 per cent in the urban area of Omdurman) had worked formerly in the formal sector, some in highly qualified positions.

Altogether, there appears to be a high occupational mobility. The social origin of the entrepreneurs, as measured by father's occupation varies broadly. Traditionally, artisan industries were subject to strict social segregation by sex and in some instances by ethnical group. This was considered an obstacle for rural small industry development. Currently, however, these segregations seem to be beginning to break down. Two examples will illustrate this:

1. In Darfur, blacksmithing was traditionally a 'despised' trade practised only by West Africans, but now indigenous people are beginning to take up the trade.
2. Weaving of mats, baskets and food covers was traditionally womens' work, but more recently it is reported that in some areas men are taking up the craft.

The ability to use local raw materials is only true to a high degree for traditional rural industries. For all modern activities, for example, carpentry, metal work, tailoring, mechanical services, and electrical workshops, the supply of raw materials and/or equipment appears to be the main constraint. Even the supply of raw materials such as wood, wool, leather, and palm leaves, for traditional industries is presently endangered as a result of environmental degradation. On the other hand, some trades are now using waste material such as scrap. The market for raw materials is often dominated by traders and shortages are aggravated by them to increase profits. Raw material shortages often result in work stoppages and lead to high excess capacities.

Very few small industrialists have access to institutionalized credit. Most rely almost exclusively on self-finance, but some trade is dependent on merchants, money or raw materials being borrowed at very high interest rates. Lack of capital appears to be a barrier to growth, but this view is not undisputed (Curtis 1980, p. 24). In any case, programmes focusing on credit supply run the risk of increasing capital intensity inappropriately—the ability to work efficiently under conditions of capital scarcity is one of the strong points of rural small industries. If increased market demand is not ensured, the result will be a displacement of many entrepreneurs by the few who get credit.

The technological level of many activities is low. Generally, they are labour-intensive and use little capital. To develop rural small industries, a modernization is necessary (except for artistic crafts). But the appropriate technologies will be intermediate, their introduction has to be carefully prepared, and due regard must be given to the conditions under which rural small industries work.

Possession of licenses and tax paying are characteristics of formal enterprises. For industries in rural areas, especially those of a traditional character, they appear to be almost irrelevant. Proposals to 'tap the broad tax base provided' by this sector (Anand and Nur 1985, p. 159; Bilal 1985, p. 30) have to be met with scepticism at this stage of development. The internal economics of the enterprises have to be fully understood in order not to restrict or even prevent their evolutionary growth.

References

Anand, V. and Nur, T. M. (1985). The role of small scale enterprises in developing countries—a case study of Wad Medani in Sudan. *Asian Economic Review*, **27**, 137–59.

Babiker, A. A. (1982). The role of rural industries in the arid and semi-arid areas of the Sudan. *GEO Journal, Wiesbaden*, **6**, 49–55.

Bakhit, A. H., Ibrahim F., and Rheingans F. (1986). *Small-scale industries in Darfur, Sudan*. Hannover, State of Lower Saxony, F.R. Germany.

Bank of Sudan (1983–8). *Twenty-third to twenty-eighth annual reports 1982–1987*. Bank of Sudan, Khartoum.

Bilal, G. K. (1985). *On the role of the informal sector in development: a case study of Medani Town*, Economic and Social Research Council Bulletin, No. 128. ESRC, Khartoum.

Curtis, D. (1980). *Small scale industry promotion. Report on a field project*, Development Administration Group Occasional Paper No. 7. University of Birmingham.

Department of Statistics, MFEP (1974). *Handicraft industries survey 1970–71*. Department of Statistics, Khartoum (in arabic).

Hansohm, D. (1986). Promotion of rural handicrafts. In *Reassessment rehabilitation programme, Kordofan and Darfur*, final report to the Ministry of Finance and Economic Planning (ed. Euroconsult), pp. 241–56. Euroconsult, Arnhem.

Hansohm, D. (1988). *Preliminary report on the research project: the potential of small industries in Sudan*. Nyala, Labour Office, 20 April.

Hansohm, D. and Wohlmuth, K. (1987). Promotion of rural handicrafts as a means of structural adjustment in Sudan. *Scandinavian Journal of Development Alternatives*, **6**, 170–90.

Hansohm, D. and Wohlmuth, K. (1989). *Sudan's small industry development. Structures, failures and perspectives*. Berichte aus dem Weltwirtschaftlichen Colloquium der Universitaet Bremen, No. 16. University of Bremen.

IBS (Industrial Bank of Sudan) (1967–87). *Annual reports 1962–1986*. IBS, Khartoum.

ILO (Industrial Labour Organisation) (1972). *Employment, incomes and equity. A strategy for increasing productive employment in Kenya*. ILO, Geneva.

ILO (1987). *Employment and economic reform: towards a strategy for the Sudan*. Report of the ILO/JASPA mission to the Sudan, August–September 1986. ILO, Geneva.

Jenkins, S. J. (1981). *Aspects of the informal economic sector of Juba, Southern Sudan*, Population and Manpower Unit, Research Paper No. 1. University of Juba.

MNP (Ministry of National Planning) (1977). *The six year plan of economic and social development, 1977/78–1982/83, Vol. 1*. MNP, Khartoum.

Oesterdiekhoff, P. (1983). Industrial development. Structural deficiencies, agro-industrial prospects and alternatives. In *The development perspectives of the Democratic Republic of Sudan* (ed. P. Oesterdiekhoff and K. Wohlmuth), pp. 164–94. Weltforum Verlag, Munich.

Oesterdiekhoff, P. (1984). *Hand werk im Sudan—technische und soziookonomische aspekte*, Sudan Economy Research Group Discussion Paper, No. 2. University of Bremen.

OAU (Organisation of African Unity) (1982). *Lagos plan of action for the economic development of Africa 1980–2000*. International Institute for Labour Studies, Geneva.

Page, J. M. and Steel W. F. (1984). *Small enterprises development. Economic issues from African experience*, World Bank Technical Paper, No. 26. World Bank, Washington.

Sen, A. R. (1985). *The development of small industries in the Sudan: policies, strategies and programme*. UNIDO, Vienna.

Silsoe College (1986). Animal drawn implements. A study to assess feasibility of local manufacture in Sudan. Unpublished report for EEC Delegation, printed in Khartoum.

SRDC (Sudan Rural Development Company Ltd.) (1988). *Annual Reports and Accounts 1987*. SRDC, Khartoum.

UNIDO (United Nations Industrial Development Organisation) (1985). The Democratic Republic of the Sudan. Unpublished paper, Industrial Development Review Series, UNIDO, Vienna.

Wohlmuth, K. and Hansohm D. (1986). *Sudan: a case for structural adjustment policies*, Sudan Economy Research Group Discussion Paper, No. 8. University of Bremen.

12

Historical development of agriculture

A. TRILSBACH

The present agricultural landscape is the product of a long evolution. The contributing forces include environmental changes of the past 5000 years; the ebbing and flowing of empires from the early days of Ancient Egypt to the Anglo-Egyptian Condominium of the present century; advancing technologies from the earliest introduction of the *shaduf* to the sophistication of the Kenana sugar complex; and the position of Sudan in a world of international trade.

This chapter examines these themes and traces the development of Sudanese agriculture from the earliest known records to the present. Particular emphasis is placed on periods when agricultural development progressed in a series of rapid transformations, beginning in the late Palaeolithic.

The beginnings of agriculture

Although vague, evidence suggests that human beings appeared in the Sudan at a very early stage of Man's existence. More certain, however, is that from the end of the Palaeolithic there were three main population groups. To the south of modern Khartoum were the most northerly of the East African hunters. To the west the principal group was Negroid, of West African origin; these people were also hunters although gathering supported their basic subsistence. In northern Sudan the peoples were dominantly of Caucasoid stock and had had some experience of grain harvesting (Wendorf 1968) and domesticating livestock. The Nilotics were mainly in the Ethiopian Highlands at this time and appeared on the Sudanese plains after the coming of the Prenilotes (below).

By about 4000 BC a considerable amount of interaction had taken place, most especially in the central Nile Valley, and various types of agriculture were practised. Archaeological evidence from this time (Arkell 1949) indicates that a complex agricultural mix existed including fishing, hunting, and the domestication of sheep, goats, and dogs. There was also a well established harvesting of wild grains in Nubia, especially during wet climatic secessions. Additional evidence points to some rain-fed cultivation away from the banks of the Nile, probably also linked to the rearing of cattle (Krzyaniak 1978).

A new agriculture emerged with the Prenilotes from about 3000 BC. These were Negroes moving in from the Sudanic belt of West Africa and they brought their experiences of simple tilling and cultivation techniques to the area of modern Kordofan. Cultivation also spread to the Ethiopian Highlands and this encouraged the Nilotes to come down to the more environmentally-suitable plains of the present Sudd region.

From about 2800 BC there was growing contact with the emerging kingdom in Egypt and this facilitated the spread of the Beja into eastern Sudan. These were the first proper pastoralists to inhabit the Sudan and products such as milk and butter were added to their diet for the first time. By 2150 BC, pastoralism had expanded with the arrival of long-horned cattle. These were introduced by the ill-defined *C Group* of Caucasian, possibly Mediterranean, origin. Despite earlier thoughts to the contrary (Firth 1910; Emery and Kirwan 1935) they had probably lived in the area for many centuries, but made their presence more obvious by moving towards the Nile in response to a gradual desiccation of their traditional grazing areas.

By the time that the kingdoms of Egypt began to exert political, economic, and military influence in the Sudan, the Sudanese had already had some experience of hunting, gathering, fishing, grain harvesting, tilling, cultivation, domestication of livestock, and dairying.

Surprisingly, Ancient Egyptians had little impact on Sudanese agricultural development. It is true that many of the existing techniques and practices were consolidated and spread over greater areas. Pastoralism, in particular, began to dominate agricultural activity in the vast grasslands to the west of the Nile, in areas of modern Kordofan and Darfur. Perhaps the greatest new contribution from Egypt was the *shaduf* (water lever) which facilitated some artificial irrigation in areas which did not require a very high lift of water.

By about 1000 BC the Egyptian Empire began to collapse and power began to return to the Sudanese themselves, especially the southern Nubians (Kushites) who had challenged the Egyptians for two millennia within the so-called Kingdom of Kush. This kingdom managed to reassert its strength and a new Sudanese empire emerged, subsequently known as the Empire of Meroe.

At last the opportunities presented by the existing agricultural diversity became manifest and a new technological and agricultural revolution began.

The Meroitic Empire and the first agricultural revolution

The Empire of Meroe lasted between about 750 BC and AD 350 and marked a new era in the development of Sudanese agriculture. It was located approximately between the third and fifth cataracts, but the focal point of the Empire gradually shifted from its first capital at Napata (near modern Karima) to Meroe (near Kabushiya in the vicinity of Shendi). This was significant, because the shift is thought to have been stimulated, at least in part, by the restrictions of agriculture in the Napata area.

Cattle, in particular, need reliable pastures to survive and when the shift began about 400 BC, the Napata area began to suffer a sequence of droughts and sand invasion on to cultivable land. The search for better grazing necessitated a move to the south and Meroe town proved to be an ideal location. It had been an important settlement during Napatan times and its location offered distinct environmental advantages for agriculture.

At that time, Meroe was located within a vast area of rain-fed grassland, both to the west in Kordofan and to the east on the Butana Plain (also known as the Island of Meroe). Agriculture was no longer restricted to the Nile; in particular, the fertile clays of the Butana offered new opportunities for cultivation. Seasonal nomadic pastoral routes were established and these were still evident when Sudan became independent in 1956. On the plain *dura* (*Sorghum vulgare*) was grown for the first time to feed the urban population.

Furthermore, engineering skills which were developed during the cultural and intellectual expansion enabled the construction of canals and sophisticated irrigation networks, together with water storage pools, which were effectively the origins of the *hafir*.

These technical developments facilitated a new crop diversity. Simple irrigation techniques, combined with the use of the plough, allowed wheat, barley, lentils, cucumbers, and melons to be grown to feed the urban population especially where *wadis* provided fertile and well suited soils (Hakem 1981). Even fruits and grapes could be introduced to the most fertile locations. Of special importance was the introduction of cotton in about 400 BC. Huge plantations were grown around Meroe, sustaining a flourishing textiles industry for several centuries. It was only in the second half of the first millennium AD that cotton yields began to fall, and the cultivation of this crop on a large-scale disappeared until its reintroduction from Egypt in the nineteenth century.

The engineering developments mentioned above were only the beginning of several technological advances of importance to agriculture. Of particular note was the development of the ox-powered *saqqiya* (water wheel), which allowed irrigation water to be lifted over greater distances than the *shaduf*. For the first time, water could rise sufficiently to allow the artificial irrigation of *wadis* and basins (Shinnie 1978) and its use soon spread down the Nile to the area of modern Dongola. Although they are declining in importance, the *saqqiya* and especially the *shaduf* can be found in present day Sudan.

By about the third century AD the Meroitic Empire was at its zenith, but it soon began to break up, partly as a result of new demands from its agriculture. Overgrazing on the Butana was in evidence before the coming of Christ and, as time progressed, the pastoral peoples were forced to migrate farther to the east and west. Gradually the Empire broke down into a series of independent cells.

Consolidation and stagnation

From the end of the Meroitic Empire until the arrival of the Egyptian armies in the early nineteenth century and despite a complex and fascinating political history, there were few major agricultural advances.

Christianity entered the Sudan as early as AD 558 when a number of Nubians were converted. A series of independent Christian kingdoms began to control the northern half of Sudan, the most dominant being Makuria, in Nubia, and Soba, which was centred on Soba near Khartoum and which was partly influenced by the coptic Christians in Ethiopia. This was mainly a time of spiritual rather than agricultural advancement, but the period up to the middle of the thirteenth century witnessed the consolidation of many pastoralist groupings and the continuation of irrigated cultivation techniques using the *saqqiya* and *shaduf* along the valley of the Nile and its principal tributaries.

A major reorganization of the population groupings in the Sudan began with the first large-scale invasion of the Arabs in the thirteenth century. Muslim Arabs had begun arriving since the time of Muhammad, but most settled in the extreme north of Sudan where they were tolerated peacefully. In 1275, however, the Egyptian mamelukes invaded Makuria and Sudan was opened up to large numbers of Arab settlers. Population groups moved down the Nile and some of its principal *wadi* tributaries, such as the Wadi el Milk and the Wadi Muqqadam, to graze the grassland pastures of Kordofan and even Darfur. This consolidated the pre-existing pastoral practices in this area by encouraging trade in livestock and livestock products with Egypt. In particular, the principal tribal groupings of western Sudan were established, together with the complex mixture of nomadic practices involving a combination of camels, sheep, goats, and cattle. To a considerable extent these divisions remain, albeit linked more closely with semi-sedentary cultivation, but the principal nomadic groups are still dominated by the Jammala (camel owners) and Baggara (cattle owners).

Around 1500 a new kingdom appeared along the Nile Valley, based at

Sennar. Described loosely as the Black Sultanate with an affiliation to Islam, it is now referred to as the Funj Kingdom. Comparatively little is known about this Kingdom, but it had influence over much of what is now the northern Sudan; the main exception was Darfur which was an independent kingdom until the beginning of the twentieth century.

What little evidence there is of agriculture during the Funj suggests that few advances had taken place since the Meroitic period, although there was a widespread practice of slash and burn/shifting cultivation in the vast rainland areas of Kordofan and the Butana (O'Fahey and Spaulding 1974). There is also evidence of low-investment subsistence grain cultivation as it is known that the *sheil* system of food/money lending was in operation. Merchants stored grain to lend in times of shortage and to receive payment in times of abundance. The effect was to bring considerable profits to the merchants and great misery to the peasants; this system is still known to operate in parts of Sudan today.

Therefore, for about one and a half millennia, Sudanese agriculture advanced very little. Certainly some agricultural practices were consolidated, especially pastoralism, but in many aspects, most notably the irrigation techniques along the northern reaches of the Sudanese Nile, stagnation was apparent. However, a new impetus to growth and development occurred from about 1820 and this can be identified as the next major revolution in Sudanese agriculture.

The Egyptian conquest and the internationalization of agriculture

In 1820/1 the Sudan was invaded from the north by Egyptian forces acting in the name of Muhammad Ali, the ruler of Egypt. The conquest had many objectives such as the exploitation of Sudanese resources, including minerals and ivory, and the acquisition of manpower. However, there was also a scientific dimension and scientists, many of them French, accompanied the expeditions. Through the next few decades the Egyptians and Sudanese exchanged a number of agricultural ideas.

At the time of the invasion, the area around the Funj capital of Sennar was best documented. As the centre of the region it received a sequence of ambassadors and visitors and a series of interesting pictures about life in that region were created (Hill 1970). It is clear that by the mid-1820s little modernization had occurred. Away from the Nile, large amounts of land were still owned by the merchants and were worked by slave labourers, who tended the animals and were responsible for most of the day-to-day cultivation. *Dura* was the most important crop and it was usually cultivated by a team of three: one man to prepare the ground with a pronged stick; one to sow the seed in the hole; and one to cover the seed. Together they also constructed the simple bunds which identify their practices as a form of *terus* cultivation. This system trapped rainfall and increased the amount of water infiltration into the soil; it is still practised in many areas of the central Sudan. This type of activity, despite being simple and effective, was probably amongst the most advanced anywhere away from the banks of the Nile. With hindsight it can be seen that new changes were only likely to be externally induced:

> The economic transformation was beyond the power of the Sudanese. Export guidance must come from Egypt, for in general the Egyptians farmed better and worked harder than the capricious natives who alternated between tribal war and primitive agriculture and cattle-raising (Hill 1959, p. 49).

The scientists working for the Egyptians were keen to learn and to experiment since they were under pressure to boost the agricultural yields from the newly-acquired territories. New innovations were combined with renovation of old neglected practices. The transformation of agriculture could be seen with respect to irrigation, improved utilization of the rainlands, and new exotic crop species.

The sophisticated basin-irrigation techniques of the Meroitic Empire were resurrected. These had lapsed as a result of poor management and lack of maintenance. New canals and ditch digging began in Berber Province in the north and a large new canal was cut, organized by experienced engineers from Egypt, to open up the great Basin of Kerma, near Dongola, for intensive cultivation. Elsewhere, the *saqqiya* was introduced into Upper Nubia for the first time, allowing the irrigated cultivation of new exotic fruits and vegetables (below).

Perhaps the most innovative of all the new ideas came in the rainlands, especially in Kordofan. Despite its long history, the *saqqiya* was still little used in the Sudan, but the Egyptians saw that it could be used to provide a sequence of wells along the roads from Sennar to Kordofan. Once water was guaranteed, cattle could follow this route and thence to the markets of Egypt (Hill 1959).

The Egyptian markets were responsible for other accelerations in agricultural production, notably gum arabic for the paper trade. The best sources were around Sennar and in Kordofan and, from 1825 onwards, the Egyptians sent concessionaries to both these areas. In 1827/8 no less than 1270 tonnes of gum arabic were transported down the Nile (Hill 1959), most of it reaching Alexandria for export.

Although gum had been sold for centuries (the French imposed a special tax on it as early as 1349 (Beshai 1976)) its markets remained small. The supplies which were channelled through Egypt prompted a more widespread demand for the product and it was soon transported to the main gum trading centre in Trieste, by way of the Saharan caravans and boats from Tripoli, thus avoiding Egypt completely.

Perhaps the most extraordinary and far-sighted introduction in this period was a form of early locust control. Designed mainly to protect Egyptian territory from invasion by 'Sudanese' locusts, the system consisted of a series of trench fires set up along likely locust routes to divert the insects to other areas. Despite its obvious political implications the idea has remained and is still practised in parts of the Sudan.

Without doubt the most important agricultural transformation between 1820 and the mid-1860s was the introduction of new crop species. The invading Egyptians found cotton of exceptional quality in the area around Sennar and samples were sent back to Egypt. As a result cotton seed was shipped to Egypt and elsewhere and good quality cotton was produced in Sudan, particularly in the Dongola area of northern Sudan. In the 1860s the trend was reversed when the hardy Egyptian cotton was introduced into the Tokar region and elsewhere in eastern Sudan (Beshai 1976). The main stimulus was commercial gain. Shrewd Egyptian entrepreneurs saw a vast gap in the world cotton market at a time when Western Europe was industrializing on a backbone of textile manufacture and the main supplier hitherto, the southern states of the USA, was decimated by civil war. Although cotton cultivation stagnated during the Mahdiya (below), it soon recovered and has continued at the forefront of Sudanese commercial cultivation ever since.

Despite the long-term success of cotton, the most dramatic short-term innovation came with the introduction of sugar-cane. The Viceroys had tried to introduce a number of new crops from Egypt, the most notable failures being coffee, opium, and rice. With sugar-cane there were few problems, although it is a thirsty crop and put great pressure on the *saqqiya* irrigation systems (Hill 1959). Cane production expanded rapidly, especially in Berber and Sennar Provinces and by mid-century a large state plantation had been established at Kamlin, a centre for the rapidly expanding cultivation of indigo, another crop introduced successfully. There is no real market for crop dyes now, but at the time indigo was important in the textile and printing industries.

In addition to importing tree products, most notably gum, the Egyptians saw the value of importing whole trees for their wood content. *Sunt* (*Acacia arabica*) was brought up from the southern regions to be used for boat building, whilst ebony (*Dalberqia melanoxylon*) was carried down the Nile. Trade was not just one way, and the Egyptians introduced many new fruit trees and vegetables to areas of the Sudan where they were hitherto unknown:

> He (Ahmad Pasha) had also introduced the culture of the sugarcane, refining it and making rum. . . . lemons, apples, bananas, figs, sweet lemons, water melons, melons, cucumbers and a variety of useful vegetables have also been introduced (Anon. 13 September 1847, quoted in Santi and Hill 1980).

In addition to this impressive list, records show that grapes, pomegranate trees and others were sent to the administrators in Sennar as early as 1833.

By the time that the Egyptian Empire in Sudan began to be challenged by a new, Sudanese Islamic uprising in the 1880s, few could argue that the Egyptian influence on Sudanese agriculture had been enormous. The stagnating systems of the Funj had been transformed by a variety of new innovations and, perhaps most significantly, the production of crops and animals was no longer for subsistence and local barter, but for commercial reasons with vast new markets in the growing industrializing world of the mid- to late-nineteenth century.

The period of the Mahdi and his successor, the Khalifa, between 1881 and 1896, was a period of extreme turmoil in Sudan. Agriculture, beyond that of a basic subsistence necessity, was abandoned (Holt 1970) as populations fled the Islamic forces (Abd el Ati 1976) or flocked to support them. One simple expression of this was the decline in gum arabic exports, which fell from 7543 tonnes in 1881 to a mere 7 tonnes in 1890. By 1901, after the successful reconquest led by Kitchener (1896/8), exports had recovered to 7695 tonnes (Beshai 1976).

The only major agricultural development during the Islamic period took place in the territories of eastern Sudan beyond the effective influence of the Khalifa. In 1892, Sir William Garstin, of the Egyptian Irrigation Service, visited Tokar in the Ethiopian foothills (Mackinnon 1948). Whilst there he conducted a number of studies of the floods and silts which came down from the Ethiopian Highlands during the wet season. As a result, the first proposals were made to utilize the floodwaters for flush irrigation, a system in which the planting of crops could by synchronized with the availability of water and fertile soils. Although this project did not come to fruition for many decades, the proposal was the only major agricultural development during the last quarter of the nineteenth century. The flush system of irrigation still operates in several tributary deltas in the Sudanese territories close to the Ethiopian Highlands.

The Anglo-Egyptian Condominium and the regionalization of agriculture

The defeat of the Khalifa in 1898 was followed by the establishment of the Anglo-Egyptian Condominium. Despite the formal title, Sudan effectively became a British colony and for the next half-century the British influenced most of the major advances in Sudanese agriculture. No longer were these developments piecemeal. Grand developments were initiated, including one encompassing the whole Nile Basin. Sudan's first large-scale agricultural projects were established at this time, and it is possible to classify the developments into large-scale irrigated agriculture, small-scale irrigation, and mechanization.

Although the concept was first proposed as early as 1839 (Barnett 1977), the first attempts to develop a large-scale irrigated cotton project came in the wake of some experimental cotton growing at Zeidab, in northern Sudan, in 1904. Supported by further successful experiments on the clays of the Gezira, most notably at Haj Abdullah, the potential for cotton production was acknowledged. This was extremely important to the British, as there was an almost insatiable demand for cotton from the growing cotton industry in Lancashire, especially when supplies from Egypt and the USA were proving to be unreliable (Gaitskell 1959). Feasibility studies were conducted and the Gezira area, the clay plain between the Blue and the White Niles south of Khartoum, was found to be ideal. Not only were the soils adequate, but the indigenous population were keen to experiment with new ideas following a series of bad harvests for their traditional grain crops. Above all, however, the gradient of the plain was such that if the Blue Nile could be dammed at Sennar, water could be distributed by gravity over more than a million feddans. The geology of the Nile Valley at Sennar was suited for the construction of a dam and so the project, perhaps the most ambitious agricultural project ever contemplated in the world at that time, was agreed.

The First World War delayed the project, but the Gezira Scheme was opened finally in 1925 and, despite initial teething troubles, proved a great success. Not only was it the right project in the right place at the right time, but the social and financial organization was acceptable to the local inhabitants, together with adequate provision for the production of food and grain products in addition to cotton. The project was well-managed and had good supporting infrastructure such as a custom-built railway and scientific laboratories. Additional advantages accrued from the reservoir above the dam, where *saqqiya* and small pumps facilitated a proliferation of small pump schemes, producing grain crops, fodder crops, and vegetables.

Whilst the cultivation of cotton was expanding rapidly and efficiently along the Blue Nile, the governing authorities were sceptical whether the same could be done along the White Nile. In 1927, however, one of the Mahdi's descendents living on Aba Island near Kosti, persuaded the administration to allow him to grow cotton over 250 feddans there (Trilsbach 1987*a*). Yields were surprisingly good and this encouraged the Sudanese Government to set up its own pump scheme the following year at nearby Ed Dueim (Ali undated). In 1934 cotton production expanded further along the White Nile when the Mahdi family obtained four more cotton-growing licenses (Dirar 1970).

Agricultural production along the White Nile was transformed in 1937, with the completion of a dam at Jebel Aulia, 50 km south of Khartoum (Trilsbach 1987*b*). It was built, principally, to help even out the discharge of the Nile system in Egypt. Along the White Nile, however, a vast reservoir was created some 300 km long. The Egyptian Government paid compensation for social and economic disruption, but also financed the creation of seven new 'alternative livelihood' irrigated cotton schemes (Sudan Government 1955). Above the dam, the raised level of the water for much of the year facilitated efficient irrigation by diesel pumps, which were operated alongside more traditional *saqqiya* and *shaduf*. The growth of pump irrigation continued to expand through to Independence and was encouraged by the shortages of cotton supplies generated by political crises in the competing areas of Korea and Egypt.

Mechanization of the rainlands was the third arm of agricultural modernization during the Condominium period. Small-scale experiments took place during the 1930s for the mechanized production of grain crops in the Butana Plain, but the real acceleration came in the wake of the Second World War when there was a noticeable shortfall in grain production in eastern Sudan, especially in the Gedaref area. Originally, the plan was to concentrate on the production of sesame (*Sesamum orientale*) and *dura*, using a mixture of unskilled labour and imported machinery. Little consideration was given to the environmental suitability of the method (Habashi 1968), nor was much thought given to the management and profit-sharing system. Experiments began at Ghadambaliya in 1945. The nearby railway line was to be the principal route for export and quick profits were anticipated. Initially, the scheme was a disaster. The technical planning was minimal and there were insufficient incentives for the majority of the participants. The whole project was reappraised in the mid-1950s. Consequently, it was handed over to private investment and infrastructural improvements, most notably in domestic water supply, were made (Davies 1964). Quick profits were gained by exploiting the soils and the mechanization programme was soon expanded into neighbouring areas, plus a small area in the Nuba Mountains region in the heart of the country. However, in retrospect, it can be seen that the regions have benefited little. The land has been decimated of natural vegetation and is prone to rapid soil erosion (Cloudsley-Thompson 1978), whilst the workers have gained little reward; the managers were the principal profiteers. Nevertheless, the mechanization of rain-fed lands still dominates the southern Butana.

It can be seen that nearly all of the main agricultural transformations during the Condominium were located in the central Nile Valley, especially the Gezira, and in eastern Sudan. This is significant because it increased the economic and developmental gulf between northern and eastern Sudan, on the one hand, and between western and southern Sudan, on the other.

The grassland areas of Kordofan and Darfur received little investment in agriculture, although in many localized areas water supplies were improved. Perhaps the most important development was the construction of the railway line to El Obeid, which facilitated a more efficient marketing of rainland produce, especially gum arabic.

However, if the west received little investment in agriculture, the southern areas received virtually none. Very little has been said about the South in this chapter. It was not until the mid-nineteenth century that there was any real contact between the northern Sudan and the lands beyond the Sudd marshes. Agricultural diffusion was minimal and in many areas the southern economy was still based on a hunting and gathering subsistence lifestyle, supported in the tsetse-free areas by pastoralism, especially cattle rearing by the Nilotics. Throughout the Condominium, the South was treated as a quaint backwater of the 'Empire', where the most ambitious product was the recounting of adventures. No attempt was made to modernize production. Indeed, some were keen to preserve a part of the world which many felt had not emerged from Neolithic times. It was only after some of the southerners provided valuable service in the King's African Rifles during the Second World War that any attempt at modernization was made. The ensuing project, the Zande Scheme, was ill-conceived and a poor attempt to stave-off some of the political turmoil expected with the granting of Sudanese Independence, and caused by the divisions between the backward economy (agricultural and otherwise) of the South and the rapidly expanding economy of the northeast.

Despite various attempts since Independence to modernize agriculture in southern and western Sudan, the projects have received but a fraction of the investment received elsewhere in the country. The regional divisions in agricultural practices and efficiency remain virtually unchanged and contribute to the political and regional problems which exist in Sudan today.

Agricultural modernization in independent Sudan

The Sudan became independent at the beginning of 1956 and inherited a complex of traditional agriculture, both arable and pastoral, together with a variety of irrigated systems and mechanized cultivation. Ironically, it also inherited a number of technical reports on potential developments in Sudanese agriculture. The irony was that the granting of Independence coincided with a long sequence of lower than average rainfall; what was possible in the wetter 1930s and 1940s was much less so in the drier 1960s and in recent years (Hulme and Trilsbach 1986).

The principal agricultural developments since 1956 are discussed in other chapters of this book, so this section does little more than highlight the principal themes of the continuing story of agricultural development.

Following the successful coup of Colonel Nasser in Egypt, the victory over the Anglo-French forces in the Suez in 1956, and economic links with the USSR, Egypt proposed to build a dam within its own territory to protect its water supplies. Partly financed by the USSR and confirmed by negotiation with the Sudan during the 1959 Nile Waters Agreement, the High Dam was constructed at Aswan.

The principal international effect of the construction of the dam was that the reservoir above the dam flooded parts of northern Sudan, most especially large parts of the Sudanese town of Wadi Halfa. The Egyptians agreed to pay compensation for the flooding and for the resettlement of the Nubians. As part of the 1959 Nile Waters Agreement, Sudan's allocation of Nile waters was also increased substantially. The combined effects of the financial compensation and the greater availability of water led to the development of new large-scale irrigation schemes. The Gezira Scheme was expanded with the Managil Extension, whilst many of the displaced Nubians were resettled on a new irrigation scheme on the river Atbara at Khashm el Girba. The Khashm el Girba Scheme, recently renamed the New Halfa Scheme, was an attempt to grow cotton away from the Blue and White Niles and to combine it with the production of other crops such as sugar.

These developments overshadowed other agricultural changes in the 1960s, but other projects were initiated. Small pump irrigation schemes were set up along the White Nile with both private and government capital, whilst a series of irrigated sugar schemes were developed along both the Blue and White Niles. Elsewhere in the country, a number of agro-industrial schemes were established with Eastern European funding to help boost specialized agriculture. Examples included fruit and vegetable canning factories at Karima and Wau, an onion processing plant at Kassala, and a milk dehydration factory at Babanusa. These proved to be failures, mainly due to insufficient planning (Williams 1987), but they were attempts to diversify agricultural investment. Perhaps of greater importance was the extension of the railway to Nyala, in Darfur, and to Wau, in the south. This facilitated the easier transportation of rain-fed crops and livestock produce to the capital and the other major cities.

The one other major investment linked to the 1959 Nile Waters Agreement was the construction of a new dam at Roseires on the Blue Nile, south of Sennar. Although it was built for a number of reasons, its real value emerged in the 1970s with the next major spurt in agricultural investment.

Since Independence, the principal themes in the Sudan have been the continued civil war between the North and the South, which led to a collapse rather than development of the meagre southern agricultural economy, and a succession of governments, both military and civil. In the short-term, the military coup of Numeiri in 1969 helped to eliminate these two problems. Numeiri retained power for 16 years, albeit with major shifts of policy and political/religious allegiance and the civil war was brought to a conclusion.

After giving concessions to the southern peoples, Numeiri helped to placate the Muslim northerners by drawing Sudan more firmly into the Arab economic sphere. Coinciding with the vast increase in oil revenues achieved by neighbouring petroleum states, the Sudan was deemed to be the 'bread basket' of the Arab world. Many Arab States, especially Kuwait and Saudi Arabia, invested large sums in Sudan to help finance the expansion of agriculture. The Rahad Scheme was constructed on the Butana Plain, utilizing water drawn from the new Roseires Dam and transported by a complicated network of canals, pipes, and siphons. Further east, a new modern high-technology sugar refinery was the centre piece of the Kenana Scheme, built close to Rabak.

The success of these schemes is discussed elsewhere in this book, but it is worth considering the rain-fed and traditional economies during the period since Independence. Very little money was invested in the southern or western Sudan during the whole period. The major investments, ostensibly to bolster the national economy, were in prestige industrial or infrastructural schemes, such as the Jonglei Canal project and the development of oil reserves. These were perceived by many to be for the benefit of the population of the north-east of the country. Other investments have often been ill-conceived, with the most notable major project, the Habila Mechanization Scheme in Southern Kordofan, being a potential environmental disaster. The small-scale subsistence farmers, nomads, and semi-nomads received very little assistance, financial or otherwise. Methods developed little but, stimulated in part by a rapidly growing population, pressure on the land increased. Combined with a quarter-century of relatively dry years (Trilsbach and Hulme 1984) the environmental consequences have been enormous, with a decimation of the natural vegetation and spectacular erosion. Because of this and together with inadequate communications and aggravated by the deterioration of the railway system, production became marginal with only rare surpluses to avert potential crises. Lack of central organization from Khartoum or the main regional centres, plus a series of economic disincentives for grain production, made famine inevitable. The first emerged in the late-1960s and has continued more-or-less ever since, reaching a crisis of staggering proportions in 1984/5.

The most spectacular feature of the agricultural disaster in western Sudan and also in parts of eastern Sudan, has been the contrasting lack of investment and interest at a time when enormous sums were being expended on large-scale irrigation projects. Now, more than ever, Sudan's agricultural profile is characterized by a gulf between the two Sudans, not the North and the South, as political scientists would argue, but between the irrigated, primarily cash-crop production areas and those where production is still largely subsistence-based.

Conclusions

It is possible to identify a number of important phases in Sudan's agricultural history.

Despite the sophistication of the Ancient Egyptians and their political dominance in parts of northern Sudan, they had little influence on agricultural development. Simple cultivation techniques and pastoral routes were already established before their arrival and, although they helped to expand many of the practices, often helping to refine them a little, they contributed little in the way of new ideas. The one major exception was the *shaduf*, which can still be seen operating successfully in many areas of Sudan.

The first major agricultural revolution occurred during the Meroitic Empire. New crop production techniques were developed and, in particular, advanced irrigation systems were introduced to provide water for crops grown away from the immediate vicinity of the Nile. In many ways the scientific advances during this time can be compared with those of its contemporary empire based in Rome.

From around AD 300 there was little agricultural advance for one-and-a-half millennia. Indeed, as in the Dark Ages of post-Roman Europe, many of the advanced systems, developed during the Meroitic Empire, fell into disrepair and were forgotten. Apart from the invasion of the Arabs in the thirteenth century and after, which established many of the traditional pastoral areas of western and eastern Sudan, very little of note occurred before the coming of the Egyptians in the 1820s.

The Egyptian invasion inaugurated a half-century of rapid agricultural development, with the introduction of many new crops and ideas. This was the time when development focused on the production of crops for export and international trading patterns have played a key role on the fortunes of Sudanese agriculture ever since. The advances of this time suffered some setbacks during the brief rise of the Islamic forces in the last quarter of the century, but the arrival of the Anglo-Egyptian armies at Omdurman in 1898 heralded a new phase of agricultural development.

The half-century or so of 'effective' British influence witnessed the development of large-scale agricultural schemes, most notably the Gezira Scheme. It was also a time when the Nile was controlled more fully with dams constructed at Sennar and Jebel Aulia. However, with hindsight, it is possible to speculate that the most long-lasting agricultural consequence of the period was the focusing of investment in the north and east of the country, with virtually no significant projects in the west and south. This was to exaggerate regional divisions when Independence was achieved.

The granting of Independence in 1956 saw the beginning of a series of political upheavals and the full scale savagery of civil war. Whilst this was occurring, agricultural development concentrated further on the north and east, thereby exacerbating regional trends. The large-scale Managil Extension and the Khashm el Girba Schemes of the 1960s were followed by the Rahad and Kenana Schemes of the 1970s. In each case there were managerial problems, insufficient planning, and worse than expected yields.

One effect was to bankrupt the country and to reduce further any possibility of effective agricultural investment in the west and south. The resurrection of civil war and regional disputes, entailing massive defence expenditure will do little to assist agricultural development in the foreseeable future.

References

Abd el Ati, H. A. (1976). Dueim town growth and functions. Unpublished B.A. thesis. University of Khartoum.

Ali, S. M. (undated). The Ed Dueim pump scheme. Unpublished B.A. thesis. University of Khartoum.

Arkell, A. J. (1949). *Early Khartoum*. Oxford University Press, London.

Barnett, A. (1977). *The Gezira Scheme. An illusion of development*. Frank Cass, London.

Beshai, A. A. (1976). *Export performance and economic development in Sudan, 1900–67*. Ithaca Press, London.

Cloudsley-Thompson, J. L. (1978). Human activities and desert expansion. *Geographical Journal*, **144**, 416–23.

Davies, H. R. J. (1964). An agricultural revolution in the African tropics. *Tijdschrift voor Economische en Sociale Geografie*, **55**, 101–8.

Dirar, A. R. M. A. (1970). The economics of agricultural production in the private pump schemes—Kosti and Ed Dueim districts. Unpublished M.A. thesis. University of Khartoum.

Emery, W. B. and Kirwan, L. P. (1935). *Mission archeologique en Nubie, 1929–1934. The excavations and survey between Wadi es-Sebua and Adindan, 1929–1931*. 2 Volumes. Government Press, Cairo.

Firth, C. M. (1910). *The archaeological survey of Nubia*. National Print Department, Cairo.

Gaitskell, A. (1959). *Gezira: a story of development in the Sudan*. Faber and Faber, London.

Habashi, W. (1968). *The development of agricultural production in the Sudan*. Lectures presented in arabic to the Department of Extra-Mural Studies. University of Khartoum.

Hakem, A. A. (1981). The civilization of Napata and Meroe. In *Ancient civilizations in Africa. UNESCO general history of Africa volume 2* (ed. G. Mokhtar), pp. 298–321. Heinemann, California.

Hill, R. (1959). *Egypt in Sudan, 1820–1881*. Oxford University Press, London.

Hill, R. (1970). *On the frontiers of Islam*. Clarendon Press, Oxford.

Holt, P. M. (1970). *The Mahdist state in the Sudan*. Clarendon Press, Oxford.

Hulme, M. and Trilsbach, A. (1986). Rainfall trends and rural changes in Sudan since Nimeiri: some thoughts on the relative significance of environmental changes and political control. In *Sudan since Nimeiri* (ed. P. Woodward), pp. 3–17. School of Oriental and African Studies, University of London, London.

Krzyaniak, L. (1978). New light on early food production in the central Sudan. *Journal of African History*, **19**, 1959–72.

Mackinnon, E. (1948). Kassala Province. In *Agriculture in the Sudan* (1st edn) (ed. J. D. Tothill), pp. 699–735. Oxford University Press, London.

O'Fahey, R. S. and Spaulding, J. L. (1974). *Kingdoms of the Sudan*. Methuen, London.

Santi, P. and Hill, R. (1980). *The Europeans in the Sudan*. Clarendon Press, Oxford.

Shinnie, P. L. (1978). The Nilotic Sudan and Ethiopia, *c*.660 BC to AD 600. In *The Cambridge history of Africa* (ed. J. D. Fage), Vol. 2, pp. 210–71. Cambridge University Press, Cambridge.

Sudan Government (1955). *Irrigation by pumps from the Nile in the Sudan*. Ministry of Irrigation, Khartoum.

Trilsbach, A. (1987*a*). Agriculture west of the White Nile. In *Rural development in White Nile Province, Sudan* (ed. H. R. J. Davies), pp. 46–64. United Nations University Press, Tokyo.

Trilsbach, A. (1987*b*). Environmental change and village societies west of the White Nile. In *The Middle Eastern village. Changing economic and social relations* (ed. R. I. Lawless), pp. 13–50. Croom Helm, London.

Trilsbach, A. and Hulme, M. (1984). Recent rainfall changes in central Sudan and their physical and human implications. *Transactions of the Institute of British Geographers* (NS), **9**, 280–9.

Wendorf, F. (1968). *The prehistory of Nubia*. Fort Burgwin Research Center and Southern Methodist University Press, Dallas.

Williams, J. (1987). Babanusa 'milk' factory. *Sudan Studies*, **2**, 4–6.

13

Systems of agricultural production in northern Sudan

R. T. WILSON

Introduction

For the purposes of this chapter, northern Sudan comprises that part of Northern Darfur Province lying north of approximately 16°N latitude and the whole of Nile and Northern Provinces. The total area is about 650 000 km^2. With the exception of narrow strips of land along the Atbara and Nile Rivers it is wholly desert.

Excluding the largely unoccupied area of Northern Darfur, the population of the two provinces along the Nile is of the order of 6–700 000 people. About 500 000 of these are in Nile Province, the remainder in Northern Province. Extrapolated to an average density, this is somewhere in the region of 3.5 persons per km^2 in Nile Province and 0.5 persons per km^2 in Northern Province. In both provinces, however, the population is concentrated in the major towns along the main Nile (Wadi Halfa, Dongola, Merowe/Karima, Berbera, Atbara/Ed Damar, Shendi) and the Atbara Rivers in the fertile river valleys and along their irrigable terraces. In the valleys and along the terraces, population densities are as high as 150 persons per km^2 along a stretch north of Atbara and along the Merowe–Dongola reach in the great bend of the Nile between 18° and 19°N and 30° and 32°E. Away from these areas, the permanent population is effectively zero. The two or three small oases, including the famous Selima and Laqiya Arba'ín oases on the Darb el Arba'ín, in Northern and Nile Provinces which have long been considered havens of productivity in the surrounding hostile desert environment, have very small numbers of permanent residents and are only occupied seasonally. Northern Darfur also boasts two famous oases, Nukheila and El Atrun (from which salt is mined) which are not permanently occupied.

The Nile has for a long time been the principal means of transport in the north–south axis. East–west movements have been by pack animal, mainly the camel. The famous, or perhaps more correctly notorious, Forty Days

Road or Darb el Arba'in traverses the area in a north-east direction from El Fasher in Northern Darfur to Assiut in Middle Egypt. Part of this route is probably still used for the illegal (i.e. not officially recorded) export of camels from Sudan to Egypt. Nowadays river traffic is largely seasonal and on a small scale. The railway line follows the river as far north as Abu Hamed and then continues directly across the desert in a NNW direction to Wadi Halfa. A branch of the railway strikes off in a south-westerly direction from Station Number 10, following the course of the Nile in the same direction, as far as Merowe. There are no roads and transport other than by river or rail (air services are irregular to non-existent) is accomplished across desert dune or rattling rock in *suq* lorries or 4-wheel drive cross-country vehicles. Wheeled transport averages not more than 30 k.p.h. and is very uncomfortable at that speed.

Population

The whole of the Nile Valley area, from the second cataract near Wadi Halfa in the north to upstream of the sixth cataract at Sabaloka Gorge near Khartoum in the south, is known as Nubia. Ethnic groups in this area are now very diverse and differ considerably in origin. The original Nubians were Negroid, probably occupied the whole of the valley and undoubtedly spoke a Nubian language of the Eastern subfamily of the Sudanic group (Murdock 1959, p. 161). The Nubians are now scattered as far west as Darfur and as far south as Dilling in the Nuba Mountains in Southern Kordofan. In the Nile Valley they are now essentially confined to north of the fourth cataract at Merowe. Many Nubians, who had survived for thousands of years without being displaced from their original home, were forced to migrate from their ancient land between the first and second cataracts during the 1960s. The rising waters behind the Aswan Dam had reached Wadi Halfa by 1964 and Nubians (including 11 000 from old Wadi Halfa town) were transported to New Halfa and the Khashm el Girba Irrigation Scheme on the Atbara River (Sørbø 1985).

Southwards from the fourth cataract, there is now a kaleidoscope of ethnic groups mainly of Arab descent. In addition to the ubiquitous and largely detribalized *jallaba*, or traders, of the towns the major groups along the river are the Gaalin, Hasania, and Shayqíya. East of the river, in the desert and from south to north, the major pastoral groupings are the Arabic-speaking Shukriya and the Beja-speaking Hadendowa and Busharayen. West of the river the livestock owning tribes are Kababish and Bedeyat with smaller incursions of many other groups.

History of settlement

There is archaeological evidence of settlement in the valley and on the terraces dating from the Lower Palaeolithic. Radiocarbon dates from the Middle Palaeolithic provide dates in the region of 36 000 BP. From this time an increase in the intensity of occupation can possibly be related to a longer duration of settlement by formerly more nomadic groups. The Late Palaeolithic in the Nile Valley seems to date from 18 000 to 7000 BP. At that period, as at present, settlement on the terraces provided access to resources both from the river and from the surrounding desert. No agriculture was practised. There is evidence that fish were a major source of food and wild game included hippopotamus from the river, and hartebeest, African wild ass, and hares from its hinterland. Gathering of wild grains at appropriate times of the year would also have been a seasonal occupation (Hassan 1980).

The Mesolithic period and the beginning of a transition to farming probably started about 7000–5000 BP. It is probable that at that time rainfall was probably about 500 mm per year and the major economic activities were still hunting, some gathering, and the exploitation of aquatic resources. Three dates obtained by ^{14}C analysis for a site at Abu Darbein, 2 km east of Atbara, yield estimates of 8500 ± 100 BP, 7860 ± 190 BP and 7700 ± 140 BP (Haland 1988, p. 49). Bone and lithic artefacts were common at the Abu Darbein site, mostly usable for hunting, although a few small grinders and hammerstones indicated at least some use of grains as food. The faunal remains show that a wide range of aquatic resources including fish, shellfish, and crocodiles was exploited, as well as antelope and buffalo. In some areas there were numerous seeds of the wild plants *Ziziphus* and *Celtis*.

The Neolithic and the transformation to a more settled existence obviously overlaps the Mesolithic. From about 6100 BP an increasing percentage of domestic animal bones, particularly of cattle, goats, and sheep, indicates a move towards more intensive food production but with the emphasis still on animal husbandry rather than on farming. The little evidence that exists for the utilization of grain suggests that gathering of wild seeds, particularly of *Sorghum*, was the predominant agricultural activity. Animal husbandry would be at least seasonally migratory, the river being used as a dry season refuge.

There is an unexplainable gap in the archaeological evidence (Haland 1988, p. 49) from about 5000 BP to about 3000 BP, which is the beginning of the Meroitic period. It is possible that this gap occurs because the main economic activities were still related to nomadic animal husbandry. From this period *Sorghum* was clearly cultivated and millet had also been domesticated and was being used as a food crop. In addition to the Sudanic staples of sorghum and millet, other crops which were introduced for cultivation were water-melon, gourds, okra (ladies fingers), sesame, and cotton.

Egyptian complex crops introduced over varying periods of time, particularly during the Pharaonic period, included barley, dates, peas, lentils, figs, carob, wheat, and grapes. Much more recently American crops such as maize, groundnuts, peppers, and varieties of squash have been introduced.

Following the Pharaonic period, Nubia was independent for a long time, with its seat of government first at Napata and subsequently at Meroe (east of Shendi and not Merowe in the Nile bend). It was at Meroe that iron working was first developed, from which it spread to the rest of Africa (Mair 1977). Following a period of anarchy and Beja incursions, Christianity arrived in Nubia during the sixth century AD. Isolated from other Christians by the Arab spread into Egypt in the eleventh century, Nubia remained Christian until AD 1315 when a Moslem king acceded to power (Murdock 1959, p. 160).

The northern Sudan environment

Climate

The climate of northern Sudan is one of the harshest in the world (Table 13.1). A brief and unreliable wet season produces rain (in some years and in some months) from July to September, but the coefficient of variation exceeds 100 per cent. Low rainfall in the south of the area decreases rapidly northwards so that it is effectively zero at the Egyptian border.

TABLE 13.1. Northern Sudan: climatic data for Dongola (19°10′N; 30°29′E; 225 m alt.)

Month	Mean temperature (°C)		Rainfall (mm)		Evaporation (mm/day)
	Maximum	Minimum	Mean	Maximum	
January	28	9	0	0	7.9
February	29	10	tr	1	9.5
March	34	14	0	tr	11.8
April	38	18	0	tr	14.5
May	42	22	1	20	17.0
June	43	24	tr	tr	17.8
July	42	25	9	50	15.7
August	42	26	13	56	14.5
September	41	25	tr	3	15.1
October	39	21	tr	3	13.7
November	33	15	0	2	10.0
December	29	11	0	tr	7.8
Annual	37	19	23	60	13.0

Source: Griffiths and Soliman (1972).

Summers are extremely hot with a mean maximum of about 41 °C (and an absolute maximum of more than 50 °C) and a minimum of 25 °C. The large diurnal variation and the low relative humidity provide some relief from the enervating process in the early mornings. Winters can be described as cool, with temperatures below 0 °C occurring in the north and minimum air temperatures of 5 or 6 °C in the south of the area: in winter mean maximum and minimum temperatures are about 30 °C and 15 °C.

Sunshine averages 80–90 per cent of possible hours on an annual basis Wadi Halfa is one of the sunniest places on earth with about 4000 hours per year. Even during the wet season the average daily sunshine is about 8 hours, equivalent to 60 per cent of the maximum possible.

Potential evaporation is very high, being more than 4 metres at Merowe and almost 5 metres at Dongola. Strong northerly winds blowing for much of the long dry season contribute to the evaporation potential but, conversely and because they also transport sand and reduce the sun's effects, they also assist to some extent in reducing evaporation. The strong winds are the sources of the duststorms, known in Sudan as *haboob* (Arabic, to blow), which are the bane of northern Sudan, leading to sudden temperature drops and greatly reduced visibility. As many as 18 *haboob* may occur in a year and an average of five at Atbara during August.

Soil

The soils of northern Sudan are basically of two types: imported soils brought down by the river and deposited in its valley and on its terraces; and resident soils which have evolved *in situ*.

It is the Nile, originating in the highlands of Ethiopia and of central and eastern Africa up to 4000 km to the south, which is the life-force of northern Sudan. Without its depositions of soil and its constant flooding, hardly any agricultural activity would be possible in the region. Valley and terrace soils are alluvial and fluviatile, fairly heavy, with a large percentage of fine clay particles. They are often difficult to cultivate. Nitrogen is deficient and there are apparently poor responses to applications of phosphorus and potassium. Areas which are naturally flooded in most years, in the valley and on the lower terraces, pose fewer problems than the higher terraces.

The higher terraces are irrigated by lifting water from the river. Formerly this was done by hand, by the traditional *shaduf*, or by the Persian water wheel or *saqqiya* which is powered by camel or ox. In later years mechanical pumps have replaced human and animal power. The *shaduf* can irrigate little more than a few tens or at most a few hundreds of square metres. The *saqqiya* can irrigate as much as 2 hectares at high water but somewhat less than 1 hectare at low flood. By contrast, some of the so-called pump-schemes with large mechanical engines are capable of irrigating several hundreds to several thousands of hectares. Irrigation imposes a severe penalty on river soils through the attendant risks of salinity and alkalinity. These risks were established as early as 1904 when some of the first soil analyses were done. Good drainage can reduce the salinity hazard and, theoretically, alkalinity can be reduced by applications of gypsum but leaching following further irrigation is an additional problem.

The river flows over Basement Complex rocks of Precambrian age and those of the Nubian sandstone series. These geological formations are the parent material of the soils of northern Sudan away from the river. Much of the desert is of gravel and stone with only thin layers of soil. For thousands of years aeolian action has shifted sand to form and reform massive dune complexes.

Water

The main Nile and its major tributary, the Atbara, are virtually the only sources of permanent surface water. The few oases provide water at reasonably shallow depth and there are some deep bores in the Nubian sandstone series. The ratio of low water to high water in the main river on an annual basis is more than 1000 per cent. While the Nile itself is very flashy, the Atbara is even more so with water flowing along it for a maximum of 4 months and a maximum mean monthly discharge (1912–62) in excess of 2100 m^3 per second in August. The Atbara's lowest August discharge in that period was 1450 m^3 per second, the highest being 4100 m^3 per second. The mean long-term contribution of the Atbara amounted to 22 per cent of the Nile water north of its confluence with the main river, but in recent years this has been much reduced due to prolonged drought in the northern Ethiopian Highlands.

Occasional water sources can be created in some of the fossil river beds such as the Wadi Howa in Northern Darfur or the Wadi el Melik in the western part of Northern Province but, in general, the desert is dry. The vast reservoir of Lake Nuba remains little exploited.

Normal annual fluctuations in river levels create several problems for agricultural production. A poor high flood, amounting to only a few centimetres below an average level, means that large areas of potentially irrigable land can not be watered. The situation is further aggravated if a low level is compounded by a short duration. In many years, more than 50 per cent of potentially irrigable land is not used, but the frequency of such years has increased considerably since the recent series of drought years began in the late 1960s.

Vegetation

Slightly over 200 species of plants are known to occur in the Nubian region. Only nine (4 per cent of these species) are considered to be endemic

(Ghabbour 1972). Most plants (about 69 per cent) are not even indigenous to Africa but are widespread aliens that are distributed through the activities of man. The riverain flora can be assigned to four habitats (El Hadidi 1976).

The water plants habitat is characterized by species such as *Potamogeton* spp. which grow in the slow-flowing river and irrigation canals. In the Nile bank habitat annual species, whose seedlings appear after the retreat of the annual flood, occupy the lower silt terraces: typical genera include rushes such as *Cyperus*, and *Senecio, Potentilla*, and *Alternanthera*, but there are also a few perennial shrubs, including *Dichrostachys cinerea*, and an introduced and noxious American weed which is now naturalized along much of the Nile and elsewhere in Africa, *Argemone mexicana*. The cultivated land habitat has a flora mostly composed of weeds and is restricted to the higher levels of the river terraces.

The scrub-palm habitat provides the widest variety of plant life and consists predominantly of indigenous trees, shrubs, and tall grasses which were originally found on the middle terraces. The most characteristic trees are *Acacia nilotica, Tamarix nilotica*, and *Acacia seyal*, while *Acacia albida* and the doum palm, *Hyphaene thebaica*, occur on slightly drier ground. The date palm, *Phoenix dactylifera*, is the main cultivated tree and typical grasses and weeds include *Saccharum spontaneum, Phragmites australis*, and (on slightly more sandy silts) *Desmostachys bipinnata* and *Sorghum virgatum*.

Many more xerophytic plants such as *Calotropis procera* and *Leptadenia pyrotechnica*, as well as some ruderals such as *Solanum incanum, Tephrosia apollinea*, and *Abutilon pannosum* have invaded the cultivated ground.

The transition from river-influenced vegetation to the desert is very sharp. Northern Sudan, away from the river, has two main vegetation types: the desert where rainfall is less than 75 mm per year; and the semi-desert where rainfall is greater than this amount. The desert proper has little or no vegetation except in seasonal or fossil water courses. Vegetation of the semi-desert scrub and grassland type is influenced as much by soil type as by rainfall. On lithosols, *Acacia tortilis* and *Salvadora persica* are perhaps the commonest woody species and *Aristida* spp. are the commonest grasses. On the more sandy areas *Leptadenia pyrotechnica* and clumps of *Panicum turgidum* are the most conspicuous plants. On clay soils, shrubs include *Acacia mellifera, Capparis decidua, Ziziphus spina-christi*, and *Balanites aegyptiaca. Aristida* spp., *Chloris virgata*, and *Schoenefeldia gracilis* are the commonest grasses.

The once rich scrub-grassland of the Butana in the south-east of northern Sudan is now a sorry sight as a result of prolonged drought and severe over-use by pastoralists. Driving through this area one has the impression that the only living things are the plastic bags flapping lazily in the wind and festooned on every morbid stump.

Major ecotypes

The classification of northern Sudan into different ecotypes (Watson *et al.* 1977) provides an overall view of the relationship between the productive river system and its desert hinterland. The river system, on which life in this area depends almost entirely, covers about 2 per cent of the area while degraded and denuded desert systems cover more than 90 per cent (Table 13.2; Fig. 13.1).

Much has been said in recent years about desert encroachment, desertification, and desertization. A widely cited but unpublished document written following a short aerial reconnaissance and some ground work in 1975 concluded that the desert was advancing southwards at an alarming annual rate and that 'unless very extensive control measures are introduced in the near future, the whole of the north and north-east-facing banks of the Nile will be covered by sand within one or two decades' (Lamprey 1975). The Nile banks are not yet covered by sand and, according to a very recent article based on analysis of satellite imagery (Hellden 1988), there is currently no systematic desertification through encroachment of sand dunes, there is no systematic desertification around villages, and variations in crop yield are not caused by an expansion of cultivation into marginal and vulnerable lands.

Farming systems and land use

Northern Sudanese farming systems are characterized by intensive use close to the river and very extensive use away from it. Crops are produced mainly for subsistence although there is a flourishing market garden trade, particularly on land close to the main urban centres. Some of the pump schemes produce winter crops such as wheat and barley on a more commercial basis. Livestock are kept mainly for subsistence or in support of agricultural operations, as primary sources of power or for transport of crops, in addition to being used for personal transport. The fisheries potential of the Nile is not fully exploited—offtake is very low when compared, for example, with the intensive fishing of the Niger in West Africa—and the commercial possibilities offered by Lake Nuba are not yet utilized adequately.

The economy of the whole of northern Sudan is based on the Nile. Almost all its agricultural activities as well as the government administrative centres are crowded in its valley and on its terraces. About 90 per cent of the region's human population is compressed into less than 5 per cent of its area. Away from the river valleys, there is little permanent settlement.

The desert population is described as nomadic although in reality very little true nomadism is practised. Livestock owners are, with few

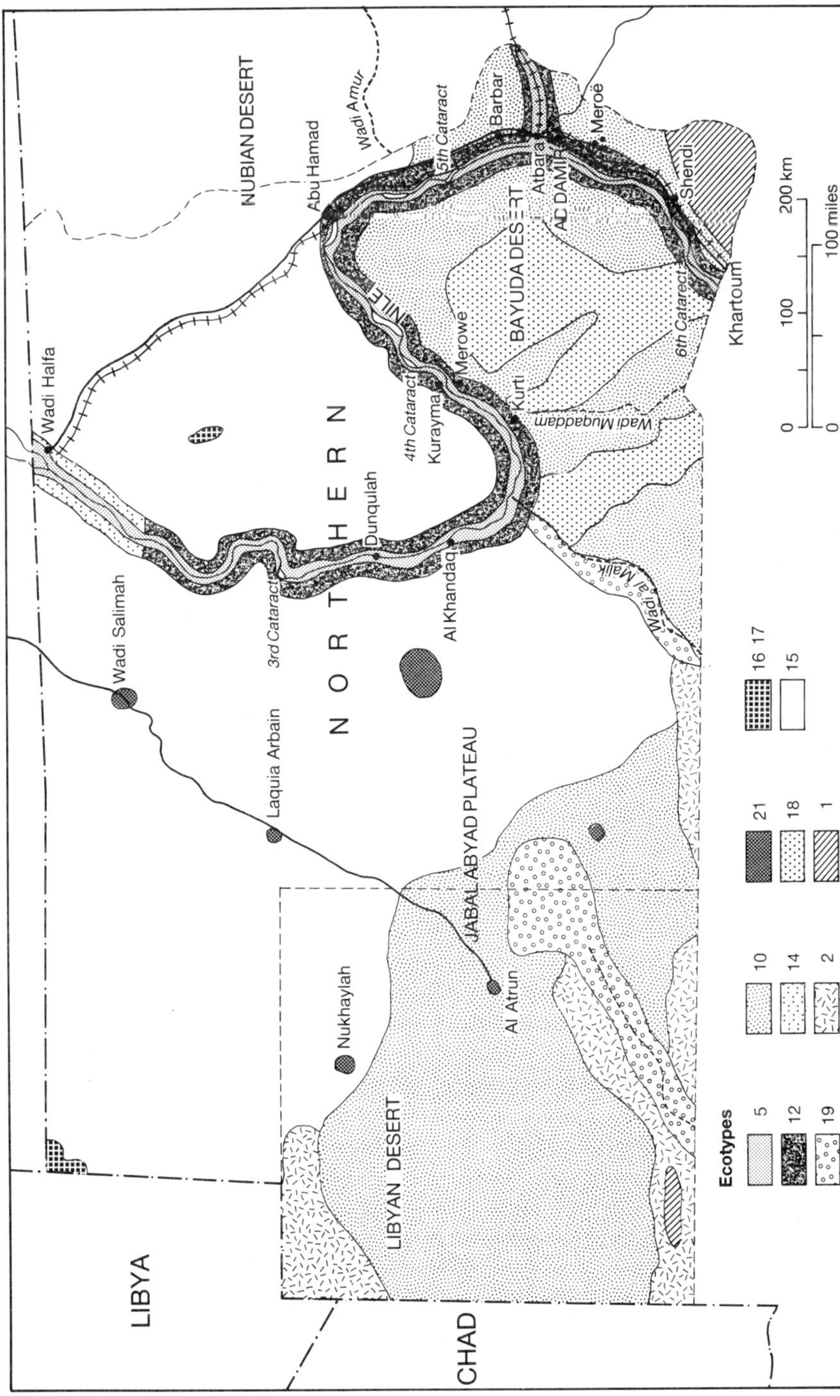

FIG. 13.1. Northern Sudan: principal ecotypes.

TABLE 13.2. Northern Sudan: major characteristics of the principal ecotypes

Eco-type	Area (km^2)	Vegetation	Geomorphology	Soil	Drainage	Topography	Use
1	3200	Sparse woody bush	Tertiary volcanic	Very stony	Deeply incised	Steep and mountainous	Browsing by goats and camels
2	20000	Very sparse bush, sparse grass	Pediments and jebels of Basement Complex	Thin, stony brown soil and sand	Deeply incised	Steep to hilly	As ecotype 1
5	13000	Medium woody cover close to river, sparse on bank and fans	Alluvial deposits and fans	Brown alluvium	Well developed fans and incised valleys	Flat	Grazing, browsing, flood irrigation
10	150000	Very sparse bush in drainage lines	Sand and stone desert	Unstable sands and stones	Poorly developed and incised	Undulating	Nil
12	26500	Sparse bush, very sparse grass in drainage lines	Sand and stone desert	Unstable sands and stones	Poorly developed and incised	Undulating	Very occasional pastoral use from river valley

14	5000	Medium woody cover close to river, sparse bush in drainage lines	Sand and stone desert	Unstable sands and stones	Poorly developed and incised	Undulating	Flooded valley
15	385000	Nil	Sand desert with rock outcrops	Unstable sands and rock outcrops	Poorly developed and incised	Undulating	Flooded valley
16	500	Very sparse woody cover	Jebels	Stone, rock, some sand	Well developed and incised	Steep	Very occasional browsing
17	1000	Very sparse woody cover	Mountains	Stone, rock	Well developed	Steep	Very occasional browsing
18	25000	Sparse bushes and grass in drainage lines	Jebels	Stone, rock, some sand	Well developed and incised	Steep	Very occasional browsing
19	20000	Sparse woody, some grass	Fossil rivers and fans	Alluvium and stony fans	Well developed	Flat	Limited pastoral: *gizu*
21	800	Trees around water sources	Oases	Unstable sands and stones	Poorly developed	Flat	Permanent pastoralism at low densities, some cultivation

Source: Watson *et al*. (1977).

exceptions, rarely true pastoralists who rely on their animals for their livelihood. Most economic units have at least part of the family living in one of the towns along the river. Livestock movements are generally over short distances and of short duration. Exceptions to this general statement are the Shukriya in the south-east of Nile Province in the northern Butana, and the camel-owning groups of Northern Kordofan and Northern Darfur who occasionally make opportunistic use of the northern desert areas.

People, crops, and livestock are intimately linked in a complex system. In Northern Province it has been estimated that 91 per cent of the people, almost 100 per cent of cattle and horses, 70 per cent of sheep, 67 per cent of goats, and 95 per cent of donkeys occupy the river valley and terraces where the cultivation intensity is about 23 per cent of the land available (Watson *et al.* 1977). Camels provide the only exception to the almost absolute dependence on the Nile. Only about 7 per cent of camel numbers are found within the river confines although most of the remainder are in strata adjacent to it. The picture in Nile Province is somewhat similar but even more exaggerated, as the Nile Valley south of Atbara has a cultivation intensity of about 67 per cent.

Dryland farming is practised to a very limited extent on account of the highly unreliable rainfall, the principal crop is bulrush millet, *Pennisetum americanum* and the cultivated area is mainly in the south of Nile Province. Even in the south of Nile Province, the rainfall is such that less than one-tenth of 1 per cent of the area is under cultivation and 40 per cent of that is left fallow in any year. The common agricultural practice is to sow millet at very wide spacing as a speculation: if rain comes a crop of sorts might be harvested, if there is no rain the crop is simply abandoned.

The Nubian agricultural system

The Nubian agricultural system is characterized by irrigated farming of annual crops, traditionally mainly for subsistence, together with the production of some perennial tree crops, particularly dates, for sale. There has been pressure on land for a long period and subsistence cropping has been complemented more and more by migration of labour. In the old Wadi Halfa area in 1960, for example, 34 per cent of the population (and 53 per cent of the rural population) were absent, and 43 per cent of households were dependent, at least in part, on cash remittances from migrant labour. Many households were chiefly of women, children, and aged people. Of the residents, 65 per cent were engaged in agriculture but most younger male members of the community were aiming at a clerical or skilled manual occupation. Literacy was higher than in any other part of the country, with 75 per cent of boys and 40 per cent of girls in the age group 6–15 being literate (Sudan Government 1960).

The date palm was a cornerstone of Nubian life. Trees were individually owned and were indeed at one time used as a basis of taxation (Leach 1919). Due to the high level of migration and the basis of inheritance, holdings in land as well as in dates were split many times. Eventually individual holdings became minute, several dozens of people often having a share in one date tree. Nubian women had an important role in agriculture and often outnumbered men in the labour force (Dafalla 1975). Share-cropping also became a part of the system because of absenteeism.

Nubian society is not markedly structured on class lines but nor is it egalitarian. Among rural Nubians there were differences in access to land and to other resources which were largely determined by political processes, starting in the sixteenth century when Turkish and Bosnian soldiers imposed themselves on the local population to become rulers and wealthy landowners. Differentiation has also occurred through trade. Farmers who were able to acquire dates from their fellows for sale could then use the money to buy consumer goods to sell back to other farmers. Money-lending became an important function, the proceeds of which were often reinvested in land. This was particularly so after 1920 when mechanical pumps arrived in the area, making it possible to irrigate land that was too far away from the river or too high to be cultivated using traditional animal-powered lifting devices. As money-lenders, traders provided goods such as tea and sugar against the credit of the next date crop and a vicious cycle of dependence set in. Following the forced Nubian exodus of the 1960s to the Khashm el Girba Scheme in eastern Sudan, little of the traditional system remains on the agricultural lands of the Nile. Social differentiation and dependency continue on the irrigated schemes and in the large urban areas where many Nubians now reside.

Irrigated cropping in the Nile Valley and on the terraces

The microcosm of the Nile Valley provides a fine example of an integrated farming system. Crops are grown for subsistence and sale. Residues and by-products of crop processing provide feed for livestock. Livestock provide transport for crops, still provide some power for lifting water for irrigation, and some power for crop processing (milling of groundnuts or sesame for oil, for example), and they also return manure to the land. Irrigated fodder, particularly berseem (lucerne, *Medicago sativa*), is grown to supplement the natural protein-rich feed provided to animals by the fruits of such trees as *Acacia albida, Ziziphus spina-christi*, and *Balanites aegyptiaca*. Cattle, sheep, and goats provide meat and milk directly to their owners or are sold for cash.

The mixed farming system and the mixed species of livestock kept—which include equines and poultry—demand a range of skills and provide employment throughout the year for all ages and sex categories constituting the labour force. Land preparation starts as the annual flood begins to recede during late September or early October. Sorghum may be the first crop planted on the higher terraces, as many varieties will germinate and mature on the residual moisture in the soil and require no irrigation. As the flood recedes farther, the lower terraces and the river bed are worked to a seedbed and irrigation channels are made or repaired. Annual weeds, particularly grasses and sedges, are a major problem. The first requirement of whatever crop is grown is water but water also encourages weeds. For a reasonable crop to be obtained, water has to be applied in adequate quantities at regular intervals. Watering and weeding are both time consuming and labour intensive. Some crops, particularly the vegetables and salads such as tomatoes, peppers, and onions, are grown in small seedbeds and transplanted later, imposing yet further demands on the labour force. There are conflicting demands for labour: in addition to that required for agricultural operations, there are the needs of the livestock which also have to be kept away from the cropped areas during the growing season. In some cases animals may be totally confined so that all feed has to be carried to them. A part of the labour force may also be wage-earning, either on another agricultural property or as day labour, part- or full-time, in a nearby town.

Maize is grown in small quantities as a vegetable and also to produce winter fodder for livestock. It is normally planted early, in November if possible. Some sorghum, particularly the short-season variety Abu Sabaeen, is also grown as forage.

Wheat is a traditional winter crop but since the early 1960s and in tune with the changing tastes of the population for white flour, its production has been encouraged by the Sudanese Government, as an import-substitution measure. It has been grown on the major irrigation schemes in central and eastern Sudan, as well as on the pump schemes in northern Sudan. Most is grown for home consumption and any surplus is supposed to be sold through official channels at controlled prices. However, there is a thriving parallel system on which open market prices are paid. The area planted to wheat varies each year. Estimates for Northern and Nile Provinces are from 20 000 to about 40 000 feddans (8000 to 16 000 ha) which represents, in different years, from 5 to about 25 per cent of Sudan's total wheat area. Yields fluctuate widely from field to field and from year to year, a major determinant of yield being the date of sowing. Late October to early November sowing usually assures a good crop but there are severe yield penalties of as much as a 1 per cent reduction per day for later sowings. The potential yield loss is as high as 60 per cent if sowing is delayed until December. Farmers are not all convinced that yield is a function of sowing date which accounts for many of the late sowings observed. Competition for water by other crops is another contribution to loss of yield—irrigation every 10–14 days is the ideal. Weeds, insect, and mammal pests and various bacterial and viral diseases also reduce yields well below the potential.

Winter pulses and legumes are other major crops whose production is confined almost solely to northern Sudan. It has been estimated that more than 60 per cent of the area planted to faba beans (*Vicia faba*) is in small areas around Shendi in Nile Province and Dongola in Northern Province. In many years, faba beans are the most important crop, in terms of area sown, in northern Sudan. Dry beans (*Phaseolus vulgaris*) and chickpeas (*Cicer arietinum*) are other important irrigated legumes. Minor crops are field peas, lentils, hyacinth beans, pigeon peas, and cowpeas. About 90 per cent of the faba bean area in Sudan is on small scale private farms and this crop, eaten as *foul masri*, is an important item in the Sudanese diet being, in particular, a rich source of protein. The beans are prepared for eating by soaking for about 12 hours then simmering for 6–8 hours before oil (sesame oil or ghee—liquefied butter) and spices are added. Faba beans are usually sown in mid-November after the wheat has been planted and they are often sown in rotation with wheat. Yields, as for wheat, vary widely and are influenced by the same factors. Lack of inoculation by the correct species of symbiotic *Rhizobia* may be a further yield constraint, particularly for chickpeas sown on new ground or ground which has not been used for growing legumes for some time. Groundnuts are a relatively minor crop in northern Sudan.

Other crops grown include cotton, an increasing area of Irish potatoes, and perennial tree crops, the most important of which are dates, mangoes, and citrus. Northern Sudan is also one of the world's largest suppliers of senna pods, from *Cassia angustifolia*, which are used in the pharmaceutical trade as a laxative. Yields of up to $7\,t\,ha^{-1}$ of air dried pods have been recorded.

Harvesting of crops, following the intensive watering and weeding activities of the growing period starts as early as January and continues in a leisurely fashion until the river rises again in July. Indeed, harvesting is often too leisurely and losses are incurred both in the standing crop and in the temporary field and permanent home stacks.

The flood season is a more relaxing time as, unless rain-fed cropping is attempted, herding and tending of livestock are the main activities. Herding may require more or less constant presence to avoid loss or theft but is not physically demanding, at least where surface water is available.

Pastoral nomadism

'Nomadism and transhumance are ecological adaptations to the rigours of a stern environment' (Phillips 1961). Nowhere on earth, as we have seen, is the environment more stern than in northern Sudan. Migration by pastoralists is a logical response to the ever present problem of finding sufficient nourishment. Pastoral systems in eastern Sudan described for the Shukriya in Chapter 14 have relevance to the south-eastern part of northern Sudan in the Butana area near Atbara and Shendi. The Baggara system described in Chapter 15 is also relevant to the western area of northern Sudan.

In the vicinity of the Nile, nomadism is not very extensive in terms of distance covered nor indeed in terms of the number of people engaged in it. Farther to the west, however, more extensive movements involving large numbers of people and animals, particularly camels, occur. These movements are erratic, both temporally and spatially, and are related to the occurrence of the *gizu* in approximately two out of every five years (Wilson 1978). *Gizu* is a Sudanese term for an ephemeral type of succulent winter grazing which occurs mainly in Northern Darfur and in Northern Province just north of the border with Northern Kordofan (Fig. 13.2). It has ecological equivalents in a number of other desert areas. There is no free water in the *gizu* area, but the lack of water is no impediment to its use. A major advantage of the *gizu* is that it provides feed for animals and can also (because of the succulent nature of the plants which are present) provide sufficient water to cover stock maintenance and production requirements for several months. The longest documented period of a group of animals (and their human herders) living without access to free water is

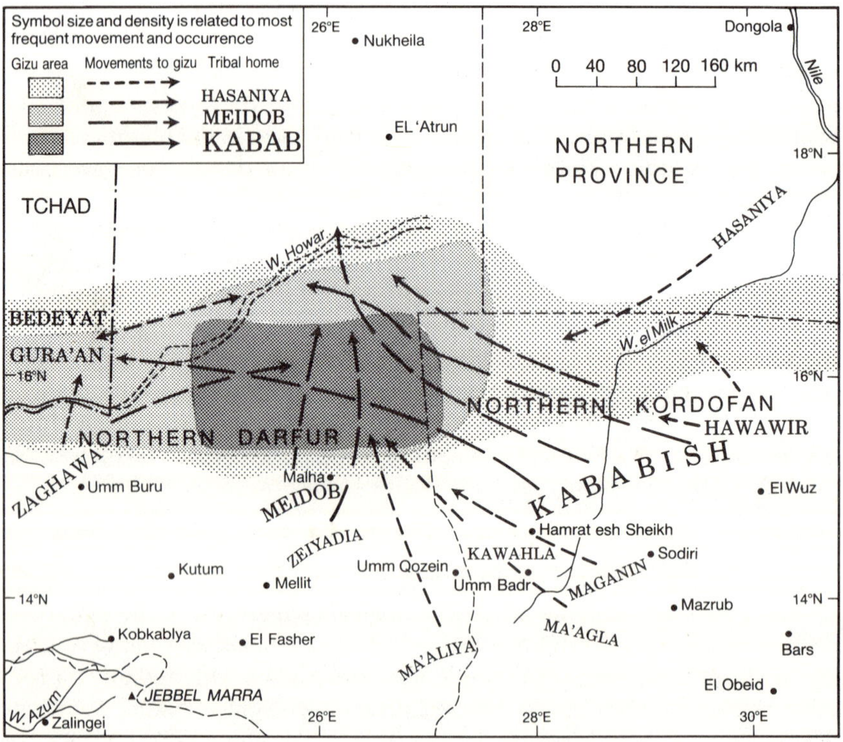

FIG. 13.2. The location and tribal use of *gizu*.

probably about eight months: a one-time Government Inspector noted in his diary for 11 May 1917 that a large herd of camels was 'just back on its wells and having its first drink since the rainpools dried up' (Davies 1957).

May 1917 was the first chronological mention of the *gizu* but over the next 60 years, '*gizu* years' occurred about 20 times. Sometimes, as in 1920–4, they appeared in a run of consecutive years (Newbold 1924); at other times there were none for as long as 8–10 years. In arid zones livestock carrying capacity is usually limited by lack of feed and water during the dry season. The *gizu* is probably unique in that it provides, at the same time and from the same resource, both feed and water. Shortly after the *gizu* was first recorded, its value was recognized by Sudan's administrators who maintained that pastoral tribes whose livestock produce and reproduce are unlikely to get into other forms of mischief. This is perhaps why there are constant references to the presence or absence of the *gizu* in monthly and annual reports of the western and northern Sudan administrations for the whole period that records have been kept.

The Kababish have historically been the most important exploiters of the *gizu* but many other groups have used it with more or less intensity (Fig. 13.2). Camels have been the main beneficiaries of the *gizu* and the Kababish have often made round-trip migrations of 1200 km in search of it. Other domestic species, as well as wildlife (particularly the addax, the scimitar-horned oryx, and the Addra or dama gazelle), also benefit from it on occasions. Sheep travel fairly regularly to the *gizu* although probably only when it is not too far north or west, and goats have also been recorded to use it (Maydon 1923).

Gizu is welcomed by livestock owners and administrators alike and in *gizu* years there are seldom problems among different tribes in search of grazing for their animals. Although herdsmen can get their liquid requirements from the milk of their animals, the nights can be bitterly cold in the far north in December and January and at this time people suffer from bronchitis, rheumatism, and pneumonia. Their concern for the welfare of their stock is paramount, however, and for this reason they are prepared to put up with almost any hardship.

Crops and livestock

Crop research for northern Sudan is carried out at Hudeiba Research Station and at Dongola Substation. Hudeiba serves mainly smallholder agriculture along the Nile and carries out research on vegetables, winter pulses and legumes, wheat, barley, fruit trees, and date palm. Some research on soils and salinity problems is also undertaken.

Livestock research is confined to the Butana breed of cattle at Atbara, a station originally established about 80 years ago to supply Atbara town and its garrison with milk.

Crops

Maize areas in northern Sudan are small but even so fluctuate widely from as little as 30 ha to as much as 1364 ha. Yields are generally low, from about 850 to 1600 kg ha^{-1}. Most maize is grown as a garden crop and is eaten green. As a forage, maize planted during November, has a fresh matter yield of 33 t ha^{-1}. Insect pests, especially earworms (*Heliothis* spp.), army worms (*Spodoptera* spp.), and sucking bugs can reduce yields significantly.

Although yields of wheat at Hudeiba have reached almost 5 t ha^{-1}, the long-term average on smallholder plots is about 1.2 t ha^{-1} but fluctuates widely between years from 0.9 to 1.6 t ha^{-1}. Some indications of the problems associated with wheat growing have been discussed, but the gap in yield gives some indication of the potential for increasing wheat output. Research concentrates on introducing new varieties with short straw and with resistance to rust. Short-season varieties are also required, for yields are reduced when excessive temperatures are met during the maturity period: the need for early sowing until such varieties are developed is paramount. Stem rust (*Puccinia* spp.) is the major disease problem. Insect pests include aphids and the wheat stem borer (*Sesamia cretica*). The multimammate rat (*Mastomys natalensis*) and the Nile rat (*Arvicanthis niloticus*) consume large amounts of wheat especially in plague years. Birds also cause considerable damage.

The area planted and the yields of pulses also vary widely. Average smallholder yields for faba beans are about 1.5 t ha^{-1} ranging from 1.2 to 2.0 t ha^{-1}. Dry beans have similar yields but chickpeas yield about 30 per cent less because only a proportion of the crop is irrigated, the remainder being grown under flood conditions. Root rot and wilting diseases (especially *Fusarium solani*) reduce yields of winter pulses significantly: mildews are also a problem and may cause up to 50 per cent loss of yield potential. Sudanese Broadbean Mosaic Virus, transmitted by aphids, can also be very severe, affecting 100 per cent of the crop. Insect pests include aphids, army worms, the mung moth, the American bollworm, *Heliothis armigera*, and thrips and leafhoppers. Bruchid beetles and weevils cause considerable losses in the stored product. Research on pulse improvement has concentrated on improving local (*baladi*) varieties and crossing these with Egyptian stock. Agronomic research concentrates on sowing dates, seed rates, and plant spacing.

Livestock

In view of the role of livestock in the northern Sudan economy little research is done on them and little is known of their productivity. Research on camels is undertaken at the University of Khartoum but this is almost entirely on its physiology: research on production might be more useful, particularly as racing camels are now being exported to Saudi Arabia at individual prices as high as US$40 000 (Abu Sin 1986).

The Butana breed of cattle is one of two types of North Sudan Short-horned Zebu—the other being the Kenana—which are of considerable interest as milk producers. There is concern for their conservation (Wilson *et al.* 1987) and they are in danger of being absorbed through crossing with exotic breeds. At Atbara, production characteristics of the Butana breed are an age at first calving of 44 months, a calving interval of 416 days, a lactation length of 253 days, and a lactation yield of about 1450 kg of milk. It is possible that some of these parameters are exceeded in traditionally managed herds.

No research on small ruminants is carried out in northern Sudan and there is relatively little elsewhere in the country. The various types of Sudan Desert sheep are known to be productive elsewhere in Sudan under both traditional rain-fed (Wilson 1976*a*) and station irrigated (Sulieman and Wilson 1989) management. Their potential remains to be fully exploited. Goats along the Nile are of the Nubian type which are reputed to be good milk producers and are, indeed, exploited for their milk, but there is little firm information on them. Away from the river, the Sudan Desert goat is the characteristic type and this breed is also known for its reproductive performance, lack of seasonality in breeding, fast growth rate, and good milking ability (Wilson 1976*b*).

Conclusions

Northern Sudan provides a harsh and difficult living environment. Existence is rendered possible by the ribbon of the Nile flowing from south to north at the eastern extremity of the area. Integrated and mixed crop and livestock farming is possible in the Nile Valley and along its terraces because of the water the river provides. Away from the river the vast and forbidding desert is exploited ephemerally by hardy animals and their owners. Communications are generally poor, the area is neglected by central Government, and more investment and more research into improving crop and livestock output is required if northern Sudan is to contribute its full potential to the national economy.

References

Abu Sin, M. E. (1986). *Breeding and marketing of racing camels in eastern Sudan: the case of Butana district*, Camel Forum Working Paper No. 15. Somali Academy of Sciences and Arts, Mogadishu.

Dafalla, H. (1975). *The Nubian exodus*. Hurst, London.

Davies, R. (1957). *The camel's back: service in the rural Sudan*. Murray, London.

El Hadidi, M. N. (1976). The riverain flora in Nubia. In *The Nile. Biology of an ancient river* (ed. J. Rzoska), pp. 87–91. Junk Publishers, The Hague.

Ghabbour, S. I. (1972). Flora of the Nile region at the Dongola Reach, Sudanese Nubia. *Revue de Zoologie et de Botanique Africaines*, **85**, 1–29.

Griffiths, J. F. and Soliman, K. H. (1972). The northern desert (Sahara). In *World survey of climatology. Climates of Africa* (ed. J. F. Griffiths), Volume 10, pp. 55–131. Elsevier, Amsterdam.

Haland, R. (1988). Problems in the Mesolithic and Neolithic culture-history in the Central Nile Valley, Sudan. In *Nubian culture: past and present*, pp. 47–74. University of Bergen, Bergen.

Hassan, F. A. (1980). Prehistoric settlement along the Main Nile. In *The Sahara and the Nile. Quarternary environments and prehistoric occupation in northern Africa* (ed. M. A. J. Williams and H. Faure), pp. 421–50. Balkema, Rotterdam.

Hellden, U. (1988). Desertification monitoring: is the desert encroaching? *Desertification Control Bulletin*, **17**, 8–12.

Lamprey, H. F. (1975). Report on the desert encroachment reconnaissance in Northern Sudan. Unpublished mimeograph. United Nations Environmental Programme, Nairobi.

Leach, T. A. (1919). Date trees in Halfa province. *Sudan Notes and Records*, **2**, 25–31.

Mair, C. (1977). *African kingdoms*. Clarendon Press, Oxford.

Maydon, H. C. (1923). North Kordofan to South Dongola. *Geographical Journal*, **61**, 34–41.

Murdock, G. P. (1959). *Africa. Its peoples and their culture history*. McGraw Hill Book Company, New York.

Newbold, D. (1924). A desert Odyssey of a thousand miles. *Sudan Notes and Records*, **7**, 43–92.

Phillips, J. (1961). *The development of agriculture and forestry in the tropics: patterns, problems and promise*. Faber and Faber, London.

Sørbø, G. M. (1985). *Tenants and nomads in Eastern Sudan. A study of economic adaptations in the New Halfa scheme*. Scandinavian Institute of African Studies, Uppsala.

Sudan Government (1960). *Population census in Wadi Halfa rural area and town*. Department of Statistics, Khartoum.

Sulieman, A. H. and Wilson, R. T. (1989). Productivity of three subtypes of Sudan Desert sheep at El Huda research station, Sudan. In *African small ruminant research and development* (ed. R. T. Wilson and M. Azeb), pp. 552–64. International Livestock Centre for Africa, Addis Ababa.

Watson, R. M., Tippett, C. I., Rizk, F., Jolly, F., Beckett, J., Scholes, V., and Casbon, F. (1977). *Sudan national livestock census and resource inventory*. Veterinary Research Administration, Khartoum.

Wilson, R. T. (1976*a*). Studies on the livestock of Southern Darfur. III. Production traits in sheep. *Tropical Animal Health and Production*, **8**, 103–14.

Wilson, R. T. (1976*b*). Studies on the livestock of Southern Darfur. IV. Production traits in goats. *Tropical Animal Health and Production*, **8**, 221–32.

Wilson, R. T. (1978). The '*gizu*': winter grazing in the South Libyan desert. *Journal of Arid Environments*, **1**, 325–42.

Wilson, R. T., Ward, P. N., Saeed, A. M., and Light D. (1987). Milk production characteristics of the Kenana breed of *Bos indicus* cattle in Sudan. *Journal of Dairy Science*, **70**, 2673–9.

14

Systems of pastoral and agricultural production in eastern Sudan

G. M. SØRBØ

Introduction

Eastern Sudan is defined for the purposes of this chapter as the country lying to the east of the Blue Nile and to the north of the line of permanent villages and regular cultivation that runs approximately from Sennar through Gedaref to the valley of the Settit River. To the north it includes the whole of the Red Sea Hills area as far as the Egyptian frontier. In administrative terms, the whole of Red Sea Province and the major part of Kassala Province are included.

There are some distinct differences between the two provinces. In Kassala Province, the presence of fertile clay soils has led to the development of highly profitable grain cultivation (Gedaref), a relatively extensive river system to large-scale cash-crop production in irrigated schemes (New Halfa and Gash), and groundwater reserves to commercial horticultural production (Kassala).

In Red Sea Province, no parallel agricultural development has taken place except for the cotton-growing that is carried out by way of flood irrigation in the Tokar Delta. Its rural areas are predominantly inhabited by people who base their living on various forms of pastoralism and agro-pastoralism.

In ethnic terms, the Red Sea Province is also less heterogeneous than Kassala. Except for the city of Port Sudan which offers a kaleidoscope of ethnic types, the rural areas are dominated by the Beja, a group of pastoral tribes speaking a single, unwritten Kushitic language (Tu Bedawie). Besides the Beja, the pastoralist Beni Amer and Rashaida are also represented in the rural parts of the province.

The situation is different in Kassala Province where the various irrigated and rain-fed agricultural areas have attracted people who belong to many different ethnic categories. Thus the Butana which is the flat clay plain between the Nile and Atbara Rivers is inhabited by almost twenty different groups during the whole or part of the year.

Writing in 1961, the British geographer K. M. Barbour commented on the major contrast between the south-west and the north-east of the region, this roughly corresponding with the present provincial boundary between Kassala and Red Sea. He predicted a gradual shift in the south-west from nomadism to settlement and a 'general participation in the growing prosperity of the whole country' (Barbour 1961, p. 233). In the north-east, on the other hand, Barbour saw the rural areas as being much closer to 'a primitive, self-sufficient economy'. He saw little reason to expect extensive cultivation by way of irrigation, nor did he see Port Sudan emerging as a major centre for industrial growth due to the aridity and harsh climate of the coastal plain. In his eyes, 'the picturesque poverty of the inhabitants of the Red Sea Hills' was likely to persist.

However, while the contrast between the two areas still remains and while the Red Sea Hills are still among the poorest, most depressed and least well served areas in the Sudan, its inhabitants are being affected increasingly by developments such as the growing importance of Port Sudan, which is the country's only port, and the road and railways that connect it to Khartoum; by the influx of refugees from war-torn Ethiopia; and by the sequence of droughts starting in the late 1960s, continuing in the early 1970s, and culminating in the drought and famine years of 1979–85.

No analysis of pastoral and agricultural systems of production in eastern Sudan can be adequately carried out without taking such developments into account. Thus, one important effect has been a shift towards the settlement of nomads in large parts of the region—related to migration, famine, and the growth of settlements along the roads and on the schemes. Even in Red Sea Province, nomads are estimated to be no more than about one-third of the total population, 240 000 out of a total provincial population of 730 700 (Swift 1986, p. 13).

However, the distinction between nomads and sedentary rural people is not a particularly meaningful one, since most of the nomads do not move very far and, conversely, most of the settlements are inhabited by people who, to a large extent, still depend on livestock for their livelihood (Swift 1986, p. 11).

Rather, the decline in the proportion of nomads in the rural parts of eastern Sudan is closely related to the emergence and development of growing linkages between livestock and crop cultivation through new patterns of land use. While such developments have expanded the subsistence base of a substantial number of people, they also seem to have a number of other, less fortunate implications. Thus, increasing evidence suggests that the relatively successful adjustments between farming and pastoral societies to specific environmental conditions are under increasing strain from a number of separate, but related causes. In the following, I shall discuss these trends and some of their implications for those who presently struggle to make a living in eastern Sudan. I shall draw primarily upon examples from the Red Sea area and the Butana.

Production systems in the Red Sea area

Environment

The main landscape features of the Red Sea Province are (a) a narrow coastal plain, 20–50 km wide, (b) a line of hills, parallel to the coast and up to 100 km wide from east to west (rising in altitude to 1500 m above sea level), with (c) a plateau to the west at about 1000 m above sea level, scattered with isolated hills and cut by shallow *wadi* (*khor*) drainage systems flowing west towards the Nile, 200–400 km away.

The climate varies by location, with three main types. In the Red Sea coastal plain, the mainly dry climate is modified by the sea; rain may occur at any time, but is most common in winter, from October to January. North of around latitude 19 °N there is a typical arid climate with little rain at any time. The land south of this latitude has a dry winter, with a monsoon, summer rainfall regime.

Neither regime brings much rain. It is estimated that 98 per cent of the province receives less than 200 mm of rain annually. As a result, apart from the Tokar River in the south which is fed by higher rainfall in the Eritrean Highlands, the main surface drainage in the province is in seasonal *khors*, some of which form extensive networks carrying large quantities of runoff water for short periods towards the Nile or the Red Sea. These are cultivated with the help of water harvesting bunds to divert the *khor* flow (Swift 1986, p. 8).

Soils in the province are mostly poor, much of the terrain consisting of rocky, mountainous land, with gravel, stony or sandy *khor* beds. More fertile soils have developed in the silty flood plains of some of the largest *khors*.

The vegetation is typical of such an arid area and consists of drought-resistant species. It is extremely sparse in the north and is composed mainly of semi-desert *Acacia* scrub which increases in density as one moves southwards. Generally, three types can be distinguished:

(1) open *Acacia* communities in hill-valleys, *khors*, and catchment areas of inland and coastal plains, with bushes and trees of *Acacia tortilis, A. radiana, A. etbaica, Salvadora persica*, and *Maerua crassifolia*. Annuals and ephemerals include *Cenchrus* spp., *Panicum, Euphorbia* spp., *Aristida*, etc.;

(2) saltmarsh vegetation on the Red Sea coast consisting of succulent halophytes;

(3) hill vegetation with a unique type of open woodland formations, controlled by relatively cool and moist conditions due to the altitude and off-sea wind.

Population and production

As mentioned above, most of the rural areas are inhabited by the Beja who are divided into three main tribal categories: the most numerous are the Hadendowa, who live mainly in the south of the Red Sea Province, spilling over into Kassala Province and especially the Gash Delta; the Bushareyen in the central and northern part of the province; and the Amara'r, who occupy coastal areas north and west of Port Sudan.

Although the Beja engage in agriculture and wage-labour, their main economic activity is herding. Pastoralism among the Beja lacks uniformity and ranges from the camel and goat pastoralism in the arid north to the mixed farming practised by the southern Beja.

In the north, among the Bushareyen camel pastoralism is the main source of subsistence and area of specialization. The Bushareyen are more commercialized in camel trade than the neighbouring Amara'r. Goats and sheep are reared but are of less importance. The Um-Ali Bushareyen in the north breed a fine and light riding type camel which is very famous. The southern Bushareyen near Atbara are also camel herders but camel herding is combined with cultivation and small stock rearing.

The pastoral movements of the Bushareyen are rather systematic, vertical and naturally constricted to *wadi* routes. Some coastal rainfall in the winter and streams from the hills produce some annual pasture which is exploited between November and January when the Um-Ali Bushareyen move towards the coast. During the late dry season, from April onwards, they retreat to the hills. Other Bushareyen are based more or less permanently on River Atbara and move inland only after rains (Morton and Fre 1986, p. 23).

The Amara'r are situated south of the Bushareyen in a much smaller, compact territory (from Port Sudan to Musmar stretching north-east). Rainfall during summer is scanty and unreliable. But the coastal Amara'r benefit from winter rainfall in the eastern slopes of the hills north of Port Sudan.

The Amara'r keep camels, sheep, and goats and have been largely sedentary for decades. Khor Arbaat, north of Port Sudan is their main cultivation area. However, the inland Amara'r move irregularly between the hills and the coastal plain depending on the incidence of rain. The Amara'r also provide much of the labour for the Port Sudan harbour (Morton and Fre 1986, p. 24). The Hadendowa are the most numerous group of Beja occupying areas south of the Amara'r. The inland Hadendowa south of the Erkowit Hills from Musmar to Haiya to Gebeit and up to the Gash Delta are mainly hillmen owning camels, goats, and sheep. The southern Hadendowa have been partly integrated in the Gash Scheme since 1920 (see below), and they own cattle, sheep, goats, and some camels, in addition to having tenancies on the scheme.

Some Hadendowa who cultivate in the Gash Delta move to their open grazing land outside the delta during summer rains. They return to their permanent villages and camps during October. Northern Hadendowa move southwards in search of grazing after the rains and during the late dry season they return north to their bases and wells. There are also Hadendowa tribal sections in the Red Sea Hills and on the eastern slopes around Erkowit which receives winter rainfall and Hadendowa grazing extends as far as the Beni Amer territory towards Eritrea in the east.

A rather different type of pastoralism is practised by the Rashaida, Arabic-speaking, camel and sheep herders, who are relatively recent arrivals from Arabia. They move much longer distances than the Beja, ranging from Kassala Province or even Eritrea in the south, to Egypt in the north. Within Red Sea Province they use the coastal plain and in the mid-1970s, there were estimated to be about 30 000 Rashaida in the province (Salih 1976, p. 98).

In the south of the province straddling the border with Eritrea, there are also unknown numbers of Beni Amer, who are mainly cattle pastoralists.

The Beja supplement their livestock economy with flood-retreat farming, using the natural *khor* flow which is channelled on to fields by putting down brushwood and some sand and letting the flood pile up and consolidate the bund. A number of the wide, flat, silted and better-watered *wadis* are suitable for cultivation and sorghum and some millet are grown on these fields, as well as small quantities of vegetables, for household consumption. The staple Beja food is a thick sorghum porridge eaten with milk or clarified butter. Local cereal production has never covered requirements and most grain and other consumption goods are bought, traditionally on credit, in small shops throughout the province. Livestock are marketed to provide income, with camels often going in caravans to Egypt or by boat to Saudi Arabia, and smallstock to Port Sudan for consumption there or export to Saudi Arabia. Subsidiary household income comes from the making and sale of charcoal, *birish* matting, and woollen rugs in local markets. There are also substantial wage-labour opportunities in Port Sudan and on the Tokar and Gash agricultural schemes (Swift 1986, p. 17).

Social organization and land rights

Each of the three main Beja tribes are subdivided into sections which, in turn, are divided into smaller kin units. The most cohesive of such kin groups and also the basic land rights-owing unit, is the small lineage segment known in Tu-Bedawie as *diwab*, composed of men who trace descent in the male line from a common ancestor five to eight generations back (Salih 1976, p. 98).

The *diwab* units are territorial units in the sense that each of them has collective rights to land, vegetation, and water in a limited length of a *wadi* bed and the barren upland countryside surrounding it. The territories of closely-related *diwabs* are not necessarily placed next to each other but may be widely scattered. Within the *diwab* territory, individual households may live together in a single large camp, or may be divided into several smaller groups which move and camp together. Although households belonging to a *diwab* may leave *diwab* territory in the course of regular seasonal migration or because of an event such as the famine of the early 1980s, individual households identify closely with the territory of their *diwab* and generally stay within the limits of their territory. Shared ownership of this land is highly valued by the Beja and is a defining feature of Beja culture (Swift 1986, p. 20).

Traditional Beja law recognizes two types of rights: (a) collective ownership rights (*asl*) in a territory and its natural resources; and (b) usufruct rights (*a'mara*) to grazing and cultivated land, to water and to camp sites.

Ownership rights (*asl*) are claimed by original occupation, that is by the *diwab* whose ancestors are believed to have been the first occupants of the territory. Once obtained, these rights are the permanent property of the lineage and are inherited collectively by male lineage members in the patrilineal descent line; women are excluded from this collective inheritance of territorial rights.

Although collective *diwab* land ownership cannot be subdivided, individual members can hold special rights—permanent inheritable usufruct rights without rights of alienation—in particular resources: wells, cultivated plots, camp sites, and possibly individual large *Acacia* trees (Swift 1986, p. 21).

The more general category of usufruct rights (*a'mara*) gives non-lineage members (as individual families or as a whole *diwab*) the right to graze and water their animals, cultivate, and pitch camp on *asl* land of another lineage. This is subject to negotiation and to payment of annual dues to the *diwab* with ownership rights. The dues are not fixed and depend more on the wealth of the person or group seeking usufruct rights: they are paid in kind, commonly an agreed numbers of animals, a fixed proportion of a harvest, or a specified portion of any animal slaughtered. However, these dues are considered largely symbolic, a recognition of ownership of the territory by another *diwab*. Even closely-related *diwabs* are expected to seek permission before using the territory of another *diwab*.

The Gash Scheme

The Gash Delta spreads out to the north of Kassala and has been built up by the annual flood of the Gash River. While it is part of Kassala Province, it is also traditionally part of Hadendowa territory. It is, therefore, dealt with in this part of the chapter.

In its natural state the Gash Delta provides an area of exceptionally good grazing, the flooded silt soils yielding a heavy growth of grass, which continues to be green and palatable long after the rains have ceased and the surrounding country has dried out. The region has always been highly valued by the Hadendowa, both for grazing and sorghum cultivation.

The commercial exploitation of the Gash Delta dates back to the nineteenth century, but it was not until 1926 that the Gash Scheme was established and tenants asked to cultivate cotton under a joint account system. For each 10 feddans of cotton they were allowed to cultivate one feddan of sorghum free of all charges for their own benefit. It was decided that the Hadendowa should have a prior right to hold about 65 per cent of the total irrigated area every year. In 1959, castor was introduced in the rotation, and cotton was abandoned in 1970. Although castor cultivation has expanded, yields have declined and average incomes have decreased since 1962 (Salih 1976, p. 98).

The majority of Hadendowa tenants have not settled down in the Gash Delta. Hassan Mohamed Salih (1976, p. 98) writes that there were 16 000 registered tenants in 1972, but that only 5000 (30 per cent) cultivated regularly and the remaining 70 per cent visited the scheme only occasionally to see to it that they still maintained their rights in the area. The major reasons for continued priority being given to pastoralism are the following:

(1) The size of the tenancy is too small to establish profitable farming in the Gash Delta. Due to the fact that stretches of land are allotted to lineages and then distributed to individuals, a process of fragmentation has reduced the average size of a tenancy to about 3 feddans only, although the Sudanese Government has fixed the minimum size to 5 feddans.

(2) The system of land allocation and the varying floods which determine the size of areas to be cultivated from year to year do not encourage sedentarization. The individual tenant does not have the same tenancy, nor the same amount of land, allocated to him from season to season and, in years when the floods are low, the majority of tenants can only be accommodated on the scheme by accepting very small pieces of land which, in turn, reduce their incomes.

(3) The Hadendowa household units do not generally command sufficient labour to allocate on both pastoralism and agriculture. The time of sowing (September) conflicts with the labour requirements for herding during a period when the herds are away from the delta, and the time of harvesting (January–April) coincides with the annual rush of animals towards the delta, when there is need for many herders if animal trespass and conflicts with other nomads are to be avoided. Hadendowa culture, which excludes women from taking part in herding or agricultural activities, is also a main factor explaining the general scarcity of household labour.

(4) The administrative organization of the Gash Delta depends entirely on the Hadendowa sheikhs for their contact with the tenant farmers.

Tenancies are registered when the tenants are away in the Red Sea Hills and the sheikhs are able to take sizeable holdings for themselves (the average holding of a sheikh is 30 feddans) and also to appropriate a considerable proportion of the loans and income from the castor cultivation (Salih 1976, p. 98).

As the Hadendowa give priority to pastoralism, the peculiar situation arises that when the rains are good and the floods high, there is a high rate of absenteeism as tenants move with their animals outside the delta; when the rains are poor and the prospects for agriculture are equally poor, there is a high pressure on the delta as tenants seek the added income from agriculture to take them through a year of drought.

A large number of tenants in the Gash Scheme are poor and stick to cultivation mainly because they have few animals. The tenants claim that the loans provided by the Sudanese Government are not sufficient to cover the high costs of production and those who are unable to finance cultivation, for example, by selling animals, are forced to borrow money and often pledge their crop or tenancies as security for accumulated debts. In view of the prevailing low incomes, most of the tenants are unable to settle these debts and this may result in secret agreements between Hadendowa tenants and moneylenders/shopkeepers to whom they concede their rights of cultivation. The latter tend to belong to ethnic categories from the Nile Valley (*jallaba*) and this is a process whereby riverain people are able to control and run large allotments of castor successfully because they can afford the high expenses of establishing and maintaining the crop and because they have no animals to distract them from attending to their cultivation regularly. So the benefits of the Gash Scheme are passing to non-Hadendowa farmers who neglect the cultivation of food crops on an area which, in the past, provided the chief area for grazing and the main supply of sorghum to the local population.

The Beja and their future

Many aspects of livestock production among the Beja are still rather poorly understood as there has been very little research. This is critical at a time when a growing number of relief and development agencies are implementing their projects, generally based on insufficient knowledge. It seems that a basic concern with *asl* rights and a flexibility over actual land use, form basic principles which are adapted to suit different environments and activities among the Beja (Morton and Fre 1986, p. 35). Thus, by means of seasonal migrations and by herding several species of domestic animals together, the Beja have achieved considerable flexibility in their exploitation of a natural environment characterized by great fluctuations both in the amount and the nature of available feed.

However, the Beja have clearly suffered major shocks from the sequence of droughts over the last two decades. Physical recovery—both in terms of herd and pasture reconstitution and careful land use to encourage environmental rehabilitation—may prove difficult. While there is insufficient evidence to suggest that Beja territory is severely damaged ecologically, observations made by different authors indicate that the long-term ecological viability of Beja pastoralism may be threatened. Thus the tendency for an increasing number of people to settle in and around the small, urban centres (particularly on the Port Sudan–Kassala road) has created local overgrazing and deforestation. Households have come to rely more on sheep and goats rather than on camels, which are increasingly scarce and they tend, therefore, to become less mobile. The growth of Port Sudan and other centres has created a growing market for wood and charcoal and, at the same time, the Beja themselves depend more on a crop-based diet and, therefore, on cooking by wood or charcoal.

The Beja represent a loose tribal confederacy and have never formed a politically cohesive society. As a consequence, increasing ecological pressures are often being felt as local conflicts between neighbouring groups. On a larger scale, there have been movements southwards by Amara'r and Bushareyen on to areas traditionally held by Hadendowa, while an increasing number of Beni Amer have also moved on to Hadendowa territory due to the war in Eritrea. Ecological pressures thus seem to be mounting in the central and southern part of the Red Sea Province and, particularly, in areas of better winter rainfall such as Erkowit.

The Butana Rangelands

Environment

The Butana, lying between the Nile and Atbara Rivers, is a gently undulating plain where land use patterns and population distribution have been determined to a large extent by the combined effects of erratic rainfall and a geological structure which contains virtually no water-bearing rocks.

Rainfall shows a remarkable variation in incidence, amount, time of fall, and annual as well as area distribution. Rains are concentrated during a three-month period from approximately mid-June to mid-September, but in the northern part, mean and annual rainfall variability is 45 per cent and further south about 20 per cent. Total rainfall varies from 75 mm in the north to about 600 mm along the railway line in the south (Abu Sin 1970, pp. 7–15)

The compact nature of the clay soils covering most of the Butana accelerates runoff which, in turn, makes for fast erosion and a high ratio of annual plants compared with perennials. Most of the Butana is completely open without bush or tree. In the northern part, the sparse vegetation consists of semi-desert grasses and *Acacia* shrubs, the latter limited to the soils around the few inselbergs and to narrow belts along seasonal watercourses. Moving southwards, the vegetation gradually changes into *Acacia* trees, bushes, and savanna-type grasses. In the wet season, the Butana becomes green and covered with different grasses and herbs.

The Basement Complex which forms about two-thirds of the underlying solid geology of the area, bears no water except in joints. In the northern part, water-bearing rocks exist (the Nubian formation), but their extension northward into the desert has limited their value as a basis for permanent settlements. The only permanent natural sources of water are provided by the large rivers bordering on the Butana.

Production systems

In the north, due to sandy soil and low rainfall, nomadism is the main form of land use and camels are the dominant livestock. In central Butana, permanent settlements have been established, based on a combination of animal husbandry and rain-fed sorghum cultivation. Camels, cattle, sheep, and goats are all kept in this area. Wells have been dug by hand in many places, particularly where there are cracks in the underground foundation or at small local catchments of the inselbergs referred to above. Although some of them are quite shallow and thus dry up before the end of the dry season (April/May), most of them are quite deep (more than 200 feet) and allow for animals to graze around them for a long period in the year. The semi-nomadic adaptation which has characterized the central part of the Butana, is based on its dependence on such wells as well as other sources of water which exist because of the light, non-cracking clay soils in the area. This allows water to be stored in natural depressions (*maia't*) and in the numerous deltas being formed by seasonal water *khors*. A considerable number of reservoirs (*hafirs*) have also been dug by machines, particularly during the 1950s. In the Butana, all permanent settlements have either been established around wells or near the rivers.

The Butana is the traditional homeland of the Shukriya who attained dominance over the plain after a series of battles with other ethnic groups during the eighteenth century. They number between 300 000 and 500 000 and belong to a group of Arab tribes which seems to have drifted into the Sudan between the fourteenth and fifteenth centuries.

The diet consumed in Shukriya tents and huts (*qatáti*) is composed of both agricultural and pastoral products. A porridge (*assida* or *lugma*) and thin unleavened bread (*kissra*) are made out of sorghum, and the stalks serve as food for their animals. Sorghum cultivation starts in mid-July when the rains have begun and takes place mostly in the larger *wadis* where waterlogging tends to retard the growth of weeds and creates good conditions for the grain. A common practice is to construct small earth banks (*terus*) to hold storm-waters and allow better percolation, but this is not always necessary. Sorghum is stored in pits, but as there is a great variability in the amount and distribution of annual rainfall, very little produce enters the market even in good years. It is never certain whether crops will be sown, so land rotation cannot be planned and some fields suffer from over-cultivation.

Holdings range in size a great deal, all land is unregistered and belongs to the Sudanese Government. In some areas, ownership is claimed by lineages or sub-lineages and rights are inherited collectively by members of such units. In these areas, land is subdivided between households which claim rights to their plots after cultivating them regularly for a period of ten years. Such claims ultimately depend on local recognition.

Access to resources and social organization

Shukriya territorial rights are over (a) pastures, (b) water points, and (c) agricultural land (El Tayeb 1980, p. 105ff.). These rights have been organized formally and allotted according to the tribal structure which is based upon a patrilineal segmentary system. Whereas grazing is said to be communal for all the members of the tribe (although rights are claimed in certain territories by different subdivisions of the tribe), wells are considered private, or are in the hands of groups of households sharing their descent from a common ancestor. Thus grazing is only communal, in the dry season, for those who have access to water.

The four major animal types on the Butana (camels, cattle, sheep, and goats) have different husbandry requirements and both large and small ruminants are milked. Therefore, households which rely on a combination of different animal types rather than on a single species, stand a better chance of surviving during years of drought and epidemic. However, a multi-species economy, combined with sorghum cultivation, is particularly demanding in terms of labour and skills. It also requires that activities are properly co-ordinated in economic units with the appropriate labour resources (Sørbø 1977, p. 138ff.). Generally, the father and all or some of his married sons work together. The extended family is regarded as the basic social and economic unit within Shukriya society.

New patterns of land use

Residence patterns and patterns of pastoral movements among some of the Shukriya underwent important changes when mechanized farming operations were established in the Gedaref from the 1950s. Some of the most fertile agricultural and grazing lands were lost through a land-grabbing process whereby the Government issued licences to private enterprises for undertaking sorghum and sesame cultivation on a large scale. The establishment of the New Halfa Scheme in the 1960s further encroached upon and largely served to replace important dry season pastures traditionally used not only by the Shukriya but also by other ethnic categories on the Butana.

By 1970, it was also decided by the Sudanese Government that security and guarantees for rights to production resources be maintained by the administrative apparatus of the Government rather than by particular local groups. The decision was accompanied by the partial abolishment of the native administration and the subsequent collapse of a general grazing agreement which recognized Shukriya dominance over the central parts of the Butana. This resulted in wider access to land formerly retained by Shukriya pastoralists for their own exclusive use during the dry season and other ethnic categories (Rashaida, Beja, Kawahla, etc.) started to move into the Butana heartlands, encouraged by the Government's decision to create a land use system based on common tenure.

These events occurred during a relatively short time-span (1969–72) and had serious implications on land use patterns and grazing resources, particularly as they were accompanied by a number of years with insufficient rains (1969–73).

The Shukriya loss of exclusive use of dry season pastures also coincided in time with their attempts to settle as tenant farmers on the New Halfa Scheme. Attracted by amenities of civilization (schools, health services, etc.) as well as the availability of water and grazing (more than the agricultural opportunities), they arrived on the scheme in large numbers.

The New Halfa Scheme

The New Halfa Scheme is located west of the Atbara River and north of the village of Khashm el Girba and has served not only as a settlement area for local nomads, but also for the resettlement of about 50 000 Nubians from northern Sudan, due to the construction of the High Dam at Aswan.

The farming pattern is influenced by the Gezira experience. The size of the settler's tenancy is 15 feddans (6.3 ha), and a three-year rotation of cotton, wheat, and groundnuts, without a fallow year is prescribed for the tenancies. Recently, sorghum was introduced in the rotation.

Like the Gezira Scheme, the New Halfa Scheme is highly mechanized and production is closely supervised and scheduled by management according to tight timetables. Sanctioned by a tenancy agreement, a system of division of responsibilities exists between management (the New Halfa Agricultural Production Corporation) and tenants, particularly with respect to cotton production. A joint account system for settling costs existed until recently, but has been replaced with an individual account system for each crop. In principle, the system was a rather complicated share-cropping agreement for the production of cotton, although it did extend beyond this to embrace debt and cost recovery.

Since the establishment of the scheme, the New Halfa tenants have been struggling to cope with low yields and low revenues, due to factors such as insufficient irrigation water, lack of machinery, fuel and spare parts, and rising production costs. As a result, the following pattern of tenant responses emerged.

In the Nubian communities, a large number of tenancies came to be cultivated through caretaker (*wakeel*), share-cropping, or subletting arrangements while the owner–tenants were primarily engaged in other occupations or they were aged and had retired from an active life, leaving the management of their tenancies to others. As a consequence, some tenants were able to expand their own holdings and an increasing number of tenants came to combine agriculture with other sources of income, almost invariably according lowest priority to irrigated agriculture which provided them with such low and irregular incomes. Thus, while the yield figures and their development clearly told the story of a non-viable cultivation system, the final collapse of which was likely to be imminent, mechanisms existed which made it possible for the majority of Nubian tenants to maintain their position as producers.

In the 'nomadic' communities, including those of the Shukriya, off-scheme activities were equally prevalent, with the majority of households remaining dependent on animal husbandry and rain-fed grain cultivation. This resulted in a number of conflicts with management as animal trespass caused great losses to agriculture every year and as a number of tenants came to care more for their animals than for their fields. Social and economic differentiation were strengthened in the new communities: tenancies were being accumulated by resourceful tenants while others lost their lands or struggled hard to supplement their incomes by other economic activities.

The picture was quite similar to what has been reported from the Gezira Scheme. There was out-migration of youngsters who found agriculture unattractive and unprofitable and who had no tenancies of their own; poor tenants were often in urgent need of cash and sought employment wherever they could find it in order to secure their survival on the scheme; and well-off tenants pursued off-scheme interests because the possibilities for security and growth were perceived to be better outside irrigated agriculture. Therefore, although such activities kept a number of tenant households viable and enabled the New Halfa Scheme to continue, they provided the basis for increasing economic differentiation within the various communities on the scheme.

Impact on Shukriya herding activities Among the impacts of the New Halfa Scheme on herding activities, the following seem to be most important:

(1) For tenant farmers, the establishment of the scheme has resulted in a new type of transhumance. Animals spend the wet season and part of the

dry (*darat*) season on the Butana and are driven back to the scheme to depend on crop remains (mainly groundnut hay) and other grazing from December until the official opening of the scheme for grazing in late March or April. In this way the New Halfa Scheme provides an additional source of grazing during the dry season for the Shukriya.

(2) The alternative use of natural pastures and the scheme area for grazing, combined with the fact that the scheme places considerable importance on the managerial functions of tenants, has affected the composition of many Shukriya herds. Management has come to accept that tenants may keep sheep, goats, and cattle in small numbers at all times and, after harvesting (April–June) they may keep any number of sheep, goats, and cattle. Camels, which are mainly browsers, are not accepted at any time and, therefore, tend to be kept by the Shukriya outside the scheme. An increasing number of tenant households are keeping sheep, goats, and cattle in different combinations and there has been a tendency for tenants to engage in more selective selling and investment in livestock (Sørbø 1977, p. 146; 1985; Salem-Murdock 1984). The popularity of sheep, goats, and cattle in different combinations is due to (a) their growing markets, (b) the importance of milk in Shukriya diets, (c) the relative similarity in pasture and water requirements, and (c) their high response to supplementary feeding (Abu Sin 1982, p. 98).

(3) The alternative use of range pastures and scheme grazing has made herd diversification a paying practice for a growing number of households. It has also increased the replacement rate of small animals (sheep, goats, and cattle), reducing losses and improving the general health conditions of animals and their disease viability. Although reliable figures do not exist, it seems clear that the scheme has contributed to an increase in animal numbers on the Butana. This area provides important grazing during the severe conditions at the end of the dry season for animals which have to be kept and herded outside the scheme during the cultivation season (June–April). Tenant households seem to be less sensitive to ecological pressure than 'nomadic' households because they can maintain their consumption level with agricultural income (including cultivation advances) in times of shortage rather than by slaughtering and consuming their herds. Any attempts to reduce the number of livestock seem impossible in the absence of any institutional mechanisms that can control the exploitation of pastures.

(4) The new conditions in the New Halfa Scheme have had a significant effect on the direction of movements and periods of stay of the Shukriya who have not joined the scheme. Because of the deterioration of *hafirs*, mainly due to the break-down of the grazing agreement (cf. p. 225), they are forced to move early to the permanent wells. This increases the pressure on the wells and decreases their efficiency. Severe water shortages tend to force many Shukriya to move towards the scheme or to make short,

southward movements to el Adeid, el Suki, and el Fao where the wells are more reliable, but where they encroach upon areas traditionally claimed by other groups. Those who stay near the wells, join their fellow tribesmen at New Halfa after the opening of the scheme, until late June when they normally start their wet season (*nushug*) movements again (El Tayeb 1980, p. 207).

While tenants breed cattle, sheep, and goats increasingly, the direction of development has been towards an increasing density of livestock in central Butana and the most serious damage to the vegetation seems not to occur at the end of the dry season when so many animals are in the scheme, but during the wet or growing season during seed formation. As one result, the vulnerable *Blepharis* (*siha*), a highly nutritious grass which has afforded much of the best dry-season grazing on the Butana, has almost completely disappeared.

Conclusions

When eastern Sudan is considered as a whole, the general situation is not encouraging. Developments in the rangelands give cause for concern not only because of the sequence of droughts (which have been very severe in Red Sea Province), but also because of the deterioration of the conditions in the rangelands caused by changing patterns of land use and the wider access to land formerly retained by different groups of pastoralists for their own exclusive use. Environmental degradation caused by the removal of plant cover seems much more rapid and more serious in Kassala Province than among the Beja in Red Sea Province. There have been dramatic changes in land use patterns in Kassala Province, large tracts of natural rangeland have been cleared, and the linkages between cultivation and livestock rearing, already discussed for the Butana, clearly contribute to declining production capacity as a result of over-utilization.

The typical approach to development in such degraded areas is still to treat symptoms, often over-emphasizing the easily visible, but often superficial phenomena at the expense of the root causes of these phenomena. There has been a tendency to assume that the immediate human cause of any particular symptom of ecological degradation is the significant cause, to look no further to pursue the reconstruction of a chain of causation, but to prescribe simple remedies in the form of management regimes and expect them to be implemented easily and effectively. The various attempts to establish ranch models, planting to arrest sand-dunes, and water drilling programmes are all examples. It is typical for such approaches that solutions are generally designed by focusing on the natural symptoms of the problem and by attempting to rearrange the social factors in relation to them.

This approach has often failed to bring about a satisfactory solution and has sometimes brought about undesirable and unintended consequences. If the farmers on the New Halfa Scheme invest in animals because the incomes deriving from the production of cotton and groundnuts for the world market are insufficient, and if in so doing they contribute towards the destruction of grazing areas, it seems unrealistic, both politically and ecologically, to combat desertification only by looking at management solutions for the grazing areas. Yet this has been a common approach adopted by both the Sudanese Government and by other organizations implementing development and conservation projects in the Sudan.

It is clear that we have to enlarge our perspective, not only by considering improvements in pastoral resources within the framework of a comprehensive area plan that would allow for the different land-use systems (irrigated and rain-fed agriculture, and pastoralism), but also by suggesting the political, economic, social, and ecological dimensions of any process that involves the interaction of human activities and natural processes. If the problem of low production and ecological instability in eastern Sudan is to be approached realistically, all these dimensions must be taken into account.

References

Abu Sin, M. E. (1970). The regional geography of the Butana north of the railway. Unpublished M.A. thesis, University of Khartoum.

Abu Sin, M. E. (1982). A change in strategy of animal herding among the nomads of the Butana—eastern Sudan. In *Problems of agricultural development in the Sudan* (ed. G. Heinritz). Edition Herodot: Federal Republic of Germany.

Barbour, K. M. (1961). *The Republic of the Sudan—a regional geography*. University of London Press, London.

El Tayeb, O. Y. (1980). The Shukriya of the Butana: pastoralism and problems of settlement and agriculture in New Halfa Scheme. Unpublished M.A. thesis, University of Khartoum.

Morton, J. and Fre, Z. (1986). *Red Sea Province and the Beja*. Oxfam UK, London.

Salem-Murdock, M. (1984). Nubian farmers and Arab herders in irrigated agriculture in the Sudan: from domestic to commodity production. Unpublished Ph.D. thesis, State University of New York, Binghamton.

Salih, H. M. (1976). The Hadendowa: pastoralism and problems of sedentarization. Unpublished Ph.D. thesis, University of Hull.

Swift, J. (1986). *Rehabilitation and long-term development for pastoralists in Red Sea Province, Sudan*. UNICEF, Khartoum.

Sørbø, G. M. (1977). Nomads on the scheme—a study of irrigation agriculture and pastoralism in eastern Sudan. In *Land use and development* (ed. P. O'Keefe and B. Wisner), pp. 132–48. International African Institute, London.

Sørbø, G. M. (1985). *Tenants and nomads in eastern Sudan*. Scandinavian Institute of African Studies, Uppsala.

15

Systems of agricultural production in western Sudan

G. HAALAND

Introduction

Western Sudan includes the two regions of Darfur and Kordofan, and covers a vast area of 850 000 km^2 ranging from desert in the north to high rainfall savanna in the south. Of a total population of 6.5 million, 85 per cent is classified as rural, deriving all or part of their livelihood from agricultural and pastoral production.

Since 1973, census data indicate that western Sudan has had an annual population growth rate of about 4 per cent, a dramatic increase from about 2 per cent during the period 1956 to 1973. The average population density of 8 persons per km^2 is rather meaningless since there are great variations in different localities according to environmental, technical, and marketing constraints. Most of the rural population is concentrated in a belt between 11 and 14 °N. North of this belt, towards 16°N, decreasing annual rainfall reduces the importance of agricultural activities and nomadic animal husbandry, based on camels, goats, and sheep, becomes the dominant subsistence activity. South of the belt, the heavy cracking clay soil was hardly used by cultivators before the introduction of mechanized equipment. Migratory cattle nomads, the Baggara Arabs, travelled quickly through these areas moving between dry season pastures at Bahr al Arab and the rainy season pastures in the middle belt.

The contrast between the relatively densely populated areas on the more easily cultivable soils, largely sands, in the middle belt and the almost unpopulated areas of heavy but quite fertile clay soils in the southern belt has set the scene for a major planning issue in western Sudan: should priority be given to improving conditions for the vast majority of rural producers in the middle belt, or should priority be given to maximizing agricultural output on the southern clay soils by introducing mechanized traction and establishing large scale commercial enterprises. Against this background it is understandable that some reports and planning documents emphasize the potential role of western Sudan as a major food supplier for the country as a whole and even for the Middle East, while others draw attention to accelerating environmental deterioration as well as trends towards emerging polarization between the large number of small-scale agricultural and pastoral producers in the family farm sector and the large-scale mechanized enterprises in the commercial farm sector.

The area under mechanized farming has expanded dramatically over the last twenty years while there has been little progress towards solving the economic and ecological problems of the family farms.

In this chapter we shall look at the main types of economic enterprises involved in agricultural and pastoral production in western Sudan. We shall try to characterize these enterprises as management systems and we shall explore the 'unmanaged' systems of interdependence which emerge as a result of the actions and reactions of individual enterprises in relation to each other and in relation to the natural environment. The main features of physiography, climate, and natural vegetation have been presented earlier in the book and these inputs from the natural environment have a fundamental effect on family farms and commercial mechanized enterprises.

Through cultivation, animal husbandry, and exploitation of forest resources, farming units are trying to cope with the limitations and opportunities defined by the climate, soil, and natural vegetation. Traditionally, forest exploitation (fuel-wood, gum arabic, wild fruits) involved virtually no management of natural production, animal husbandry involved management of animals, but not of pasture, and only cultivation involved management of vegetation. These activities have, over time, led to significant unintended changes both in the vegetation cover and, in many places, in the soils. This interconnection between the system which is under the management of the farming enterprise and the wider ecosystem which is affected by the activities of the farming enterprise, but not under their management, is a fundamental problem facing western Sudan.

Economic environment

The population of western Sudan is composed of a multitude of different ethnic groups. These groups have been associated with different economic activities and have been integrated to some extent in wider systems of exchange. Some groups, mostly in a middle belt on the *qoz* and around Jebel Marra and the Nuba Mountains, specialized in cultivation, others (the Baggara and some Fulani) specialized in cattle rearing, migrating between Bahr al Arab in the dry season and the *qoz* zone in the rainy season. Other groups based their livelihood primarily on camel rearing in the northern semi-desert vegetation zone (Asad 1970) (Fig. 15.1).

Today, this traditional association of ethnic identity with specific

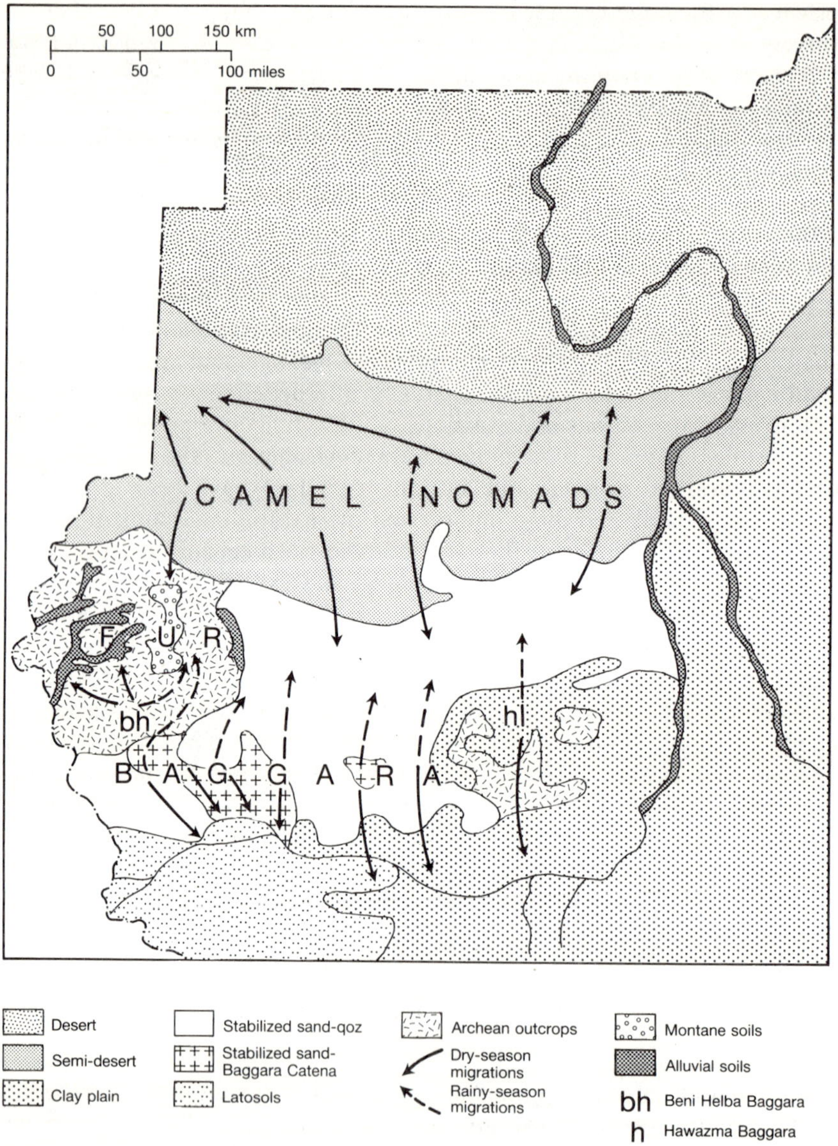

FIG. 15.1. Western Sudan: annual migration of pastoral herds.

subsistence patterns is unclear. Among the cattle-rearing groups, probably not more than 10–15 per cent of members are pure nomads; most combine animal husbandry with cultivation in the middle belt.

For centuries, an elaborate system of weekly markets in western Sudan has facilitated the exchange of goods between farm enterprises in different environments and has allowed for export of local products and purchase of imported goods such as sugar, tea, and cloth (Manger 1984). Traders from the Nile Valley, the *jallaba*, played a crucial role in the development of this marketing system and promoted long-distance trade between western Sudan and areas to the north (Egypt and Libya) and east (the Nile Valley and the Red Sea). However, until this century, the people in the Nuba Mountains area were only marginally affected by the increasing integration of western Sudan with the wider Sudanese society, except as a source of slaves in earlier centuries.

Although family farm enterprises have been affected by supply and demand forces in a wider economic environment for a very long time, markets have expanded greatly over the last 40 years from regional, to national and international. This has resulted in a reallocation of land and labour on family farms and in the establishment of a completely new type of farming enterprise—the large-scale mechanized farm. Increasingly, family farms are satisfying their consumption needs with income derived from increased cash cropping and labour migration, as well as livestock exports and commercial charcoal production. The relationships between family farms and mechanized farms are complex; competition for land has not been very important so far, except that pastoralists have faced problems of passage at the onset of the dry season when shortage of pasture and water in the north force them to migrate through the areas of the mechanized schemes on their way to Bahr el Arab; competition in the market is also limited because family farms cultivate sorghum for home consumption and rely on other crops (ochra, chilli, tomatoes, groundnuts, hibiscus, onion, citrus fruits, and mango) for market sale. The timing of agricultural activities in the traditional and modern sectors is different and this allows for a certain amount of symbiosis where family farm members work as temporary labourers in mechanized farms for short periods between the peaks of labour requirement on their own farms (Table 15.1). Despite this apparently rather harmonious picture, the exploitation of the clay areas

TABLE 15.1. Timing of activities for cotton and sorghum in the traditional and modern sectors

Activity	Traditional sector		Modern sector	
	Cotton	Sorghum	Cotton	Sorghum
Planting	May/June	May/June	July	July
1st weeding	June	Early June	July	July
2nd weeding	June/July	Late June	Early August	July/August
3rd weeding	End July	July		
4th weeding	August			
Harvest/picking	January	December/January	January/February	January

through the present type of large-scale farming may prevent the development of intermediate technologies which would make it economic for family farmers to establish medium-sized enterprises on the clay soils, thus reducing pressures in the over-exploited, more densely populated middle belt. Realization of this opportunity would depend on the Government's ability to tackle a multitude of difficult politico-administrative problems, including the introduction of new land tenure regulations, and the establishment of credit and farm inputs to serve small and medium farmers, a measure which in the short-term would have less effect on increasing agricultural output than if tailored to the needs of big farms.

To pursue a long-term strategy on such issues is very difficult for a Sudanese Government faced with the formidable short-term problems of not alienating important sections of a fragmented and heterogeneous political support basis and of mobilizing funds to serve a foreign debt which reached $US12 billion in 1987.

Family farm management systems

It is impossible to describe all types of family farm in western Sudan. However, a selected number of family farm management systems will be discussed together with the factors which have produced changes in the farmers' allocation of time and resources and changes in the organization of the management units. The effects that these changes have had in the wider context will also be considered.

The Fur agricultural production system

The management system of farm enterprises among the Fur is characteristic for non-Arab cultivators of Darfur (for the Fur see Barth (1974), for the Berti see Holy (1974), for the Masalit see Tully (1988)).

The farm management unit consists of the grown-up individual. Husband and wife operate as two distinct management units having usufruct rights in different plots of land, taking independent farm decisions, keeping the yields in separate stores and having individual control over the utilization of the crops they produce.

Through a long historical process of innovation and diffusion the Fur has developed skills and organizational capacities for tackling, with varying degrees of success, the different constraints imposed by the natural environment (variations in rainfall, soil types, topography, weeds, crop pests, livestock diseases, crop destruction by birds and locusts) on agricultural production.

Millet, grown on *qoz* soils and in the mountain ranges of Jebel Marra and Jebel Si, and sorghum, grown on alluvial soils, are staple crops among the Fur. Periodically, however, every locality can expect that crops may fail because of drought, locusts, or pests. Exploitation of a multitude of resources, primarily livestock, but also other agricultural crops as well as gathering of fruits, seeds, and roots from wild plants is, therefore, a necessary element in Fur farmers' survival strategy. Given the great variations in soil fertility between fields on *wadi* alluvials on *qoz* and on volcanic ash soils, the percentage of crop land actually cropped in a year (R value = crop years times 100/crop years plus fallow years) varies between R = 80 (on alluvial *wadi* beds) and R = 15 in marginal areas. Agricultural production is thus based on different forms of shifting cultivation. It is only in the mountain ranges of Jebel Marra and Jebel Si where the erosion danger is so great that Fur farmers have undertaken long-term investments in land by constructing terraces (primarily for soil conservation purposes on rain-fed land without irrigation). Manure was rarely used in the traditional farming system.

Where water is available from shallow wells on the alluvial *wadi* beds or from perennial streams in the mountain ranges, some irrigated farming (tomatoes, onions, melon, sugar-cane, wheat, mango, and citrus fruits) is undertaken in the dry season.

The socio-economic framework of Fur farm management Different ecological zones encourage Fur farmers to adopt different combinations of crops and different timing of cultivation activities. The timing of the different crop activities is presented in Table 15.2.

About half of a man's labour and most of a woman's labour is taken up with millet (*dukhn*) or sorghum (*dura*) production during the rainy season. This is a primary constraint on involvement in cultivation of other crops during that season. In the mountains and foothill zones in the few places where water for irrigation is available it is possible to practice a more balanced work programme.

The most important feature of the socio-economic environment which affects farming enterprises is the land administration units. Among the Fur, membership of such administrating groups is acquired by descent, by marriage, and by assimilation. Usufruct rights are held as long as there are crop remains from the last cultivation, for example, roots of millet or sorghum. Usually such remains disappear after two years, at which time rights to the field revert to the land administration group represented by its head. When a cultivator leaves his fields fallow he will request the head of the land administration unit for permission to use uncultivated land where the vegetation indicates that the soil may have regained its fertility. This land tenure system may operate quite smoothly as long as production on family farms is geared towards satisfying one's own consumption requirements and as long as there is relatively low population pressure.

Usufruct rights in land and one's own capacity to work are thus the two basic resources which are under the management of the farm units. To analyse

TABLE 15.2. Time of planting, first weeding, and harvesting adopted by Fur farmers for different crops

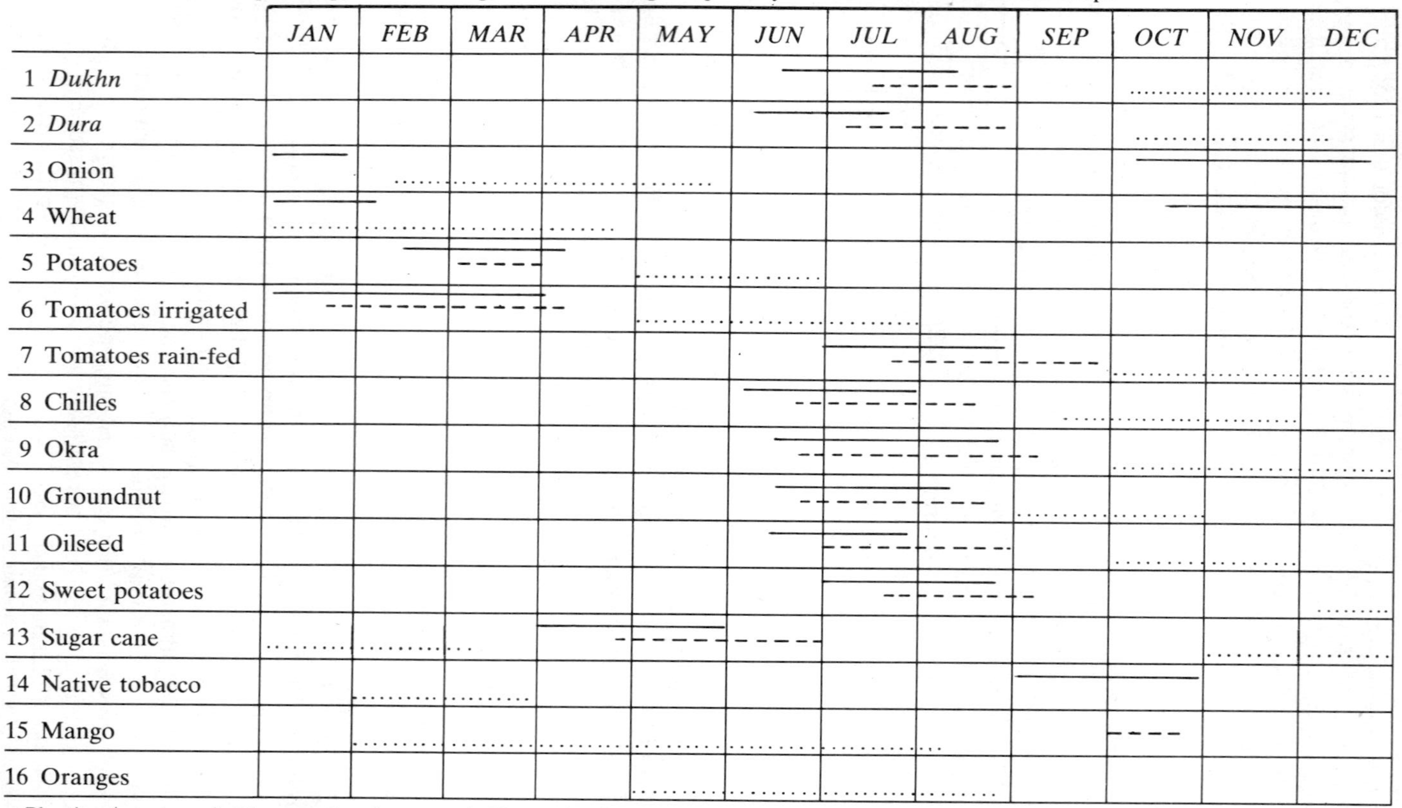

Planting (__________); First weeding (-------------); Harvesting (.............).
Source: Barth (1974).

family farmers' economic management one must identify a wider set of institutional mechanisms which affect their decision making.

The network of local markets allows for some differentiation of crop production between farms in different micro-environments, it allows for the exchange of crops for livestock products (milk products, meat, and live animals) and for the purchase of tools such as hoes, axes, and knives. At the local market, the supply and demand forces of the national and international economy impinge on the farmer. As mentioned in the discussion of the land tenure system, land is not a marketable commodity. Cultural conventions also impose limitations on transacting labour for cash—to work for one's neighbour for cash is considered a sign of inferiority and has connotations with slave identity.

However, the custom of the beer work party allows a farm unit (man or woman) to mobilize neighbours' labour in exchange for millet beer. The preparation of food such as beer or porridge is a wife's duty, but the grain required for the husband's meals comes from his store. Selling beer is considered to be an immoral activity, similar to prostitution.

Economic strategies and spheres of exchange Given the limitations and opportunities of the natural and socio-economic environment and the available technology, it is possible to illustrate the management system of the farm enterprise diagrammatically. Figure 15.2 shows the alternative allocations available for Fur farming units as flows of value brought about through production and different forms of transactions. These flows depict what is conceived to be the management system of Fur farming enterprise.

The management unit aims to steer its resources through the alternatives available in order to satisfy different objectives. Until quite recently, the primary objective of the vast majority of Fur farm units was to produce sufficient crops to satisfy their own requirements (the wife being responsible for production of food for the children). The annual millet requirement for an adult is about 250 kg, but the land and labour requirements to produce this depends on soil quality and climatic conditions. Yields per feddan vary greatly between farms on the alluvial *wadi* beds west of Jebel Marra, which receive 700 mm rainfall and on the drier sandy areas in Northern Darfur, which receive 300–400 mm. Within localities there are also great annual variations in rainfall, crop pests, locust damage, and animal diseases.

From year to year and from locality to locality, a farmer may expect great variation in the productivity of different crops and livestock species. Under such conditions of uncertainty, the farmers' chances of survival depend on their capacity to take advantage of opportunities where and when they are favourable. These situations favour farm management which is short-term and opportunistic—spreading the risk on many crop

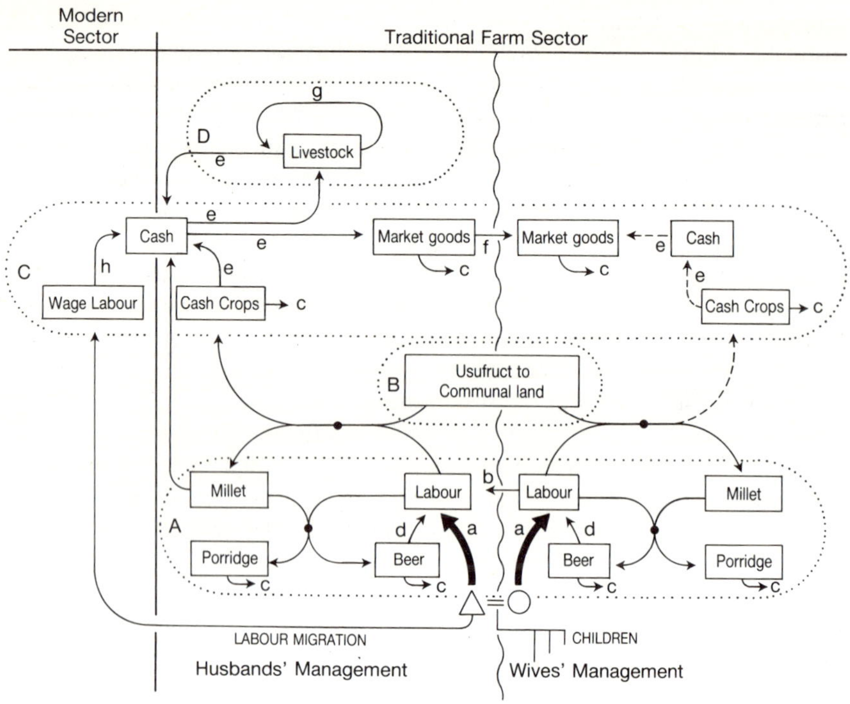

☐	Forms of value	a	Input of own time in production
(dotted)	Spheres of exchange	b	Wife's labour contribution husband (i.e. cooking)
——→	Major value transformations	c	Consumption (→C)
- - →	Minor value transformations	d	Mobilization of labour in beer party
•	Process of production	e	Market transactions
A	Labour-millet-beer sphere	f	Husband's contribution market goods to wife
B	Land sphere		
C	Cash sphere	g	Livestock reproduction
D	Livestock sphere	h	Wage employment

FIG. 15.2. Diagrammatic representation of the management system of Fur farming enterprise.

species and varieties, on cultivating in many plots in different ecological micro-environments, and investing in different livestock species.

The flows in Figure 15.2 demonstrate a feature common in many peasant economies, namely that different forms of value are not directly exchangeable (Barth 1967). The land tenure system and the cultural conventions relating to labour transactions make it virtually impossible for the farmer to invest cash from farm products in inputs of agricultural production, except via the beer work party which allows him to convert millet to beer for the mobilization of labour for further agricultural production.

Livestock investment and nomadization In the absence of investment in agricultural production, investment in livestock was virtually the only possibility for growth in the traditional farming system. Livestock investment is attractive for several reasons; the capital invested increases automatically with livestock reproduction, the animals are consumable, they can be moved between localities according to natural conditions, they are used on ceremonial occasions to establish important social relations such as marriage, and they yield income in the form of milk products. Farmers who manage to accumulate livestock have a buffer against crop disaster and can afford to risk adopting cropping patterns which yield a higher average income in the long-term, but expose them to greater yield fluctuations from year to year. However, farmers with large numbers of livestock have tended not to use this asset as a means of adopting more capital-dependent and long-term cropping patterns. On the contrary, because of natural constraints and the limited investment opportunities in cultivation, they have tended to leave agriculture and become migratory pastoralists.

This adoption of migratory animal husbandry by the most successful Fur farmers invariably leads to a change in the structure of the management unit from individual to family based enterprises. Nomadization also leads to cultural change in that the nomadized Fur accept the ethnic identity of Baggara Arabs.

In the mid-1960s, the annual rate of nomadization and recruitment to the Baggara by sedentary Fur farmers was about 1 per cent and this affected the demographic balance between the two ethnic groups. It relieved pressure on land in the Fur area, but it increased pressure on pasture in the Baggara area (Haaland 1972).

Farm management systems in Kordofan

For most of the sedentary population groups in Darfur, the management systems of family farms have a similar structure to that described for the Fur. In Kordofan, however, there is a wider variety of management systems reflecting different cultures and variations in the natural environment. There is a major cultural distinction between the Arab cultivators found in the qoz area which extends from Darfur to the Nile and the many different groups collectively labelled Nuba, who inhabit the Nuba Mountains in Southern Kordofan. Among both groups, however, the family farm unit is composed of elementary or extended families and the farm management systems are thus more complex than that described for the Fur. The rules regulating value transformations are also different.

Traditional Arab family farms Among the Arab cultivators of the middle zone, the rules regulating cultivation and grazing rights are similar to that described for the Fur (Hill 1968). In areas (mainly in the *qoz*) where the collection of gum arabic from *Acacia senegal* (*hashab*) constitutes an important source of income, rights to tap gum from *A. senegal* trees growing on virgin soil are allocated by the local chiefs (sheikhs or omdas) according to community members' needs. However, where *A. senegal* seeds naturally on land left fallow, the land is called *gineina* (garden) and is under the private control of the cultivator. It is unclear whether this control implies the right to transact trees for cash.

Trees of *A. senegal* begin to yield gum at an average yield of 1–5 lb per tree after about five years (Blunt 1926), and they remain productive for about 15–20 years. Then they are cut down and the fields are cultivated for 4–5 years. The income from gum-tapping varies according to climatic conditions and the farm-gate price, which is set by the Gum Corporation after deducting its profit from the world market price. Trees of *A. senegal* provide the farmer with a cash income, but also increase the nutrient content of the soil, provide browse for animals, and provide wood for cooking or for charcoal production and sale.

Gum-tapping, the cultivation of millet and groundnuts, and animal husbandry (sheep, goats, camels, and cattle) constitute the main elements of a family farm's multi-resource management. In Kordofan, this multi-resource management is a fundamental aspect of farmers' survival strategy.

In Kordofan, the opportunities for investing profits in agriculture are easier than in Darfur because labour is not only mobilized by work parties (*nafir*) but also by wage contracts. In areas such as En Nahud which are good for groundnut cultivation, high groundnut prices have stimulated enterprising farmers to invest in employment of seasonal wage-labour. Where there is external demand for cash crops and a local supply of wage-labour, the existing communal land tenure regulations do not function so smoothly and there is a trend towards emerging differentiation between rich farmers who employ labour and expand the area they cultivate and poor farmers who have not been able to satisfy their consumption requirements by their own production and, therefore, also have to rely on income from occasional wage work.

The Nuba In the Nuba Mountains the rainfall is higher and more regular, but the population density is relatively high due to the influx of refugees after wars and slave raids in earlier periods.

These conditions have resulted in more intensive land use practices and probably influenced the type of land tenure rules which have evolved. The unstable political environment in Southern Kordofan was probably the reason why weekly markets were never established in the Nuba Mountains. Instead, other institutions evolved to facilitate the exchange of goods and services and the beer work party was the most important mechanism for labour mobilization. Exchange of grain for livestock, or one type of livestock for another at fairly standardized rates took the form of *ad hoc* barter (Nadel 1947).

The bridewealth institution served to circulate livestock among a wide set of relatives and affines, i.e. several relatives on the groom's side contributing to the bridewealth and several relatives on the bride's side sharing it. Several ceremonial occasions, for example, funerals, circumcision, and cicatrization involved slaughtering of animals and communal consumption of the meat and this was a means of converting economic wealth to prestige and influence. Giving a feast on the strength of an accumulated grain surplus also provided high standing for the individual and served to redistribute wealth in society.

An unexpected feature of this essentially non-monetized economy is that cultivated land is individually owned and can be transacted. Land transactions have frequently taken the form of barter for animals (goats, cows, pigs). The existence of private land tenure is probably connected with the intensive cultivation which developed in areas of land scarcity.

In the Nuba Mountains, a family farmer usually operates different categories of farm land; house farm, hillside farm, and far farm. The house farm is located close to the home and is the responsibility of women who every morning collect dung, house refuse, and ashes for manuring. The practice of manuring enables continuous cultivation of plots and provides more integration between animal husbandry and crop production than is the case among the Fur. Hillside farms and far farms, situated on the clay plain using soils suitable for hoe cultivation, are rarely manured, but a rudimentary crop rotation, sesame or groundnuts followed by several years of sorghum is practised within a system of shifting cultivation, allowing at least 5 years fallow. The different types of crops are cultivated on the different types of farms: short-season sorghum, maize, and vegetables on house farms, long-season sorghum and sesame on hillside farms, and long-season sorghum, sesame, groundnuts, and cotton on far farms.

The Nuba farm management system is quite different from that described for the Fur, for land is a form of value which is completely within the management system in the sense that it can be bought and sold.

Enterprising management of resources within the rules of the Nuba economy has created economic differentiation between rich and poor farmers. However, while there was little external demand for their products, the sanctions on ceremonial redistribution of crops and livestock counteracted polarization processes within the Nuba system. With the introduction of commercial cotton cropping on the plains the situation changed as entrepreneurial farmers mobilized work party labour for cash-cropping. Nevertheless, the economic differentiation created in this way is of minor importance compared with the effects of the surrounding mechanized schemes in the Nuba Mountains area.

Baggara pastoral production system

The Baggara consist of a large number of Arab tribes inhabiting Southern Kordofan and Southern Darfur (Cunnison 1966; Haaland 1977). Baggara is the Arabic word for cattle and the Baggara Arabs are associated with cattle husbandry. The major tribes have their homeland on the Baggara Catena of the low woodland savanna where the nutritious 'salty' grasses provide important fodder and the *atamur* (sandy patches) are good for millet cultivation. The natural conditions of western Sudan favour migratory movements: most tribes move south to the green *toich* grasses of Bahr el Arab (except the Beni Helba tribe who move north to the good grasses and *Acacia albida* pods of the lower wadis of western Darfur) in the dry season and north to the *dar* (home area) as the risk of livestock disease increases with the approaching rains. During the height of the rainy season cattle are moved further north to the *qoz* of Kordofan and eastern Darfur, where they stay until shortage of pasture and water at the beginning of the dry season stimulate the movement back to the *dar*.

Baggara management units are based on family units (nuclear or extended). Ideally, in Baggara culture, the management unit should be able to satisfy its consumption needs by income from pastoral production without having to rely on cultivation. Since millet or sorghum porridge is the stable food this implies that a Baggara household must depend on cash income from the sale of livestock products (milk and animals). Where household income is totally dependent on pastoral production, the economic viability of the household is sensitive to herd size (Haaland 1977).

The viability threshold (which varies according to the exchange rate between pastoral and non-pastoral products and with the number of consumers in the household) is defined by the number of animals where herd reproduction balances the herd offtake required to satisfy the household's consumption requirements. Households falling below the viability threshold are thus forced to consume their productive capital and, unless they are able to satisfy their consumption needs in ways other than by livestock sale (e.g. agricultural production or wage-labour), they will eventually be sloughed off from the pastoral sector. For this reason and because of the vulnerability of livestock to natural hazards such as drought, livestock diseases, and theft, it is understandable that herd accumulation is a primary concern for pastoral management units. There are no limits to a pastoralist's interest in livestock accumulation unless labour availability creates diminishing returns from such investments. The relationship between aggregate stocking rate and pasture carrying capacity is not a factor affecting individual herd owners' interest in livestock investment because pasture is not under the control of the individual pastoral management units. Pastoralists either have free access to pasture or it is under some kind of tribal administrative control where access depends on the tribal identity of the pastoralists. Thus, the income from pastoral production accrues to the individual management unit while the costs in terms of overgrazing are shared by everyone. In a situation where the number of pastoral households is low and labour shortage limits increases in stocking rate this does not create any problem.

Overstocking and environmental degradation For reasons which are not well understood the Baggara used to have a low fertility rate and recruitment of nomadized farmers from other groups, like the Fur, served to replenish their numbers. However, the introduction of modern medicine has affected population growth and, for the last 50 years, the balance between stocking rate and pasture carrying capacity has become increasingly critical. Most pastoralists are quite aware that pasture deterioration is caused by overgrazing and too many animals and that increasing animal numbers in an overgrazed area will reduce the food available to them so that on average, they yield less income. However, the reduction of income caused by lower average yields per cow is compensated for by the increases in income from increased herd size, because the losses (environmental deterioration) are shared by everybody while gains are privatized.

While there are no mechanisms to co-ordinate individual management units' investment strategies, individuals continue to convert as much as possible of the communal wealth (pasture) into individual wealth (livestock). The system thus rewards individual strategies which are collectively disastrous. A sustainable balance between stocking rate and pasture carrying capacity cannot come about as a result of management because there is nobody with authority to co-ordinate individual management strategies. When a balance occurred in the past, it was a result of other unplanned processes such as low population growth, periodic decimation of livestock numbers by drought and diseases, and the expulsion of households from the pastoral niche during warfare. Planning and implementation of better health services for men and animals, improved administration of peace, and supply of international aid in drought years have made the unplanned processes ineffective as mechanisms for maintaining an ecological balance.

As pressure on pasture resources has increased, the Baggara have compensated for decreasing income from livestock husbandry by increased involvement in agricultural production, either by millet production on the *atamur* and *qoz*, or by sorghum cultivation on alluvial *wadi* beds. With increasing demand from national and international markets for agricultural products, an increasing number of Baggara households have turned to cash-cropping. Attractive prices for cotton after the Second World War encouraged many Baggara households, particularly in Kordofan, to grow cotton on the southern clay soils. With the fall in cotton prices most Baggara gave up cotton and took up groundnut production in *qoz* areas. Rich herd owners have also invested in groundnut production, by employing cheap labour from the Dinka areas of Bahr al-Ghazal. Under the communal land tenure system, this has allowed them to expand the areas they cultivate substantially. Recently, rich Baggara have even invested in mechanized equipment for large-scale sorghum production which the Mechanized Farming Corporation promoted on the clay plains between Bahr el Arab and the *qoz*. These developments imply that an increasing number of Baggara management units are becoming more dependent on income from cultivation, rather than from animal husbandry. Nevertheless, livestock is the most feasible investment for most households and this strategy serves to keep up the pressure on pasture. Periodic droughts as in 1982–4 may reduce the stocking rate in some areas by more than 50 per cent and this relieves the pressure for a few years. Today the human misery accompanying such disasters mobilizes large-scale international relief operations and, within a very short period, surprisingly rapid herd build-up takes place with accelerating degradation of the natural environment.

Enclosures In this situation, members of some local communities, particularly the Beni Helba of Southern Darfur have taken active steps to protect their communal wealth. By extending the perimeter fences around farm land to include several square kilometres of pasture they have, on a local community basis, tried to protect the land from grazing during the rainy season. Such measures are a first rudimentary form of pasture management. However, enclosures are still not recognized as legal and they are frequently ignored by members of other groups who, in this area of complex migratory patterns, find their access to pasture reduced. Nevertheless, it is this kind of control mechanism which is needed to halt environmental deterioration.

The enclosures erected by the Beni Helba Baggara have had repercussions on the nomadized Fur in that their access to rainy-season grazing in the Beni Helba area has been restricted. Investment in livestock by the Fur of the lower wadis of western Darfur is still attractive, but instead of migrating to more favourable areas further south they stay within the Fur area for the whole year, moving to the nearby elevated sandy areas between the wadis. With increasing population density there seems to be a trend towards the establishment of exclusive rights in pasture on a local basis which will lead to more confined migratory movements. It also seems likely that an increasing number of Fur management units will combine animal husbandry and cultivation.

Intensification and integration of agricultural and pastoral production

At the present these two activities remain fairly separate although the residue sorghum and millet stalks are sometimes fed to animals and there are cases reported of Baggara who have started manuring their fields, simply by camping their herds on the sown fields for a few days. It has also been reported that mixed farmers of the Hawazma Baggara of Southern Kordofan have started collecting dung and storing it in pits ready for spreading on the fields after sowing.

In western Sudan, the draft linkage is almost non-existent in cultivation systems. Only among the cultivators of the Kheiran has this linkage a long history (Manger 1981). The Kheiran is a series of depressions in Northern Kordofan, containing alluvial soils and ground water close to the surface. For centuries, the farmers in these depressions used *saqqiya* (Persian waterwheel) irrigation for intensive vegetable production in the dry season and used oxen to pull the *saqqiya* and plough the fields. During the rainy season, farmers practised extensive hoe cultivation of millet and sorghum on the surrounding sand dunes. Today the *saqqiya* have been replaced by diesel-driven pumps and ploughing is mainly by camels.

Although the use of animals as draft power has been known among the agro-pastoralists of western Sudan, only recently has it been taken up outside the Kheiran. As pressure on land increases, and if ox-drawn equipment suitable for cultivating the heavy clay plains is developed further, it is likely that the draft linkage may increase in importance, particularly among the Nuba and the Baggara close to the clay plains of Southern Kordofan. However, it is significant that the period of heaviest work for the animals is after the dry season when their condition is poorest, so their work performance will depend heavily on the availability of supplementary feed. At present, sesame cake from local oil presses is used to some extent for this purpose, so the future development of sesame cultivation and local oil pressing will have an important effect on the attractiveness of the draft linkage in the non-irrigated cultivation systems of the savanna.

The fodder linkage was developed in the Nile Valley where draft animals were fed *bersim* (*Medicago sativa*) which also played an important role in the crop rotation system on the irrigated land. To date, virtually no fodder cultivation has been introduced into rain-fed cultivation systems and with the present tenure system it seems unlikely that farmers will adopt this linkage.

In the savanna environment where there are critical constraints to the intensification of the production systems, developments are only likely to take place where there are perennial streams (Jebel Marra) or where there is a relatively high (600–700 mm) and reliable rainfall (Southern Kordofan, Southern and western Darfur). In the drier areas to the north, the polarization between rich households with sufficient numbers of animals (camels, sheep, and goats) to specialize in pastoralism and the poorer households who survive by their ability to derive income from their small herd, from cultivation in a marginal environment, and from labour migration, is likely to continue.

Family farms in the wider environment

The wider economic and politico-administrative environments are having an increasing effect on family farm management. Tapping of gum arabic took place for centuries as a result of international demand and, with the growth of urban centres, there has, over the last decades, been an increasing demand for charcoal from western Sudan. For many family farms, production of charcoal constitutes an important additional source of income. There are often community restrictions on tree cutting, but these are difficult to enforce in sparsely-populated areas and there is evidence that forest offtake for charcoal production exceeds forest regeneration. Particularly around larger settlements and along the main routes, deforestation is evident. This imbalance is caused not only by increased forest offtake for urban markets, but also by reduced forest regeneration due to over-browsing. When browsing pressure is reduced, however, as with the high livestock mortality caused by the last drought, forest regeneration can be remarkable; re-establishment of *Acacia senegal* by natural seeding during the first three years following the drought has been significant. However, the combined effects of rural impoverization, increased reliance on income from charcoal production, over-browsing, and tree cutting to clear land for cultivation are likely to accelerate ecological imbalances.

Cropping patterns are also affected increasingly by external demand. In addition to cotton and groundnuts, the demand for crops such as sesame, ochra, hibiscus, chilli, onion, mango, and oranges has brought dramatic changes to farm management units. Now, farmers may purchase their subsistence foods with income from cash crop production. This has brought new opportunities, but also new uncertainties, for the prices of subsistence crops and cash crops undergo great fluctuations from year to year.

Where perennial water is available, as in the foothills of Jebel Marra, these new opportunities have stimulated farmers to re-allocate their resources in ways which have completely transformed the production system. Cultivation of fruit trees by irrigation has led to permanent cultivation and privatization of landownership and investment in land has become more attractive than investment in livestock. The importance of natural and artificial fertilizers is well understood and farmers do invest in them. Many of the family farms along the perennial streams of Jebel Marra now operate as commercial units and on larger farms farm labour is no longer supplied by family members, but wage-labourers. The increase in cash-cropping has also brought a change in household structure, with the family of husband, wife, and children replacing the single adult.

Where the natural conditions are favourable, new opportunities have resulted in increased involvement in cash cropping. In areas where conditions are less favourable, people have satisfied their increased demand for market goods by moving to the towns and schemes of the Nile Valley for temporary wage work. Such migrations may just be seasonal, as for the cotton picking in Gezira, or they may last for several years. In many local communities more than 25 per cent of the male working population may be away as labour migrants, at any one time. On the one hand this relieves the pressure on land at home and provides increased net cash income to individual families, where the family is the household unit. On the other hand, labour migration establishes interdependence between the modern sector of business enterprises and the traditional sector of family farms which supply the labour force to the modern enterprises in agriculture and industry. This interdependence is an important part of the dualistic development trends which are emerging in Sudan's economy. Stagnation in the family farm sector stimulates the supply of labour migrants to the modern sector, but low wages force the migrant's family to continue with subsistence farming in the home area. It is by this means that the labour force to the modern sector enterprises is reproduced. Consequently, the wages paid to labour migrants working in modern enterprises do not have to cover the reproduction costs of the labour force. In a dualistic economy the profit earned in modern enterprises thus depends on underdevelopment of family farm enterprises. As the Government's revenue is derived largely from the profits of modern sector enterprises, there are strong vested interests for maintaining underdevelopment in the family farm sector.

Although the present and previous Governments have given programmatic expression of their intention to develop the family farm sector, this has not resulted in much concrete change. In fact, the major dilemma confronting development of the family farm sector is of a political nature. In order to halt overgrazing and deforestation, it is necessary to introduce a land tenure reform which exercises constraints on a production unit's opportunity to commonize the costs of production. The simplest solution would be to give livestock production units exclusive rights in pasture, in which case the same unit would carry the cost and reap the benefits of its land-use practices. In most parts of western Sudan, this would not be feasible for political and economic reasons. Instead, planning should try a stepwise approach by identifying and channelling peoples own attempts to cope (e.g. Beni Helba exclosures) in directions which can lead to sustainability (Haaland 1980). Such a policy would, in most places, imply constraints on an individual's utilization of common resources and many farm units, particularly pastoralists would, thereby, face reduced access to pasture. Therefore, unless alternative sources of employment are created, or new areas opened up, an increasing number of households could be marginalized from the agro-pastoral sector. In order to counteract such marginalization processes, it is important that reforms are introduced within the framework of a comprehensive land-use plan, with possibilities for accommodating marginalized groups in other areas or in alternative occupations (Hunting Technical Services 1977).

In western Sudan, the large Qoz Dango of Southern Darfur could accommodate a substantial number of marginalized family farmers if opened up for cultivation and animal husbandry by instalment of water yards. The large areas of heavy clay soils in Southern Kordofan provide a second possibility. Increasingly, however, these clay soils are being brought under cultivation by large mechanized schemes (1000-feddan units) operated mainly by managers with a background in trade and civil service, but sometimes by rich pastoralists.

Mechanized farming schemes and their development implications

As early as 1954 the Condominium Government was encouraging private farmers to take up mechanized farming in the Gedaref area of eastern Sudan. The private sector responded rapidly and soon expanded cultivation beyond the area demarcated by the independent Government. With the establishment of the Mechanised Farming Corporation (MFC), an autonomous agency under the Ministry of Agriculture and with loans from the International Development Association (IDA), the Government tried to exercise more control over land allocation and farming methods.

In the early 1970s, the MFC initiated mechanized farming schemes in Southern Kordofan. Here, investors were allocated farms of 1000–1500 feddans for a nominal rent and again the response was favourable. In terms of sorghum yield, mechanized farming compares favourably with traditional agriculture, but in 1984/5, with severe drought, yields were far lower for rain-fed mechanized farming than for the traditional subsector. For sesame production, yields in the traditional sector are usually lower than in the mechanized sector, but this varies from year to year.

In addition, the mechanized farming sector is an important employer of seasonal labour drawn primarily from family farms in the Nuba Mountains and poor Baggara households, but also from Nilotic agro-pastoralists from southern Sudan.

Despite its advantages, mechanized farming has considerable disadvantages. The starting-up costs of a new farm of 1000 feddans are high, including $US25 000 or more of foreign exchange for tractors and discs. After initial land clearing, yields are usually high but they decline sharply after some years. Sorghum, which accounts for about 90 per cent of the cultivated area, may yield 1000 kg per feddan initially, but the yields decline to as little at 100–200 kg per feddan after 6–8 years. Investment in mechanized farming has a high element of risk, not only because of weather conditions, but also because of price fluctuations. Nevertheless, the Sudanese Government has supported and subsidized the mechanized sector in various ways. First, by the provision of land at a nominal land rent; second, by making available capital for equipment at below market rates of interest; third, by granting an incentive exchange rate for purchases in some years; fourth, by taking over the losses of the MFC and the ABS, caused by low capital repayment ratio.

In addition to benefiting from government subsidies, mechanized farm owners are traders who market the crops themselves. The mechanized farm is rarely a specialized farming unit, but forms part of an enterprise which includes other activities (such as trade, transport, and building) spread over large regions. The owner does not reside on the farm, but has an agent who is responsible for overseeing the activities during the cultivation season. Rarely does the owner have any commitment to the land or to farming as a way of life. This involvement in mechanized farming is based on the attractions of investing in farming compared with other sectors and this depends to a large extent on the politico-economic and natural environments.

Mechanized farmers have a vested interest in maintaining conditions, such as cheap credit, cheap leases, and cheap labour, which minimize the costs of their operations, in preventing implementation of rules which might increase their costs, for example, stumpage fees for tree cutting, maintenance of tree shelter belts, and crop rotation systems which are beneficial in a long-term ecological perspective, but less rewarding in a short-term economic perspective, and in maintaining their right to move to new areas of cultivation as land is exhausted. Through the Union of Mechanized Farmers, the operators act as a political pressure group to maintain these politico-economic conditions.

Sudan desperately needs the grain supplied from the large mechanized schemes, so cannot afford to introduce mechanisms which cause the scheme operators to channel their capital into off-farm activities. On the other hand, the conditions which are attractive for the individual mechanized farmers have less desirable long-term environmental implications. Scheme expansion is accompanied by large-scale deforestation of the clay soils, by increasing conflict between the large scheme operators and Baggara herd owners who pass through the clay belt *en route* to their dry season pastures on Bahr el Arab, by reduced opportunity to accommodate the population surplus of the drier northern areas in medium scale enterprises on the clay soils, and by the opportunity costs involved in subsidizing rich operators with cheap loans. Agricultural development in western Sudan is thus shaped by a complex interplay of processes at different levels of organization, e.g. farm management in different types of economic units, energy conversions and nutrient circulation in the ecosystem, land tenure regulations and the political forces supporting them, the position of the farm units in the national and international exchange of goods and services, and the ability of different economic interest groups to influence the Sudanese Government's policy towards the agricultural sector.

The fundamental dilemma in formulating an agricultural strategy for western Sudan concerns the balance between rapid mechanized development of the relatively little exploited clay soils in order to secure Sudan's grain supply, and support for the family sector in order to reverse long-term ecological degradation and economic dualism. The main planning challenge is to link these two issues. However, there are formidable economic interests working against such a policy, primarily those having the resources to operate as scheme owners. Given the structure of Sudanese society, it is difficult to envisage that political processes might produce a government with the will and ability to tackle the fundamental strategy dilemma outlined here.

In western Sudan, the relationship between family farms and mechanized schemes is characterized by multiple duality in terms of technological development (the hoe versus the tractor), in terms of economic scale (family farms of one feddan versus a 1500-feddan mechanized scheme), and in terms of political influence (the poorly organized farmers of the periphery versus an interest group with a formidable power to influence Sudan's economic policy). These dualities are interconnected and it is difficult to see traditional family farms developing technologically unless they are able to wield the political influence required to change the 'rules of the economic game.' As the dual structure becomes more developed, the economic and social costs suffered by changing it become greater. The present political turmoil in western Sudan should also be seen in this politico–economic context and not only as an expression of ethnic conflict.

So far, development assistance has had little effect in bridging the gap between the two types of enterprise in western Sudan. In fact, there are clear indications that the conditions experienced by small farmers have deteriorated over the last twenty years despite a number of development projects aimed at them specifically. Unless one understands that the traditional farming systems of western Sudan are closely integrated in a developing, nationwide, dual economy, our assistance may just as well serve to reinforce this development rather than reverse it.

References

Asad, T. (1970. *The Kabaish arabs: power, authority and consent in a nomadic tribe*. Praeger, New York.

Barth, F. (1967). Economic spheres in Darfur. In *Themes in economic development*, ASA Monograph No. 6 (ed. R. Firth), pp. 149–77. Tavistock Publications, London.

Barth, F. (1974). *Human resources: social and cultural features of the Jebel Marra project area*. Department of Social Anthropology Occasional Paper, No. 1.

Department of Social Anthropology, University of Bergen. Distributed by Studia University Bookstore, Bergen.

Blunt, H. S. (1926). *Gum arabic*. Oxford University Press.

Cunnison, I. (1966). *Baggara arabs: power and the lineage in a Sudanese nomad tribe*. Clarendon Press, Oxford.

Haaland, G. (1972). Nomadization as an economic career among sedentaries in the Sudan savanna belt. In *Essays in Sudan ethnography* (ed. J. Cunnison and W. James), pp. 148–72. Hurst, London.

Haaland, G. (1977). Pastoral systems of production, the socio-cultural context and some of its ecological implications. In *Land use and development*, African Environment Special Report, No. 5 (ed. P. O'Keefe and B. Wisner), pp. 179–93. International African Institute, London.

Haaland, G. (ed.) (1980). *Problems of savannah development: the Sudan case*, Department of Social Anthropology Occasional Paper, No. 19, Department of Social Anthropology, University of Bergen. Distributed by Studia University Bookstore, Bergen.

Hill, L. G. (1968). Hababin village economy. *Sudan Notes and Records*, **XLIX**, 58–70.

Holy, L. (1974). *Neighbours and kinsmen: a study of the Berti people of Darfur*. Hurst, London.

Hunting Technical Services (1977). *Southern Darfur land use planning survey*. Ministry of Agriculture, Food, and Natural Resources, Sudan.

Manger, L. O. (1981). *The sand swallows of our land—over-exploitation of productive resources and the problems of household viability in the Kheiran—a Sudanese oasis*. Department of Social Anthropology Occasional Paper, No. 24, Department of Social Anthropology, University of Bergen. Distributed by Studia University Bookstore, Bergen.

Manger, L. O. (ed.) (1984). *Trade and traders in the Sudan*, Department of Social Anthropology Occasional Paper, No. 32, Department of Social Anthropology, University of Bergen. Distributed by Studia University Bookstore, Bergen.

Nadel, S. F. (1947). *The Nuba*. Oxford University Press, London.

Tully, D. (1988). *Culture and context in Sudan*. State University of New York Press.

16

Systems of agricultural production in central Sudan and Khartoum Province

I. G. SIMPSON and M. C. SIMPSON

Central Sudan is located in the country's savanna belt which extends from the fringes of the Sahara Desert to the broad-leaved forests, grassy river flood plains, and permanent swamps of the southern Sudan. This chapter is concerned with the eastern portion of the belt which is contained in the provinces of Khartoum, El Gezira, Blue Nile, the southern part of Kassala Province and the land east of the White Nile in the province of that name.

In this region, annual rainfall increases southwards away from the desert from an average of about 200 mm in the north to 800 mm in the south. The natural vegetation changes with increases in rainfall, from desert scrub and scattered grasses with a very short growing season, through short grass and bush to open woodland with longer grasses, in the south. Much of the land surface is covered by thick deposits of almost black, cracking clays, which characterize the region known as the clay plains of the Sudan. The Blue and White Niles flow through the region from south to north to their confluence at Khartoum, but away from the Niles and their tributaries, drinking water is very scarce during the long dry season. Sinking wells and boreholes does not solve the problem because of the impermeable nature of the clay and, therefore, settlement is confined largely to the riverain areas and to the bases of *jebels* (rocky hill outcrops) where water can be obtained easily from shallow wells.

Natural conditions have not favoured village settlements or arable cultivation even in the southern parts of the area where rainfall was normally sufficient for cropping. Settlement and cultivation were confined to areas with permanent water supplies and, in the past, most of the region was unused except for the seasonal migrations of nomadic flocks and herds. With economic development, the situation changed gradually. In the north of the region, irrigated agriculture, including the world famous Gezira Scheme, was introduced on a large scale in the 1920s. Food shortage in the Second World War, especially of sorghum, the staple grain, led the

Sudanese Government to experiment with mechanized rain-fed farming and the area under such schemes has spread rapidly.

In this chapter traditional nomadic pastoralism, sedentary traditional cultivation, mechanized rain-fed farming, and arable farming on gravity-fed or pump irrigation schemes are each discussed in turn. The first categories are traditional methods of agriculture, the second are modern methods—the dualism of the agricultural economy of the country. There are, of course, linkages between the separate systems and sectors. The principal agricultural schemes in central Sudan are shown in Fig. 16.1. Traditional rain-fed cultivation is confined mainly to river banks and to the vicinity of *jebel* villages with a perennial water supply. Nomadic and transhumant flocks and herds move over uncultivated areas mainly in the more southerly areas shown on the map.

Traditional nomadic pastoralism

The savanna belt of the Sudan, approximately one-third of the total area of the country, is traversed by the annual migratory cycles of nomadic flocks and herds in search of grazing and water. Accurate figures for the different categories of nomadic livestock and for the nomadic population are not available but it is agreed generally that the nomadic section accounts for a major portion of the country's livestock and that the nomads themselves represent 15–20 per cent of the human population. The severe droughts of 1983 and 1984 hit the nomadic livestock population badly, especially in western Sudan where conditions were worst. Again no firm figures are available.

Migratory cycles are peculiar to individual tribes. Most involve an annual return trek from north to south, but a select few tribes in the eastern savanna follow an east–west–east cycle. The basic reasons are the same, to secure a relatively dry grazing area, avoiding mud and flies in the wet season, and to obtain sufficient water and grazing during the extreme part of the dry season. The annual cycle represents the journey between the two destinations. The journey towards the wet season pastures is often made 'at the gallop' in advance of the thunderstorms during which the rains rapidly reduce the clay surfaces to a mire, so impeding stock movement and also providing a breeding ground for biting flies.

In the past, the nomads have been left very much to their own devices. They needed very little from the outside world and few coveted their wildernesses. Now, the situation has changed. Loss of land to arable schemes has forced nomads to modify their routes and drought, particularly in 1983–4, has forced nomads with only small numbers of remaining stock, to settle. Another undoubted motive for settlement has been the desire for social services, especially health and education. Many of the wealthier

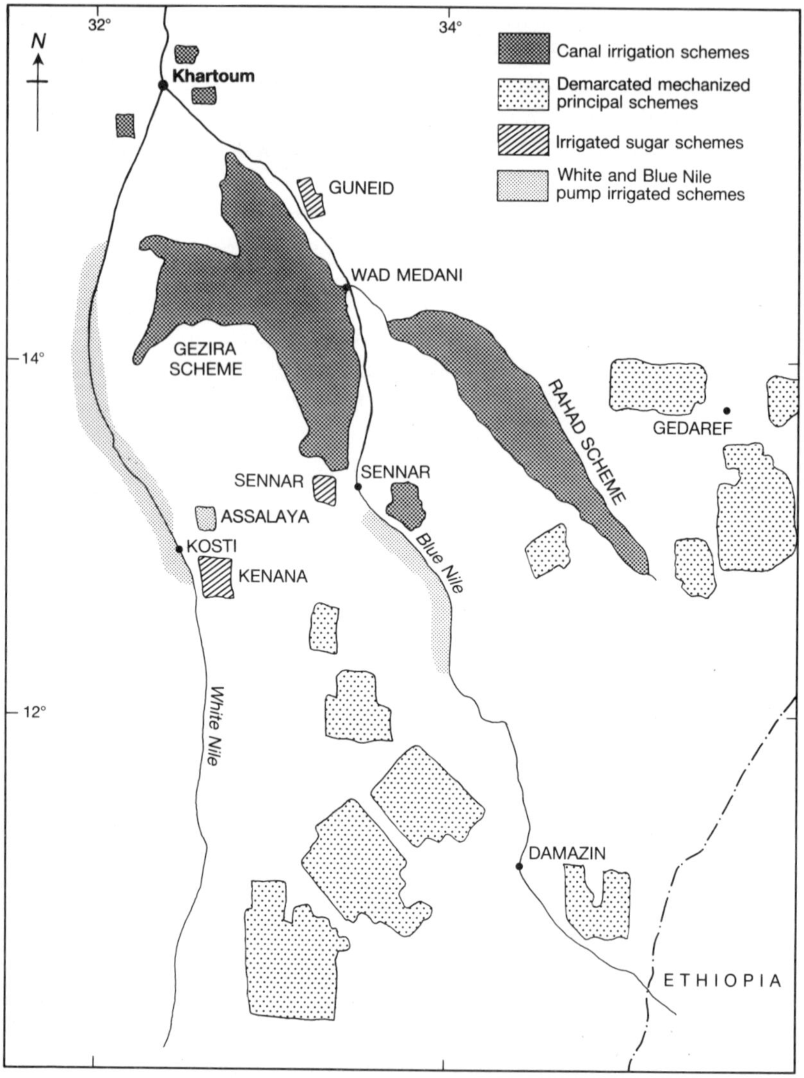

FIG. 16.1. Central Sudan: principal agricultural schemes.

nomads have had houses in towns for a long time, leaving the day-to-day running of their stock to poorer relations or herdsmen. Nomads are also market-minded, contradicting the widely held belief that nomads' cash requirements are so limited that any rise in the price of their products leads to a curtailment in the market supply—a classic case of perverse supply (Khalifa and Simpson 1972).

It was estimated (MAFNR 1977) that nearly one million cattle and over one million sheep pass through Blue Nile Province twice a year, accompanied by a smaller number of camels and goats. The movement north to escape mud and flies takes place in the southern savannas and Nile swamps. The return journey is in December and January, coinciding with crop harvesting and leading to friction between nomads and cultivators (Simpson and Simpson 1978).

Nomads on the move utilize only a very small proportion of the grasses available. Most go to seed and form a matt of old dead vegetation, the so called 'desert hay' which has little nutritive value, obstructs the movement of short-legged animals like sheep and goats, and also harbours ticks and other undesirable pests. Consequently, it is common practice for nomads to send scouts well in advance of the main party to burn the grasses, in the hope of stimulating fresh new growth. This is done particularly in December and January *en route* to dry season pastures, but some fires get out of control and large areas may be burnt out, including standing crops ready for harvesting. Nomads' grass-burning practices have also been blamed for accelerating the process of desertification.

Generally, there is little love between nomads and cultivators, whether peasants or scheme operators. Livestock trespass across cultivated crops, especially at night, trampling and feeding on the crops. Access to water is another common motive for trespass, particularly through cultivated areas alongside the Niles. The authorities have found difficulty in preventing nomads invading the Dinder Game Reserve, but the nomads did not consider it a crime. Nomads interviewed by the author maintained that the whole area was theirs traditionally and that the large scheme operators were the new colonialists. They argued that the corridors left for their passage through the cultivated areas did not last long and cultivators removed the boundary markers and then cleared the land for cropping. This was confirmed by the authorities and some of the larger farmers, who argued that they were merely safe-guarding themselves against mass trespass. Fights between farmers and nomads are common and, on occasions, both the police and army have been called in to keep the peace.

Besides trespass, small cultivators complain of theft and the spread of animal diseases and weeds. The Veterinary Commissioner for the Blue Nile Province (personal communication) confirmed that nomadic cattle were largely to blame for the spread of cattle diseases among the stock of small sedentary cultivators; mechanized scheme owners normally did not keep cattle. Infection occurred at watering points and from the practice of leaving dead cattle carcases unburied. It was almost possible to plot the routes of the nomads from reports of disease outbreaks. New varieties of weeds were believed to have been introduced by the distribution of their seeds in cattle dung and some farmers actually removed dung from their cultivated plots to prevent weed seeds germinating.

Systems of livestock husbandry and the type of stock kept differ between tribes as does the migratory cycle. However, low productivity is common to all nomadic stock. Animals are slow to mature, effective birth rates are low, death rates are high and consequently annual offtake rates are low.[1] Surplus males are usually sold, but bulls used to carry possessions during trekking may contribute to the increase in the numbers of cattle retained. Sales are either to itinerant traders, often from the same tribe, or through auction markets in towns near the nomads' route. Limited income is obtained from hiring camels for transport, collecting gum arabic, and the sale of *semn*, a clarified butter. The nomads themselves live on milk, meat from slaughtered sheep and goats and purchases of sorghum grain, together with tea, coffee and sugar. They seldom slaughter cattle.

The main tribes are the Kenana, the two branches of the Rufa'a al Hoi and the Rufa'a al Sharg, all of Arab origin, and the Umboro said to have come from West Africa. As incomers, the Umboro are blamed by other tribes for the collective sins of the nomadic community, in particular, their failure to attend temporary veterinary camps, set up along migratory routes for routine innoculations and dipping. As illegal immigrants it is believed that they wish to keep away from any Sudanese authorities.

The Kenana follow an east–west cycle, the other tribes a broadly north–south route. The Kenana are noted for their grey Zebu type cows which give average yields of about 900 kg milk over a six-month lactation. These cattle, like other variants of the Baggara breed, have excellent genetic potential for beef production. Under optimum conditions liveweights of 500–600 kg can be obtained for mature steers and bulls at four to five years old.

The Kenana practise a form of transhumance; in the dry season the young men take the cattle to the Ethiopian foothills, leaving the rest of the tribe with the sheep and goats in temporary settlements on the plains. In the past, during the course of their annual cycle, the Rufa'a al Hoi stopped to sow a crop of sorghum on wasteland. This was left and harvested on their return journey. Shortage of land and the presence of other nomads has halted this practice. The economy of the Rufa'a al Hoi is described in detail by Ahmed (1976).

The Rufa'a al Sharg are adapting to changes mainly due to the foresight of the previous chief, Sheikh Yousif Elagab, who was a member of parliament for thirteen years and served as a cabinet minister for part of the time (Mansour Elagab, personal communication). He laid great stress on education and encouraged settlement to secure it and other social services. By 1980, over half the tribe had settled in and around Dinder town with its primary and secondary schools, hospital, mains water, and electricity.

[1] No firm figures are available. Annual commercial offtake rates in the Sudan are estimated at about 10 per cent for cattle and 18 per cent for sheep (World Bank 1985). The cattle rate can be compared with an estimate of 16.9 per cent for the world average (1972–4) and 18.2 per cent for the projected world average for 1985 (Simpson 1980).

Settlement did not mean the end of nomadism, for most owners retained their livestock which was trekked for them by poor relations and hired herdsmen. Owners regarded their stock as a source of income and took a keen interest in the management and marketing of their animals. Only adult males required for breeding or transport are retained and two companies have been formed, one for the sale of fat sheep to Sudan's urban areas and for export to Saudi Arabia, the other for marketing camels to Egypt and specially-bred racing camels to wealthy Saudis.

Wealthier members of the tribe own crop schemes and include professional men such as doctors, lawyers, and civil servants. Furthermore, the late chief's policies have resulted in over 50 members of the tribe becoming University graduates. However, the old migratory routes are still followed by other nomadic members of the tribe, moving up and down the east side of the Blue Nile between the Butana and the area of the Dinder and Rahad Rivers towards the Ethiopian border.

Another category of settled nomads are those forced to abandon their way of life, usually because they have lost the bulk of their stock through disease or drought. They are generally given land on which to establish a village and cultivate crops. The authors visited two such villages inhabited by former Kenana nomads. They found crop growing much harder work than herding stock and some had saved enough to build up stock numbers again. One man had acquired 200 head of cattle which were managed under a transhumant system. The young men of the family trekked them northwards to the Butana to avoid the mud and flies of the rainy season, returning in the dry season when cattle were fed on crop residuals. Plentiful water was available within easy distance and the manure, said to increase crop yields, was prized.

Settlement does not necessarily result in a reduction of nomadic livestock numbers although it may be the result of increasing annual offtake rates, by improving standards of husbandry and securing earlier sale ages for mature and immature stock. Immature animals may go to better land or feed lots for fattening and this is already happening on a limited scale in the Gedaref area where immature, nomadic sheep are in demand for finishing on crop residuals and oilcake from local small oilseed crushing mills. Sheep are favoured as they require much less water than cattle and are controlled more easily in open unfenced fields. When fat, most are sent by truck to Khartoum on the new tarmac trunk road.

Traditional sedentary cultivation

In the present context, a sedentary cultivator refers to a farmer living in some form of permanent settlement, usually a village, and who grows his crops and manages his stock within easy travelling distance of his house. In the savanna, the main locational factor for such settlements is the proximity of year-round water supplies—a river, well, spring, or even a *hafir*. The latter is a large pit dug out of the impermeable clay soil which fills up during the rainy season and acts as a reservoir, hopefully for the duration of the dry season. Two broad categories of traditional sedentary cultivators can be distinguished, the riverain cultivators found alongside the Niles, and the *bildat*, those found away from the rivers. The term *bildat* refers to a small plot of land; not the total size of an individual farm but one of its constituent small scattered plots.

The systems of farming differ according to whether farmers are *bildat* or riverain, but traditionally they have two factors in common. The first is a method of land cultivation which involves the minimum of manual work and, secondly, a method of maintaining soil fertility which does not require much weeding and very little fertilization. The riverain cultivator waits for the river to recede and then plants his seeds in the silt left behind the retreating water, the much-prized *gerif* land. There the annual deposit of fresh silt maintains soil fertility. The *bildat* farmer clears his land by slash and burn, the *harig* system, and sows his seeds in the ashes. When yields fall to unacceptable levels, he moves on, leaving the abandoned land to rest as bush fallow. Each category of cultivator has evolved a system of farming well adapted to his natural environment, but now they are in a period of change.

Traditional riverain sedentary cultivators

Riverain cultivation is associated with the seasonal rise and fall of river levels, especially those rivers which have a high silt content when in flood. In the Sudan, this refers to the Blue Nile and its tributaries with their head waters in the Ethiopian Highlands. Heavy seasonal rains and the rapid falls in gradient lead to valley erosion by the fast flowing water followed by deposition of the eroded material at the river sides when the relatively flat plains of the Sudan are reached. The White Nile by comparison, is a sluggish river with few changes in gradient. Its direct source is Lake Victoria, so little in the way of seasonal variation in flow is experienced.

The erection of dams to check seasonal variations in river flow and divert water to major irrigation schemes is slowly modifying riverain farming systems. The amount of silt being deposited downstream of dams on the Blue Nile is said to be diminishing, while upstream of the dams, silt is accumulating rapidly. Releasing dammed water reduces levels at the back of a dam and uncovers land for cultivation, but the release of water causes flooding of riverain crops downstream. The cultivator has the risks associated with man's attempts at major river control, in addition to those arising from the natural conditions with which he is familiar.

During the course of field work, a Blue Nile riverain village near Damazin was visited in early January. The river was still receding and had not fallen to its final dry season level. Individual plots were marked by branches of cut thorns which had been placed upright in the mud and judging from the debris attached to them had survived the previous flood. The plots ran from the top of the high bank well out into the main stream. The slope was steep at first but flattened out considerably and the land yet to be uncovered also looked relatively flat judging from the shallowness of the water.

Immediately alluvial mud emerged from the river, a crop was planted literally on the water's edge. A great variety of crops was sown, in no apparent order; cereals like maize and bearded wheat, fruit including melons and pawpaw, and vegetables—tomatoes, egg-plant, okra, onion, a variety of different beans, squash, and the leafy vegetables *molokhiya* (*Corchorus olitorius*) and *rigla* (purslane, *Portulaca oleracea*). No irrigation system was used, the river was said 'to leave sufficient water'.

Most farmers also had *bildat* plots away from the river. There they grew sorghum and sesame by traditional methods described in the next section. The sorghum was mainly for their own consumption while the sesame was a cash crop. The village itself was also situated away from the river amid wooded wasteland used for grazing and the provision of firewood. Each household had a *zereiba*, a thorn enclosed compound for cattle. The stock included a few goats, sheep, and chickens, and cattle were a speciality.

Between them, the people had about 200 head of cattle which were run communally by the village cowherd, who was paid ten piastres per month per head in 1978. Each day, he collected and returned the animals to their owners' compounds and grazed them on the wastelands around the village and on the crop residuals of the *bildats*. Kenana cattle, good for milk production, could be kept all the year round; there was enough keep and mud presented no major problem on the alluvium and gravels of the riverain area. Water was plentiful. Cows calved for the first time at four years old and usually managed to produce seven live calves before becoming barren. Only three to four of the seven calves would reach maturity, the main problem being infectious diseases for which the nomads were blamed.

Nearly all the mature males were sold to local butchers at Damazin or Roseires at around four years old. In these riverain villages bulls are not used for raising water from the river or for land cultivation. Milking cows averaged about 2.3 kg of milk per day for human consumption besides suckling a calf. If the cow calved again in a year, the suckling period was for ten months, otherwise calves continued suckling until the cow was in calf again. Calves ran with their dams but had sacking tied over their mouths when not feeding. Cows were milked twice daily, but the udders were not stripped and the calves were then allowed to suckle. Cows received little supplementary feed, locally-produced oilcake being fed only when 'a cow looked weak'. Cattle dung is valued highly for its properties in binding mud for plastering and brick production. The dung in the compounds was sold, not applied to the land as a fertilizer.

In a survey of riverain cultivators in the Damazin area (Simpson and Simpson 1978, pp. 74–5) the average size of holding was about 25 feddans of which only 2.5 feddans was the prized *gerif* land, the rest consisted of rain-fed *bildats*. Average annual cash sales per feddan amounted to £S6.0 compared with a cash outlay of £S2.5 for production costs. The largest item was the cost of hired labour; none of the farmers' women folk worked on the land.

Traditional rainland sedentary cultivators

An aerial survey of the Sudan (MAFNR 1977) showed that just under 3 per cent of the land area is under traditional rain-fed crop production, by far the most important category of land use. In the area discussed in this chapter traditional rain-fed farming has been largely superceded by mechanized rain-fed farming and, as a consequence, is facing new problems which threaten its viability in areas bordering mechanized schemes.

The main problem is the loss to new schemes of the traditionally-used wastelands surrounding the villages. These wastes play an important role in the village economy, providing grazing for livestock, firewood for cooking, and, most important, the extra space needed for shifting agriculture. The wastes also allow supplementary earnings from the collection and sale of wild honey and of gum arabic.

Usually the first notification of the loss of land comes when incomers move in and start land clearance. This is legal and the villagers have neither the knowledge nor the resources to protect their position.[1] Farmers claim that crop yields fall with the reduction of the fallow period and livestock numbers have to be reduced or additional feeding expenses are incurred.

Broadly speaking, under the traditional system a farmer has two *bildats*, each 5–10 feddans in size. One is cropped with sorghum (the subsistence crop), the other with sesame (the cash crop). If another crop is grown, an additional *bildat* is needed. A simple rotation of sesame followed by sorghum is followed and after a period of four to five years of cropping, the land is rested for four to six years. In a survey of *bildat* farmers in the Damazin area in 1975–6, yields averaged 1.25 sacks per feddan for sesame and 4 sacks for sorghum.[2] These appear low but are in fact higher than

[1] Most of the non riverain land was never registered and as such belongs to the State. The 1970 Unregistered Land Act abolished traditional rights of access and usufruct so the State is free to demarcate areas for development. A village may apply to have its land demarcated to it, but only collectively either as a co-operative or a company.

[2] Crop yields are generally expressed in terms of sacks per feddan. Weights vary between varieties of a given crop, but a sack of sorghum and sesame weighs an average of 90.9 and 76.9 kg, respectively.

those achieved by a sample of mechanized farms—0.5 sack per feddan for sesame and 1.75 sacks for sorghum (Simpson and Simpson 1978).

Cultivation practices are simple. The land is prepared annually at the end of the dry season, by slashing the vegetation and burning the rubbish. After the first good rains, seeds are sown. For sorghum, this involves a man poking a hole in the ashes and a child putting a seed in the hole and covering it up. Sesame is broadcast and then scratched in using a traditional tool. Sorghum is thinned and weeded, sesame weeded only. Both crops are cut, threshed, and winnowed by hand. Crops for sale are sold directly after harvest when cultivators are short of money. Grain for home use is kept in traditional mud and thatch beehive-shaped structures raised on stilts to keep out rats. Family grain needs are about 10–25 sacks sorghum per year, according to family size and the requirements for seed and livestock feed. Any surplus is sold.

Livestock include sheep, goats, chickens, and some cattle. Water availability in the dry season is the limiting factor and often cattle are run in a transhumant system being sent away for part of the year, either by a communal arrangement or entrusted to passing nomads. In the home village, stock graze the wastes and are fed crop residues. Surplus males are sold.

Proximity to a mechanized scheme has advantages if it does not covet a village's land. Casual work is available and earnings from it are said to have freed many villages from chronic indebtedness under the notorious *sheil* system. Tractors and discs can also be hired for contract work and this has enabled families to increase the area they cultivate, provided they can acquire land and resources not only to pay for the tractor but, much more expensive, the extra labour needed for weeding and harvesting. Once they can do this and are successful, they have moved from the subsistence cultivator to the 'yeoman farmer' class and may well expand further by purchasing machinery for their own use and for contract work. The late Tom Martin, leader of the World Bank Mission to Sudan in 1977, had great hopes for these farmers and their role in the future development of the rainlands.

Mechanized, rain-fed farming

The systems discussed so far are long-standing, providing a means for survival in a harsh natural environment. They are small in scale, geared to the capabilities of the family, and using existing practices and techniques. These traditional farming systems are threatened by other large-scale schemes competing for land and water resources and backed by the State in the name of development. However, the main competitor, large-scale mechanized farming has yet to prove that it can provide a means for long-term survival.

In the 1970s, the leaders of the Sudan proclaimed that the country was to become the 'bread basket' of Africa and the Middle East through the development of the central rainlands—one of the largest reserves of cultivable land in the world. By the late 1970s, such prospects seemed much less hopeful. Sudan became a net importer of grain and the disastrous drought of 1984–5 turned the country into a 'begging bowl'.

Traditional shifting agriculture has been replaced by shifting mechanized farming and the scale of land clearance has led to the degradation of the natural environment. The worst effects are reported in the western Sudan where the predominant sandy soils are extremely vulnerable to drought and wind erosion when cleared of their natural vegetation. The clay soils in the east of the country are less vulnerable.

In the Gedaref and Damzin areas of the eastern clay plains the type of farming described in the following paragraphs has led to degradation and land abandonment (Simpson and Simpson 1978, p. 87).

The farms are standard holdings of 1000 and 1500 feddans rented from the Mechanised Farming Corporation (MFC).[1] Most of the tenants were the original occupiers who cleared the land. Gangs of workers used a variety of the traditional slash and burn method for land clearance, followed by disc harrowing two or three times after the start of the rains followed by the first crops.

Very little equipment is required; a four-wheeled tractor of about 65 hp together with a set of wide level disc harrows fitted with a seed box, and a vehicle for transport, usually a pick-up. Farms have no permanent buildings, the usual arrangement is a collection of traditional round mud-walled, thatched huts which provide a temporary camp during the cropping season, June to January. In the dead season machinery and stores are removed and the farm is left deserted.

The farms are out in the wilds, approached only by long tracks which run off unmade public roads. The whole road network is reduced to a clay quagmire once the rainy season starts, so the farms are isolated from the outside world. Consequently men, machines, and stores including food, tractor fuel, and spare parts have to be moved to the land towards the end of the dry season.

Discing starts when sufficient rain has fallen to partially close the deep cracks in the clay. Farmers would like to disc twice before sowing, because the second discing helps to control germinating weeds. However, there is pressure to sow the crops as quickly as possible and so the seed is usually sown during the first discing, using a disc-mounted seed box. Sesame yields fall off drastically if the crop is not sown by the end of July, but sorghum is more tolerant. A further complication is the likely build-up of soil moisture

[1] In 1976/7, the Damazin and Gedaref areas accounted for 86 per cent of the 3.5 million feddans of land under mechanized, rain-fed farming in the Sudan. Of the national total, 90 per cent was let in individual holdings of 1000 to 1500 feddans (Simpson and Simpson 1978).

in August and September; tractors are easily bogged down in the wet clay soil and prevented from working efficiently.

This situation leads to the neglect of fallowing and the monoculture of sorghum. Farmers believe that when soil conditions are suitable it is natural to sow as much land as possible and sorghum is far easier to grow than sesame.

Weeding and thinning are done manually by casual workers paid on piece work; sesame is weeded, but not thinned. A leading Gedaref farmer estimated that 100 workers were required over three weeks to weed a 1000-feddan holding just once. The expense involved and the logistics required to collect and transport workers to the farm as well as the food and water required for their maintenance was considerable. Most farms have neither private water supplies nor permanent water facilities nearby. Ideally, crops should be weeded twice to control early-season and late-season weed species, but the expense acts as a constraint especially if a crop looks rather poor and the farmers feel that they may not get their money back.

Harvesting starts with sesame which is cut and threshed by hand using casual workers. Timing is crucial since ripe seed heads burst and losses from shredding are serious. Labour requirements can amount to 30 man-hours per feddan.

Sorghum harvesting times are more flexible, starting near the beginning of December (after the sesame harvest is completed) and continuing until the second half of January. Usually casual workers cut off the seed heads leaving the stalks to be burnt off when the ground is prepared for next season's crops. The heads are collected and piled in large heaps for threshing by a combine harvester used as a stationary thresher. The combines are usually hired from the Agricultural Bank of Sudan (ABS), which is said to provide the service in order to monitor the crop yields of its debtors. The less popular manual alternative is to use a flail followed by winnowing to remove the chaff.

Mechanized farming is something of a misnomer, because the system is actually labour intensive. Data for 1975–6 for the most common sequence of operations including combine threshing and one weeding (Simpson and Simpson 1978, p. 47), gave averages of nearly 29 man-hours per feddan for a sorghum yield of 3 sacks per feddan compared with 47 man-hours for a sesame yield of 1 sack per feddan. Expenditure on labour accounted for 63 per cent of total expenses, as opposed to 23 per cent for machinery.

Labour costs have been rising, partly because of the increased demand for casual workers on new development projects elsewhere and partly because of the increased price of purchased food given to the workers. Mechanization of harvesting has not proved successful to date. Combining the sesame crop involves excessive losses from shredding, the popular long-strawed, multiple-headed sorghum varieties are not suited to combining and the more suitable short-strawed varieties are not popular. Lack of water for mixing sprays prevents the use of herbicides for weed control and some of the common weeds of sorghum are of the same genus, such that chemical control would damage the crop.[1]

A major problem of the present farming system is the tendency for crop yields to fall over time and stabilize at a low plateau. Typically, yields peak in the second year after clearance, then decline until the seventh year and stabilize (Simpson and Simpson 1978). Although management, rainfall, and the incidence of pests and diseases all have an effect, a scientific study of the causes of falling yields has yet to be made. Various possible reasons are cited, including the lack of scientific knowledge and risks involved in pioneering in a semi-arid wilderness. If the costs of production are greater than the returns given by the plateau yields the farm becomes financially unviable. Many farmers told the authors that they cut their expenditure on inputs such as certified seed and fertilizer because they feared that they would not get a return on their investment.

The break-even point for farmers depends not only on the level of yield but also the ratio of input prices to output prices. The price of inputs has risen much faster than that of output and the marketing system for sorghum and sesame gives farmers little incentive to increase yield. In a recent World Bank report (1985), the role of price incentives was singled out as a key factor for encouraging higher yields and output. When yields show signs of stabilizing at unacceptably low levels, the logical strategy is for the farmer to move on. Fresh land is easy to obtain and the higher yields in the initial years after clearance should leave him a profit for his efforts. The net result is a system of shifting agriculture which represents a financial gain to a few successful individuals but a social loss to the community.

The former military regime of Numeiri refused to acknowledge the failure of the 'bread basket' policy and continued to encourage the horizontal expansion of mechanized farming in order to encourage potential foreign investment.[2] His overthrow led to the acknowledgement of the problem, but in the mounting spiral of inflation and devaluation fuelled by increasing external and internal indebtedness resulting from civil war and political instability, little was done to find a solution. The adjustment and restructuring programmes of the International Monetary Fund (IMF) (1978 onwards) failed in their agricultural objectives, while the few re-

[1] Common weeds of sorghum are *adur* (*Sorghum aethiopicum*) and *aneese* (*Sorghum purpureosericeum*), both of which cross-pollinate with sorghum. A farmer using his own seed risks his crop degenerating into a mixture of sorghum and grasses.

[2] The state farms which previously made heavy losses were disposed of to joint Arab–Sudanese organizations. Major concessions of up to one million feddans were made to commercial companies. Their record has not been encouraging. Unforeseen difficulties in starting up schemes have led to delays, while one company, Triad of Luxemburg has backed out of a one million feddan concession. In 1976–7, state farms and companies accounted for just under 10 per cent of the area devoted to mechanized rain-fed agriculture (Simpson and Simpson 1978).

habilitation efforts undertaken were largely confined to irrigated cotton (United Nations Conference on Trade and Development 1989).

Arable farming on gravity-fed and pump irrigation schemes

In central Sudan, irrigated agriculture depends on water drawn from the Blue and White Niles and after their confluence in the Khartoum region, on water from the main Nile. The availability of water for irrigation is regulated by the Sennar and Roseires Dams on the Blue Nile and by the Jebel Aulia Dam on the White Nile. Total annual extraction is limited by the terms of the 1959 Second Nile Waters Agreement to 20.55 milliard (10^9) cubic metres measured at Sennar. The Sennar and Roseires Dams provide storage of water for use in the Sudan during the months of low river flow. Water is released from the Jebel Aulia Dam according to Egyptian needs but this has important repercussions for the operation of the pump schemes upstream on the White Nile. Rainfall supplements irrigation mainly in the months of July, August, and September. At Khartoum mean annual rainfall is less than 200 mm, but increases southwards reaching about 450 mm at Sennar.

Irrigated agriculture tends to be organized in relatively large schemes, each with its own irrigation regime and crop rotation. It is appropriate to discuss agricultural systems on a scheme by scheme basis, starting with the Gezira as the oldest and the largest irrigation scheme in the Sudan.

The Gezira Scheme

The Gezira Scheme is based on a gravity-operated canal network which takes water from the Sennar Dam to the southern environs of Khartoum, a distance of 200 km and serves an irrigated area of 2.1 million feddans (800 000 ha). The main Gezira Scheme dates from 1925 but the area was virtually doubled with the development of the Managil Extension between 1958 and 1962.

Crop production and irrigation practices are related to the nature of the Gezira soils. These are vertisols, dark cracking clay soils with 50–70 per cent clay content, high alkalinity (pH 8.5–9.5), low organic matter (usually less than 1 per cent), and low total nitrogen levels (0.02–0.03 per cent). Permeability is restricted with water penetration mainly through the extensive cracks which develop as the soil dries out. Ground water is at a depth of 20–30 m and with low permeability, salinization from a rising water table is not a problem. However, salt levels in restricted parts of the scheme are sufficiently high to inhibit growth. Although the sodium content of the Blue Nile water is low, it can be expected that with continued irrigation the salt content will rise to critical levels in the long term. However, after 60 years of cultivation, albeit with frequent fallows, soil deterioration is not a major factor.

The irrigation regime The Gezira Scheme is laid out in 90-feddan fields (*numbers*), each a rectangular block of 1350 m by 280 m. Each of these blocks is subdivided into plots (*howashas*) which are allocated to individual tenant farmers. Each *number* receives canal water from a field outlet pipe. The water passes into the field channel (*abu ishreen*) from which subsidiary channels (*abu sitta*) supply nine 10-feddan *howashas*. The *howashas* are separated at intervals of 40 m by low ridges (*tagnets*) with a final water channel (*gadwal*) half-way between these. The area bounded by a *tagnet* and a *gadwal* is a unit of 0.714 feddans (0.3 ha) and this is referred to as an *angaya*, often used for determining piece-work wage rates. The *angaya* is subdivided by minor ridges into small areas known as *hods*. The Gezira canals were designed for night storage with water flowing on to the fields only during the daytime. Watering was carefully controlled by the tenant or his employee who regulated the flow from *hod* to *hod* to ensure even flooding even when the ground level was uneven. This was very time-consuming, but not arduous work. The normal interval between irrigations was fifteen days with watering taking three to four days.

Nowadays, this system is widely, but unofficially, modified. The field outlet pipes remain open day and night with water flowing uncontrolled over the *angayas*. Time spent by the tenants in irrigation is greatly reduced but at the expense of uniform flooding and a longer interval between irrigations; this is about twenty days, eight for irrigation and twelve for drying out (El Bashir 1982). Land planing to remove micro-variations in surface level would be desirable but there has yet to be a general programme for this. The main canals are in use from early June until the end of March leaving a dead season for maintenance. A number of summer canals have water throughout the year primarily to maintain drinking water for both humans and livestock but affording restricted irrigation mainly for fodder crops.

Cropping system The present cropping system is a result of successive modifications to that used when the scheme began in 1925. The main objective was to produce extra-long-staple Egyptian cotton (*Gossypium barbadense*) for the British textile mills. A partnership system involving the Sudanese Government, a British Company, the Sudan Plantations Syndicate (replaced in 1952 by the Sudan Gezira Board) and local farmers was adopted. The scheme was managed by the British company but with Sudanese tenant farmers, supervised closely by the company's inspectors, responsible for all manual operations. In addition to cotton, tenants were permitted to grow *dura* (*Sorghum vulgare*) as a food crop and *lubia* (hyacinth bean, *Lablab purpureus*), primarily for animal fodder. The three crops were combined in a six-course rotation as follows: fallow–cotton–*dura*–fallow–cotton–*lubia*.

Following severe outbreaks of cotton disease in 1931–3, an eight-course rotation with more fallow was introduced: fallow–fallow–cotton–fallow–*dura*–*lubia*/fallow–fallow–cotton. The fallows helped to prevent the carry over of pests and diseases, assisted in perennial weed control, and permitted the building of soil nitrates. A tighter rotation was adopted for the Managil Extension: fallow–cotton–fallow–cotton–*lubia*/fallow–*dura*.

After the 1959 Nile Waters Agreement and with extra water storage following the construction of the Roseires Dam, an intensification and diversification programme was initiated. Groundnuts and wheat were introduced as additional cash-crops and fallows seemed less necessary as the use of nitrogenous fertilizers and insecticidal spraying, first introduced in the late 1940s, became routine. There was a measure of success in developing disease-resistant cotton varieties, although whitefly (*Bemisia tabaci*), continued to be a major cotton pest and *lubia* as a host plant was taken out of the rotation in 1975.

The standard rotations in 1987 are: main Gezira: cotton–wheat–groundnuts/*dura*–fallow; Managil: cotton–wheat–groundnuts/*dura*. Cotton, *dura*, and groundnuts are summer crops sown between June and August. Wheat is a winter crop sown in November and requiring water after *dura* and groundnuts have reached the harvesting stage. This successional water use permits 75 per cent cropping in the main Gezira although the canals have capacity to irrigate only 50 per cent of the area at one time. Similarly 100 per cent cropping is possible in the Managil although the canals permit simultaneous irrigation of only two-thirds of the area. The mean areas of crops in the Gezira Scheme, including the Managil Extension, over the 5-year period from 1979/80 to 1983/4 are shown in Table 16.1. Additionally, 30–40 000 feddans of vegetables were grown in the groundnut phase of the rotation. The standard rotations are somewhat modified in practice. Rice was grown on areas of up to 9000 feddans

TABLE 16.1. Gezira: mean areas of principal crops, 1979–84

Crop	Area (feddans)
Cotton	503 000
Dura	421 000
Groundnuts	137 000
Wheat	266 000

Source: MAFNR (1985).

TABLE 16.2. Gezira: average labour requirements of the main crops (undated)

Operation	Long staple cotton	Medium staple cotton	*Dura*	Groundnuts	Wheat
		man-days per feddan			
Cleaning field channels	2.0	2.0	2.0	2.0	1.0
Sowing/fertilizing	2.0	2.0	2.0	4.4	0.2
Gapping	0.2	0.2	0.2	0.4	
Thinning	0.8	0.8			
Weeding	6.6	6.6	5.2	9.6	1.6
Side dressing	1.8	1.8			
Irrigation	5.0	4.2	2.8	2.8	2.2
Scouting	3.6	2.4			0.2
Harvesting	31.5	21.0	10.8	27.0	0.7
Post-harvest activities	5.0	5.0			0.3
Total	58.5	46.0	23.0	46.2	6.2
Percentage of labour contributed by tenant's family (1976–81)	28		28	29	51

Source: Sudan Gezira Board Economic and Social Research Unit (1983).

between 1974 and 1980 but yields were low. Some fodder sorghum is grown outside the rotation.

Table 16.2, based on survey data collected by the Gezira Board's Economic and Social Research Unit gives details of the average labour requirements for the main crops. Comparable data for vegetables are not available.

Cotton Extra-long-staple Egyptian cotton remains the Gezira's principal crop. The original varieties came from Egypt but since the early 1930s, varieties bred locally at the Gezira Agricultural Research Station have been used. Barakat is currently the most important variety. Medium staple American cotton (*Gossypium hirsutum*) of the Acala type was introduced into the southern parts of the Gezira towards the end of the 1970s but did not account for more than a quarter of the total cotton area. It was not planted in the 1985–6 season because of marketing difficulties.

Tenants' cotton production continues to be supervised, albeit less closely, by the inspectors of the Sudan Gezira Board. Ideally, cotton is planted in late July or early August and harvested from early January to late March (December to early February for medium staple varieties). During the winter, land for cotton is cultivated dry by ridging or disc ploughing using Gezira Board tractors. Deep blading with crawler tractors is carried out periodically to control deep-rooting perennial weeds, particularly nut grass (*Cyperus rotundus*), known locally as *seid*. Immediately prior to planting, the land is ridged again (ridges are 80 cm apart), this time with tractors owned by private contractors, the Board does not have the capacity for this in the short time available. The seed is dropped by hand into holes made with a stick (*seluka*) and young plants are thinned to leave three per hole. Some resowing is needed due to waterlogging and other losses. The crop is ridged once more (green ridged) and until herbicide use became general, was hand-weeded up to four times. By 1982/3, the entire cotton area was treated with herbicides, greatly reducing the need for hand weeding. Currently, a standard application of urea at an N rate equivalent to 54 kg feddan^{-1} (3N) is applied (N = 40 lb or 18.14 kg N feddan^{-1}). It used to be distributed to tenants for manual spreading but the standard of application was poor and pilferage was common. Now, the application of urea is undertaken by the Board at 6–8 weeks after sowing using tractor-drawn distributors. Cotton is extremely vulnerable to pest attack and control by repeated aerial spraying is increasingly costly. The entire crop receives an average of seven to eight sprays.

Cotton is harvested manually and the long staple cotton requires four or more pickings on account of its indeterminate flowering habit. Around 150 000 to 200 000 extra workers come to the Gezira for the cotton harvest. It is the tenants' responsibility to have their cotton picked, packed into sacks, and delivered to cotton stations on the light railway system. After harvesting, animals are allowed to graze the old plants before they are uprooted manually and the stalks burnt or removed unofficially for fuel. This operation is primarily to prevent the carry over of blackarm or bacterial blight (*Xanthomonas malvacearum*). Machines have been developed for the uprooting but have not been generally adopted so far.

In spite of the increased use of fertilizers, herbicides, and insecticides, Gezira yields have not risen, although inter-year variability has increased. Jewitt *et al.* (1979) provide an extended time-series of seed cotton yields in the main Gezira (Table 16.3). For more recent years the introduction of medium staple cotton is a complicating factor but the mean seed cotton yields of the long staple Barakat (Table 16.4) do not indicate any improvement.

TABLE 16.3. Gezira: yield of seed cotton, 1924–75

Years	Mean yield (kg ha^{-1})	Years	Mean yield (kg ha^{-1})
1924–7	1237	1952–5	1424
1928–31	827	1956–9	1512
1932–5	1092	1960–3	1546
1936–9	1417	1964–7	1132
1940–3	1431	1968–71	1705
1944–7	1305	1972–5	1502
1948–51	1610		

Source: Jewitt *et al.* (1979).

TABLE 16.4. Gezira: yields of seed cotton for the long staple variety Barakat, 1980–5

Years	Mean yield (kg ha^{-1})
1980–1	712
1981–2	1321
1982–3	1465
1983–4	1438
1984–5	818

Source: Sudan Gezira Board Economic and Social Research Unit (unpublished).

The absence of any significant yield improvement in spite of increased expenditure on fertilizers, herbicides, and insecticides has reduced the financial value added and made cotton a less attractive crop for both the nation and for the tenants. The mean scheme yields conceal wide variations between different locations within the scheme and between the performances of individual tenants. For administrative purposes the scheme is divided into 107 blocks. Differences between locations are reflected in the mean block yields (Table 16.5).

TABLE 16.5. Seed yields of long staple cotton, 1980–1 to 1983–4

	Mean block yield (kantars $feddan^{-1}$)				
	<3	3–4	4–5	5–6	6–7
No. of blocks	5	32	34	7	2

1 kantar = 141.523 kg; 1 kantar $feddan^{-1}$ = 337 kg ha^{-1}.
Source: from Sudan Gezira Board Statistics.

Usually yields are higher in the southern than in the northern blocks of the scheme. Soils tend to be better in the south, irrigation is more reliable at the head than at the tail of the system, and higher rainfall encourages weed germination before rather than after sowing. Although cotton yields vary widely, there is little variation in input use; high- and low-yielding blocks are treated similarly.

Dura *Dura* is the staple food grain and has always been included in the Gezira rotation as the tenants' crop. It is valued not only for its grain but also for its straw (*gassab*), which can be stored for fodder. Tenants keep *dura* for their own consumption and for the cotton pickers, but considerable quantities are sold to private traders and replaced by purchases of grain from outside the scheme.

Dura is normally sown in July prior to cotton and harvested in November.

The crop is sown on ridges prepared by tractor but all subsequent operations are manual with sowing, weeding, and irrigation practices being similar to those for cotton. Fertilizer is not usually applied for there is no recognized private fertilizer distribution network and the Gezira Board gives priority to cotton and wheat. However, in 1986 there was an allocation of subsidized urea for *dura*. The crop is harvested manually and the heads separated from the *gassab*. The former may then be thrashed using a stationary combine harvester, but grain yields are low. The Ministry of Agriculture and Natural Resources (MAFNR 1985, p. 15) gives an average grain yield of 432 kg feddan^{-1} for the 5-year period, 1979–84. However, mean yields were 630 and 690 kg feddan^{-1} for 1984 and 1985, respectively, the increases being attributable to the increasing use of hybrid sorghums. However, even the higher yields are well below the potential of 2 t feddan^{-1}. Hybrid sorghums are not popular with consumers and in years of plentiful supply are likely to fetch lower prices than the traditional varieties. The prevalent parasitic weed, *buda* (*Striga hermonthica*) is a serious constraint on *dura* yields.

The *gassab* is either put into small stacks (*rakubas*) in the fields, often surrounded by a barrier of thorns, or it is stored in the villages to feed livestock during the summer months (June and July) when other forage is extremely scarce. The stubbles are grazed.

Dura is grown as a fodder as well as a grain crop. The main fodder sorghum is the native variety Abu Sabaeen, but recently the hybrid Pioneer 988 has been introduced. Fodder sorghum is grown on limited areas outside the rotation, but members of the Barakat dairy co-operative receive water for it to be grown as an early summer crop within the rotation.

Groundnuts Groundnuts were first grown commercially in the Gezira in the early 1960s and in spite of the heavy soil have proved a successful crop. Groundnuts are the first crop to be planted after the opening of the main irrigation canals around 10 June each year. Yields are higher with early planting which should take place towards the end of June, but is frequently delayed until July. As with cotton and *dura*, groundnuts are grown on ridges and are planted and weeded by hand. Harvesting should take place in November but is often delayed until December. The crop is pulled up by hand following a light irrigation to soften the soil and the pods are removed from the haulms. When dry the latter are collected and packed into old cotton sacks for use as fodder (*tibin*). Tractor-drawn lifters have been tried successfully but they are not in general use for harvesting. Decortication is usually done mechanically. Gezira tenants often sublease groundnuts to share croppers.

MAFNR estimated the average yield of shelled nuts over the 5-year period, 1979–84, to be about 773 kg feddan^{-1} compared with a potential yield as high as 3 t feddan^{-1}. Groundnuts are sold to private merchants and to agents of the Sudan Oilseeds Company. Aflatoxin, produced by the mould *Aspergillus flavus*, is often found in Gezira groundnuts and this restricts the use of groundnut meal as an animal feed.

Wheat Wheat is a winter crop in the Gezira, usually sown in November and grown during the cooler months. The rising temperatures as the crop approaches harvest in March preclude high grain yields, but the growing local consumption of wheat flour with the adoption of an urban lifestyle and the need for considerable importation has created a need for more local production. With the exception of the very restricted areas under irrigation in the Northern Region, the Gezira Scheme (together with the New Halfa Scheme) is as good a location for growing wheat as anywhere else in the Sudan. As a winter crop, wheat extends the use of the canal system by utilizing water after irrigation of groundnuts and *dura* ceases.

Wheat is the most mechanized of the Gezira crops. Land is prepared by ridging, but the ridges are subsequently flattened with a tractor-drawn steel bar (*zahafa*) and the seed broadcast by hand. The recommended sowing time is late October or early November, but sowing is often delayed until late November. An application of urea at a rate of 36 kg N per feddan is given after sowing, but very little weeding takes place and the labour requirement is restricted to irrigation. The crop is harvested with combine harvesters owned either by private contractors or by tenants' co-operative societies. The grain should be sold to authorized mills, but significant amounts are reported to be sold privately for higher prices. The straw is not usually collected but is grazed-off along with the stubble. The grain yield for the 5-year period, 1979–84, was only 482 kg feddan^{-1} (MAFNR 1985, p. 11), and in the 1984–5 season, wheat was not grown due to a shortage of water resulting from the exceptionally low Blue Nile flood of 1984.

Lubia *Lubia* is now excluded from rotations because it acts as a host for the major cotton pest whitefly (*Bemisia tabaci*). Prior to is exclusion in 1975, it was important as the sole irrigated fodder crop and also as a source of protein rich seeds for human food. It was sown in September or early October and received irrigation until January when it was grazed off. It was said that many tenants took little trouble over the crop and weed infestation was high. A fodder crop, probably clitoria (*Clitoria ternata*) is to be reintroduced into a five-course rotation initially in the Barakat block.

Vegetables In the groundnut phase of the rotations, areas usually of 2.5 feddans are allocated to certain tenants for vegetable production. Usually, these are the *howashas* closest to the field outlet pipes from the canals as these get the most reliable supply of water and can be irrigated until the end of March after the groundnuts have been harvested. The vegetable area is restricted to a maximum of one-tenth of the cotton area and in recent years has been in the range of 29–45 000 feddans. Share-cropping between a tenant and a specialist gardener (*kholi*) is a common practice. Water is available for the gardens from early June but some growers sow in nursery beds and then transplant to achieve early crops. The most important vegetables grown are tomatoes and onions together with peppers, cucumbers, and okra. Production is for the market and Gezira vegetables, notably onions, are distributed nationally.

The area of vegetables is small in relation to the total area under cultivation, but the cash output per feddan is usually high and expenditure on purchased inputs is low. Consequently, the value added per feddan is high, almost certainly higher than that from cotton and other crops. Reliable estimates of output are lacking, but in total the contribution of vegetables is probably comparable with that of the major crops.

In addition to vegetables, there is a limited area designated as village orchards and also forestry plots to supply firewood and poles for building. The trees grown are *Eucalyptus microtheca* and these are coppiced on an eight-year rotation.

Livestock Apart from the oxen used in the early years for field operations, livestock have not formed an integral part of the Gezira Scheme. However, many of the tenants own animals, the majority keep a donkey for personal transport and most households have a few goats for milk and meat. Many tenants possess a few cattle, but only a small minority possess more than twenty head. Ownership of sheep is more restricted but flocks are relatively large, often consisting of fifty or more animals. Both cattle and sheep are taken out to the fields daily, often being herded by communally employed shepherds. They are kept in the villages at night.

Numbers of livestock seem to be increasing (Table 16.6), probably due to increasing demand for their products. However, in 1984–5, when drought restricted fodder availability, losses were severe. Details of average labour requirements for livestock are not available.

TABLE 16.6. Gezira: estimates of the livestock population

Livestock	Number of head	
	1975	1981
Cattle	348 962	442 773
Sheep	597 106	641 563
Goats	657 702	410 067
Donkeys		90 411

Source: MAFNR (1977); Sudan Gezira Board (1982). *The Gezira Village Directory* (arabic) (unpublished).

The Gezira livestock depend largely on crop by-products, on the restricted grazing on the edges of the canals, and on wastes and fallows during the rains. Natural grazing is adequate in August and September, *dura* and groundnut fields are available from November to February, but then there is a gap until the wheat stubble becomes available from the end of March, and the cotton is opened for grazing in about mid-April. After the cotton is finished by the end of May livestock must utilize *gassab* and groundnut straw stored in the villages until weeds start to germinate with the rains in July. In periods of fodder shortage some use is made of concentrates, mainly cotton seed cake and wheat bran, both of which are readily available, but many stock suffer severe weight losses.

The majority of cattle are of the local Kenana and Butana breeds and the sheep are of the Sudan Desert type. Friesian blood has been introduced in some dairy herds and the cross-breeds appear to adapt to the harsh climate. There is only limited information on production levels, but for cows, yields of milk over the lactation period (often with once a day milking) are probably about 500 kg. Male cattle are slaughtered at about four-years-old and male sheep at one-year-old when their liveweights are about 30 kg.

The Animal Production Section of the Gezira Board has been concerned with the encouragement of commercial dairy and poultry production. A dairy co-operative has been established for tenants in a block adjacent to the schemes' headquarters at Barakat and its membership has increased to about a thousand. The co-operative supplies a dairy built in 1982 with Irish aid. In spite of the official support for dairy farming, most cattle herds are kept for meat with milk as a useful by-product.

The White and Blue Nile Pump Schemes

A considerable number of pump irrigation schemes, primarily for cotton production were constructed along the White and Blue Nile Rivers, mainly in the 1950s. Originally, the schemes were privately financed and operated, but since 1968 the larger schemes have been managed by public corporations. The White Nile Schemes are mostly near Ed Dueim and Kosti, those on the Blue Nile near Sennar and Singa. The soils are predominantly heavy clays.

The schemes are run on Gezira lines with a partnership system and supervised cropping by tenant farmers. Only cotton and *dura* are grown and the production methods are similar to those in the Gezira. The original rotation was: cotton–*dura*–fallow–cotton–fallow–fallow. However, the amount of fallow has been reduced and now the rotation is often cotton–*dura*–fallow or even cotton–*dura* (Magar 1986). Originally only Egyptian cotton was grown but recently the Blue Nile Schemes have grown medium staple American cotton.

Irrigation is more costly in the Blue Nile Schemes than in those on the White Nile because a higher lift is needed to raise the water to field level. Nevertheless, the extra expense has been offset by higher yields. Operation of the pumps on the White Nile Schemes is affected by the management of the Jebel Aulia Dam. Water levels are too low for pumping until the dam's gates are closed, usually in the third week of August. Consequently, irrigation is not usually available until September and cotton planting is delayed well beyond the optimum time. Cotton receives priority for water use and *dura* may be irrigated only sporadically, perhaps only twice in the season. This is of less consequence in the southern schemes where rainfall is higher. Many tenants especially those in the Blue Nile Schemes have rain-fed or riverain holdings in addition to their scheme tenancies, and they also own livestock for which the scheme cotton stalks and *gassab* provide a valuable source of dry season feed.

In recent years the area under crops in these schemes has fallen. In 1963–4, for example, there were 211 000 feddans of cotton, but in 1984–5 only 76 000 feddans. Yields are shown in Table 16.7. Over the 5-year period 1979–84, *dura* grain yields were 375 kg feddan^{-1} on the White Nile compared with 400 kg feddan^{-1} on the Blue Nile Schemes.

TABLE 16.7. White and Blue Nile Pump Schemes: comparative yields of seed cotton

Scheme		1957–65 (kantars feddan^{-1})
White Nile	Kosti	3.35
	Ed Dueim	2.93
Blue Nile	Sennar	4.68
	Singa	4.48
		1979–84 (kantars feddan^{-1})
All White Nile		2.04
Blue Nile	(Egyptian)	2.43
	(American)	4.04

Sources: Simpson *et al.* (1966); MAFNR (1985).

The reductions in cropping area and the decreasing yields reflect severe problems related to the maintenance of pumps and other machinery, shortages of diesel fuel, delayed provision of seasonal finance from the Bank of Sudan, and management inadequacies. The schemes have become unprofitable and a rehabilitation programme involving the replacement of pumps and tractors and the upgrading of maintenance facilities is underway. Pumps are being electrified and adjacent schemes amalgamated for irrigation purposes. An example of this is the Ramash Scheme of 24 000 feddans at Singa.

The Rahad Scheme

The Rahad Scheme on the east of the Blue Nile is the most recent of the Sudan's major cotton irrigation schemes commencing operation in 1977. Only Phase I of 300 000 feddans has been developed. The scheme receives water from the Rahad River in the early part of the season but as the flow diminishes, a switch is made to water pumped from the Blue Nile at Meina near Singa. There is usually some deficiency in water availability in October when this switch occurs.

Like the other schemes, a partnership operates and the scheme is divided into 22-feddan tenancies. The crop rotation has never included a fallow year; the original intention was for tenants to grow 11 feddans of cotton, 8 feddans of groundnuts, and 3 feddans of a fodder crop. However, most tenants showed little interest in the fodder and the current rotation is: cotton–*dura*/groundnuts. Only medium staple cotton has been grown in this scheme. It was intended that production would be more fully mechanized than in the Gezira, but currently, apart from the use of mechanical diggers for lifting groundnuts, operations do not differ greatly and the inputs of manual labour appear to be at least as high. There were plans for a long furrow irrigation system with plastic syphons taking water from the *abu ishreens* to the furrows but this proved to be unsatisfactory. Tenants were not present to move the syphons and ponding occurred at the far ends of the furrows. Cotton-picking machines have been used, but currently, as in the Gezira, large numbers of pickers are recruited from outside the scheme for hand picking. Tenants aim to plant groundnuts from mid-May and *dura* from mid-June.

Cotton yields have been relatively high with an average seed cotton yield of 5.77 kantars feddan^{-1} for the 5-year period 1979–84 (MFNR 1985, p. 14). These can be attributed to factors such as the use of relatively high-yielding cotton varieties and to the fertility of the virgin soils. Average groundnut yields of 794 kg feddan^{-1} are similar to those in the Gezira.

Vegetables are not grown in the main rotation, but on 400 specialized tenancies of 5 feddans. A thousand 10-feddan dairy tenancies are also being allocated for the production of Abu Sabaeen and *lubia*; two crops being grown each year to maintain a herd of twelve dairy cows plus followers.

Most Rahad tenants have a nomadic background and many still possess considerable numbers of animals which have to be kept outside the scheme for much of the year. Only household stock such as milking cows and goats are allowed to be kept in the villages until the *dura* is harvested in November. After cotton picking is finished in mid-March, the large herds and flocks can enter the scheme to feed on the cotton stalks. By this time natural fodder has become extremely scarce.

Sugar Schemes[1]

Four of the Sudan's five major functioning sugar schemes are in central Sudan. Three of these, Guneid, Sennar, and Assalaya are in public ownership. Kenana, by far the largest, is owned by an internationally funded company. Guneid and Sennar are pump irrigated from the Blue Nile, and Assalaya and Kenana are similarly irrigated from the White Nile.

The oldest scheme is Guneid, close to the town of Rufa'a. Production commenced in 1963 on 37 000 feddans on the east bank of the Blue Nile, taken over from the Gezira Scheme. About 19 000 feddans are in cultivation, with Gezira tenancies of 15 feddans retained and sugar-cane taking the place of cotton. Tenants are responsible for all manual operations including planting, weeding, watering, and fertilizer application. Harvesting is from December to May and normally two ratoon crops are taken.

The tenancy system is not very successful and is one of the reasons for the relatively low cane yields. Cane yields peaked in 1975–6 at 26.8 t feddan^{-1}, but more recently they have been below 20 t, with only 13.45 t in 1981–2. Sugar recovery rates have fallen from about 11 per cent to 8–9 per cent.

Sennar and Assalaya are twin schemes built to an identical design. The sugar-cane areas at both are centrally managed with hired labour. Sennar came into operation in 1976, Assalaya in 1979, and both schemes were plagued initially by major problems. At Sennar these were largely concerned with the factory and at Assalaya with the power supply for the irrigation pumps. However, by 1982–3, Sennar produced 40 600 t of raw sugar and Assalaya 31 700 t as against 20 300 t from Guneid. Assalaya has 27 000 feddans but the maximum area harvested so far has been 14 850 feddans in 1983–4.

The Kenana Scheme commenced operation in 1981 with 150 000 feddans of land and a potential production of 330 000 t of sugar. It is one of the largest sugar plantations in the world. The cane area is divided into twenty farms each of 4000 feddans with four sections under a section manager who supervises the paid labourers. By 1982–3 production was 229 000 t raw sugar. Each year, the company sells the first 150 000 t to the Public Sugar Trade Corporation at a fixed price with additional output sold for foreign currency at the world price. Kenana appears to have operated with some degree of success, although financial details are unavailable. High cane haulage costs due to the large area of the Scheme must have negated at least some savings associated with large-scale operation.

Molasses produced as a by-product of cane processing are exported and are also used increasingly by local owners as a livestock feed.

[1] Much of the information on sugar schemes is from Mohy El Din (1985).

Other irrigation schemes in central Sudan

In addition to cotton and sugar schemes, there are a considerable number of pump irrigation schemes for the production of vegetables, fruit, and fodder. Many of these are in the vicinity of Khartoum. With some exceptions, they are relatively small and in private ownership. There is little up-to-date and published data on these schemes and their areas are not given in the statistical reports of the MAFNR. Reference can be made to the 1963 census of pump schemes in Khartoum Province (Sudan Government 1966) when 220 functioning schemes with a total of 36 800 feddans and 50 non-functioning schemes with 10 700 feddans were recorded. Two-thirds of the schemes were of less than 50 feddans but there were ten with areas of between 1000 and 5000 feddans. The smaller schemes produced mainly vegetables, but on the larger ones fodder production was important. There were a number of irrigated fruit gardens with mango, lime, oranges, and guavas. Many of the schemes were tenanted and several large-scale fodder production schemes have been developed, including the Kuku Dairy Scheme with 3500 feddans divided into 10-feddan holdings for members of dairy co-operatives at Khartoum north on the east bank of the Blue Nile, the farm of the Arab Sudanese Dairy Company to the south of Khartoum with 5500 feddans, and the Seleit Food Production Company which was allocated 27 000 feddans to the east of the Blue and Main Niles, but this is only partially under cultivation. In 1970, the Animal Production Corporation started to develop a 12 000-feddan fodder farm at Soba, close to south Khartoum, but problems of soil salinity and livestock management caused its closure.

Fodder is grown for dairy herds and for conditioning beef animals trekked from the western Sudan and elsewhere, requiring 8–10 weeks on a high quality diet before export. Many Khartoum households also need to buy fodder in the market for domestic goats and sheep. The main fodders are the forage sorghum, Abu Sabaeen and lucerne (*Medicago sativa*), usually referred to as *bersim*. Abu Sabaeen matures in 70 days and successive crops are grown with applications of urea. A good crop will yield about 9 t feddan^{-1} of partially dry fodder. *Bersim* is expensive to establish and grow and for this reason is less popular than Abu Sabaeen. Once established *bersim* is left down for three years and with cuts at monthly intervals it can give an annual yield of green fodder reaching about 18 t feddan^{-1}.

Production of fodder for the finishing of nomadic reared cattle provides a reminder of the many interrelationships between the very diverse systems of agriculture found in central Sudan. In addition to the transfer of animals, labour, produce, and even machines move freely between the rain-fed and irrigated schemes and between traditional and modern farms. Unfortunately, space does not allow these linkages to be described in detail.

References

Ahmed, A. G. M. (1976). The Rufa'a al Hoi economy. In *Some aspects of pastoral nomadism in the Sudan*. Economic and Social Research Council, Khartoum.

El Bashir, A. S. A. (1982). Mechanization, productivity, and employment in large scale mechanized farms in the Sudan. Unpublished Ph.D. thesis. University of Kent.

Jewitt, T. N., Law, R. O., and Viga, K. J. (1979). Vertisol soils of the tropics and sub-tropics. *Outlook on Agriculture*, **10**, 33–40.

Khalifa, A. H. and Simpson, M. C. (1972). Perverse supply in nomadic societies. *Oxford Agrarian Studies*, **1**, 46–56.

Magar, W. Y. (1986). The White Nile pump schemes. In *The agricultural sector of the Sudan* (ed. A. B. Zahlan and Y. B. Magar), pp. 179–91. Ithaca Press, London.

MAFNR (Ministry of Agriculture, Food, and Natural Resources) (1977). *Sudan national livestock census and resource inventory. The results of an aerial census of resources in the Blue Nile Province*, Vol. 10. *The results of an aerial census of resources in Sudan*, Vol. 31. MAFNR. Khartoum.

MAFNR (1985). *Agricultural situation and outlook*, Monthly Report, No. 2. Department of Agricultural Economics, Khartoum.

Mohy El Din, A. El. T. (1985). Strategies to make Sudan self sufficient in the production of sugar. M.B.A. dissertation, University of Leeds.

Simpson, I. G., Yassin, M. M., and Dishoni, S. (1966). Pump schemes. In *Development of agriculture and organizational planning in the main Gezira area*, p. 113. Department of Agriculture, Ministry of Agriculture, Khartoum.

Simpson, I. G. and Simpson, M. C. (1978). *Alternative strategies for agricultural development in the central rainlands of the Sudan*, Rural Development Study, No. **3**. University of Leeds.

Simpson, M. C. (1980). *Sector appraisal manual: beef*. Overseas Development Administration, London.

Sudan Gezira Board (1982). *Gezira current statistics, season 1980/81–1981/82, season 1982/83*. Economic and Social Research Unit, Barakat.

Sudan Gezira Board (1983). *Gezira current statistics, season 1982/83*. Economic and Social Research Unit, Barakat.

Sudan Government (1966). *Report on the census of pump schemes (June–August 1963), Vol. II. Part 3: Khartoum Province*. Department of Statistics, Khartoum.

United Nation Conference on Trade and Development (1989). *Trade and Development Report, 1989*, Chapter IV. F. Sudan. United Nations, New York.

World Bank (1985). *Sudan, pricing policies and structural balances—a World Bank country study*. World Bank, Washington D.C.

17

Systems of agricultural production in southern Sudan

A. DICKIE

Introduction

The large area of southern Sudan, the wide variation in physiography and climate, and the wide variety of peoples have given rise to equally wide variations in the systems of agricultural production found within its boundaries. Underlying this diversity, however, four major factors have affected all farming systems in the south. They are:

1. A harsh natural environment.
2. A sparse population—the result of years of exposure to raiding and destruction by slavers, Turko-Egyptians, and Dervishes and, more recently, by both sides in an on-going civil war. There are few sizeable urban centres to provide local markets for agricultural surpluses.
3. Poor communications—natural disadvantages have been exacerbated by poor maintenance of the communications network or even by deliberate destruction during the periods of civil war. The fine natural corridor of the Nile has been barely exploited. This has hampered the export of agricultural products and made the importation of farm inputs difficult and costly.
4. A very low level of technological development—years of raiding, fighting, and destruction have given rise to physical and mental demoralization among indigenous southern Sudanese. Farmers have been forced to flee their farms to find refuge in the bush, or even across international borders. Even during the relatively stable and progressive period of Anglo-Egyptian rule, the development programme in the south was low-key, emphasizing disease control, famine control, and soil conservation, sometimes to the detriment of long-term development (Tothill 1948, p. 891).

The Southern Development Investigation Team (SDIT) (1955 Report) divided the southern Sudan into seven major ecological areas (Fig. 17.1).

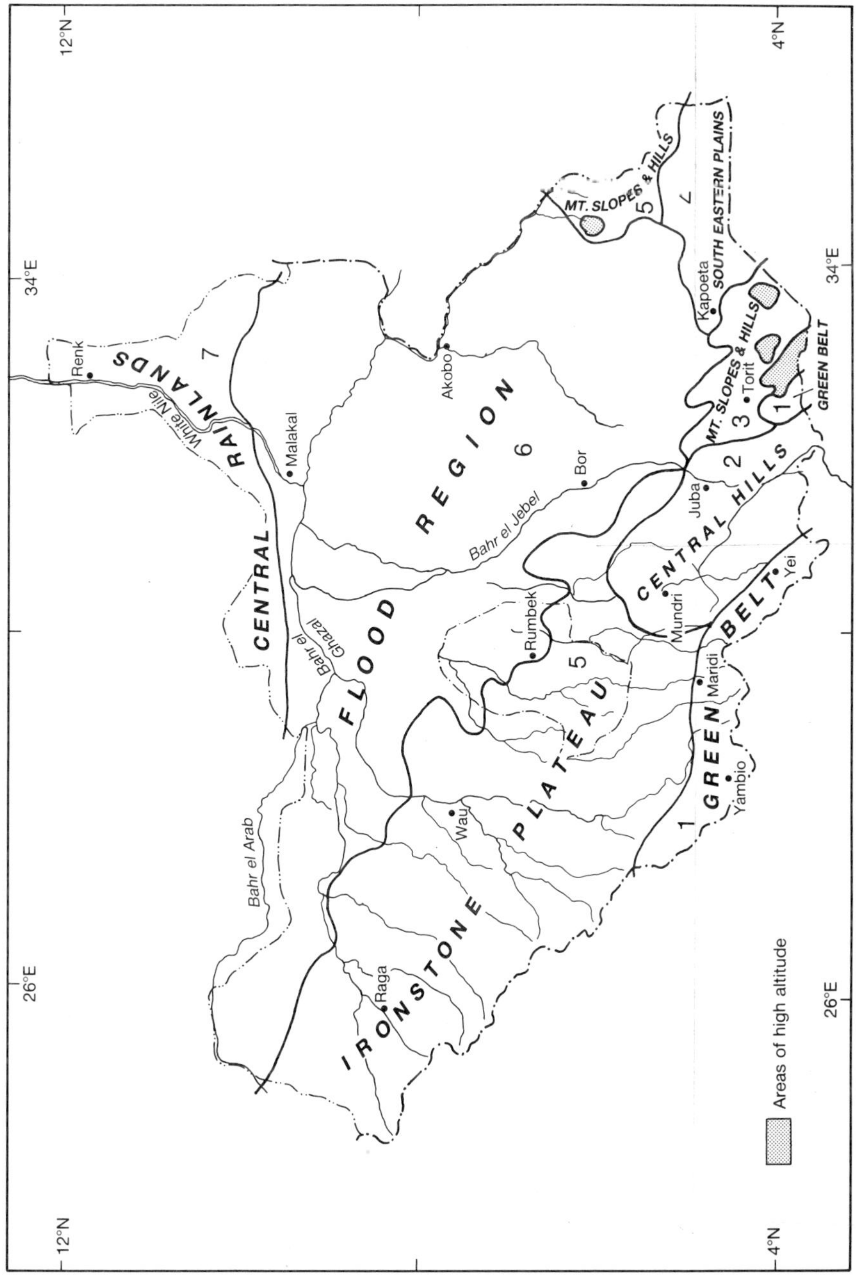

FIG. 17.1. Southern Sudan: ecological areas (After SDIT (1955)).

The most southerly areas comprise the Green Belt, the Central Hills, the Mountain Slopes and Hills and the South-Eastern Plains. Stretching north and east of the Nile–Congo Watershed (the border with the Central African Republic) lies the Ironstone Plateau which merges in turn into the vast Flood Region or Sudd, drained by the Bahr el Ghazal and Bahr el Jebel. The most northerly zone was identified as the Central Rainlands.

A general distinction can be made between the zones of the north and east, largely made up of the Flood Region, where pastoral systems of agriculture predominate and the south and west, the Ironstone Plateau area, where cropping systems are most important.

The low level of technological development in the area makes farming methods particularly sensitive to local variations in land form, soils, hydrology, and rainfall. The low population levels also mean that the majority of cultivated land is still farmed under some form of shifting cultivation, with more restricted areas, i.e. those close to towns, those with particularly fertile soils and good water supplies, or those subject to heavy refugee immigration, falling into the fallow systems category (Ruthenberg 1980, p. 70) or even that of permanent cultivation.

This discussion will concentrate on four ecological areas which illustrate the most important systems of agricultural production in the southern Sudan:

1. The Green Belt—particularly that part of it falling within Yei District. This is the area of highest rainfall and the most productive area in the southern Sudan. Cash-cropping is of increasing importance.
2. The Ironstone Plateau and Central Hills—this is easily the largest ecological zone where cropping systems predominate. Productivity is low and cash-cropping is barely developed.
3. The Colluvial Plain—in this closely settled area the pastoral systems of the Flood Plain meet the cropping systems of the Ironstone Plateau and, although not integrated, are carried out by the same farming families.
4. The Flood Region—here pastoral systems predominate and the cropping systems are least developed and most restricted in area.

The Green Belt

The term 'Green Belt' was coined by de Schlippe in his detailed study of shifting cultivation among the Azande. In using the term, he was referring to 'areas of an exceptionally rich vegetation around Yambio and Maridi, probably due to a rolling topography corresponding to a younger soil (along the Nile–Congo watershed)' (de Schlippe 1956, p. 181). The book gives a fascinating and detailed insight into the farming systems of the Azande in the last few years before Independence.

For present purposes the definition of Green Belt follows that of the SDIT Report (1954) and includes that area in the extreme south-west of the Sudan extending approximately to the 1350 mm isohyet. Besides the Zande tribal area, the Green Belt also covers the south-western half of Yei District and, in particular, the Aloma Plateau area. It excludes the Kajo Kaji Plain. The most important tribes numerically are the Kakwa and Fajulu and together with the Kaliko who inhabit the extreme south of Yei District, they have been most affected by development activities disseminating from Yei town. The less populous and less developed tribal groupings of the Baka, Avukaya, Mundu, and Makaraka inhabit the more isolated north and most of the area in question. The area has been the subject of extensive studies (Booker 1983).

Climatically, the area is wet monsoon, medium length (7–8 humid months). The higher rainfall and more extensive wet season mean that the area produces a wider variety of crops than elsewhere in southern Sudan and two successive crops (or more commonly sequential crops) can be taken from the same land in one year.

Vegetation is moist semi-deciduous forest on the Aloma Plateau, transitional to woodland savanna to the east and west of the Plateau. There is a strong tendency for much of the natural vegetation, about 65–100 per cent of the land area (Booker 1983), to become fire climax tree savanna because of extensive annual burning for hunting purposes. Soil erosion is widespread, particularly along the burned watershed areas of the Aloma Plateau and in the vicinity of Yei town. Villages are normally found along secondary roads or motorable access tracks and here vegetation has degraded to grassland savanna. The population is highly mobile, many farmers trading in Juba. Both district and international borders cut across tribal boundaries and there is considerable movement between areas. Many Sudanese fled into Uganda during the seventeen years of civil war and benefited from superior education and exposure to more advanced agricultural methods. Likewise an influx of Ugandan refugees had a major impact on agriculture in the area both through the intensification of land use, the provision of new markets for surplus crops, and through the improved methods used by the immigrants.

Since 1972 the area has undergone rapid change with cash-crops and food crops grown for sale becoming increasingly important. However, meeting subsistence needs still remains the priority of most farmers. Because of the high rainfall and relatively fertile soils, there is little direct threat of famine and a wide range of crops can be grown. Sorghum, cassava, and finger millet are the main staples with maize, sweet potatoes, and rice being grown on a more limited scale. The main relish crops are sesame and groundnuts. Cowpeas are also important as well as pigeon peas and East African beans (*Phaseolus vulgaris*). *Hyptis spicigera*, a local unimproved oilseed crop, is declining in importance. Okra, egg-plant, tomatoes, Bambara groundnuts, pumpkin, and various fruits, particularly bananas, pineapples, citrus, pawpaw, and mango, are all widely grown in this area.

The proximity of the Zande area, of Zaire and Uganda, and the steady traffic across the borders means that the region is particularly rich in different varieties of crops. For example, many more varieties of cassava, both 'bitter' and 'sweet', and varying in time to maturity are found here compared with areas to the north-east. Yields per unit area are highest for the early-maturing varieties (though not necessarily per unit of labour input) but the late-maturing varieties produce the best flour, which is sold as a cash crop and, as they are less susceptible to waterlogging they are important as a famine reserve. There are also many staple varieties of sorghum, with different season lengths, physical characteristics, storabilities, and tastes. Easily the most important are the long-season varieties (7–8 months) which ripen after the rains have finished. Other staples are important in the early part of the season.

The most popular local varieties of groundnuts are generally early-maturing sweet palatable nuts, important as early crops in relieving the hungry period. By contrast, introduced exotic varieties such as Makulu Red and Mani Pinter, are relatively late-maturing, high oil content nuts, substantially higher yielding than the early-maturing varieties, but more susceptible to unreliable rainfall and poor soils.

The most common maize variety in the area is Bazande (from the Zande area) which is fast maturing and normally eaten green.

Development efforts and lucrative ready markets, have caused the areas of both robusta coffee and tobacco to increase dramatically in recent years while cotton has all but disappeared because of marketing problems. Tobacco is grown widely by refugees as a means of earning cash quickly on relatively small areas of land. Sorghum, cassava (sold as flour or dried), and groundnuts are the main food crops grown for sale.

Poultry, goats, and sheep are kept widely throughout the area. Cattle are less common but are kept by an important minority (16 per cent) of households on the Aloma Plateau and by nearly half (41 per cent) of households further west on the Kaia–Kiju Plain (Booker 1983).

Cropping calendar

Cropping systems are designed to minimize the risk of complete crop failure and to economize on labour use rather than to maximize the yields of individual crops. Hence, a wide variety of crops are grown, normally in mixed stands and because of the predominance of hand labour, seeds tend to be broadcast or often dibbled in an apparently haphazard way, rather than line planted.

New or renewed fire lines are prepared in November and December. In January, after the harvest of the main sorghum crop, grasses and, less frequently, shrubs and trees, are cleared and left to dry until March, when they are burned. Crop residues are also burned at this time.

The first crops, maize and groundnuts are planted in the home plot. East of the Aloma Plateau this association is tending to increase to a field scale as maize increases in popularity. Optimum planting for maize is in March or April, when the rains are well established. This allows a second crop to be planted on the same land (usually relay-planted at the first or second weeding). Maize is usually line sown in clumps of three or four plants and after emergence groundnuts or beans are interplanted.

Early plantings of long-season local sorghum cultivars such as cv. Jeri are the next main sowing activity, usually taking place at the second weeding and between the rows of maize. Long-season sorghums can be planted until September but early-June plantings have given the highest yields in research trials and most farmers try to plant cv. Jeri in May or June. Planting dates are less critical than for short-season varieties and hence a much larger area can be cultivated as labour requirements are spread over several weeks.

Cassava can be planted from March through to September. Early plantings of sweet cassava are grown near the homestead and are frequently interplanted with groundnuts which form a valuable ground cover, suppressing weeds. Sometimes, sweet cassava and groundnuts are grown in association with maize and the local sorghum cv. Jeri. August plantings of cassava are often interplanted with sesame or a second crop of groundnuts, but after these are harvested, the cassava is usually left to reach maturity as a pure stand ending the rotation and often considered as a form of fallow.

While the planting of early crops goes on, men are also at work with the heavy digging and preparation of outlying fields ready for the main crops. In June, the sowing of the main cereal crops begins including the main sowings of the major long-season sorghum cv. Jeri and Godo and of a ratooning sorghum cv. Kima. These, and the later-sown cultivars, such as Ijigo (= Nyrangwa), are generally broadcast in a mixture with finger millet and sesame on newly-cleared land, if available.

The main crops of finger millet and sesame are planted in July and August. Second crops of groundnuts and maize are also planted at this time but yields are generally well below those of early plantings. Among the last crops to be planted according to current practice, are the introduced exotic, short-season sorghum cv. Serena and 9DX. Recommended time of planting is August/September.

Cowpeas can be planted from late March to August and the time of planting has little effect on yield. They are frequently grown in a pure stand, in small sub-plots, or are interplanted with any of the major crops. Pigeon peas are commonly intercropped with finger millet at the second rains.

Rice is planted between May and July. Planting later than July can reduce yields by as much as 50 per cent.

Most crops receive one or two weedings and these will often coincide with other operations, e.g. the planting of Jeri at the second weeding of maize, or the weeding of cassava at the harvesting of groundnuts, to economize on labour use. The extended period of planting means that weeding is also spread over time. Even so it demands a major input of labour and any innovation which reduced the weeding requirement (e.g. reduced spacing or interplanting with ground-cover crops) would be of major importance in releasing labour for alternative tasks, in particular for extending the cropping area planted.

Harvesting, like planting is a staggered process. In June, the first crops harvested are the ratooned sorghum cv. Kima, maize (generally eaten green) and groundnuts. Further groundnuts and maize are harvested in July, August, and September. There is then a lull in harvesting activity during October and early November before the main harvest period begins with finger millet in November, followed by the late crop of groundnuts, sorghum, and in January also sesame and *Hyptis*.

After drying on their heads, long-season sorghums are stored in specially constructed mud *gugus* (mud and dung woven baskets supported on a low platform of wooden poles with a thatched roof of elephant or other grass). Finger millet is stored likewise but can be stored for longer as it is less susceptible to weevil attack. Maize is generally stored unhusked in bunches hanging from a pole or tree. The harder, flint varieties are less susceptible to weevil attack.

Sesame flowers and ripens sequentially allowing selective picking early in the season. Dehiscence limits the time that the crop can be left standing and eventually flowering is arrested by drought. It is usually threshed in the field, the stems being heavy in relation to the seed. Groundnuts are usually picked off the vines in the field. The vines are left to rot and the nuts are stored in the shell to reduce insect attack.

Most types of seed are traditionally stored in the roof of the cooking hut where the smoke helps to protect them from insects.

Cash-crops

The tobacco calendar starts at the beginning of November with the registration or re-registration of growers with the National Tobacco Company (NTC). During December, inputs are supplied on credit and farmers are encouraged to burn over their seedbeds to sterilize them. Seeds should be sown in the nurseries by mid-January. From mid-March the seedlings are planted out into the main beds and reaping begins in mid-May. Buying starts on 1 July and all curing is finished by mid-August. Thus, tobacco growing affects mainly the early-season crops. Sesame, groundnuts, and maize are frequently planted after tobacco to take advantage of the residual compound fertilizer applied.

The main robusta coffee growing areas are to be found to the south of Yei town and the riverine terraces of the Aloma Plateau offer the best sites. Plantings to the north and east of Yei town are less successful. In areas to the west and east of the Aloma Plateau, careful site selection along river valleys is required for successful production. Two privately managed and long standing coffee estates are at Iwatoka and Kabengere. Increasingly, smallholders are producing their own coffee seedlings.

Seedlings are planted out in June when the first rains are well established and slash weeding is carried out throughout the wet season at the end of which the trees are clean weeded and mulched to prevent excessive water loss. Harvesting starts in November when the trees are 3-years old and continues as the cherries ripen, until January. Pruning begins as soon as harvesting is finished and continues into March. Suckers are removed throughout the year apart from during the main harvesting.

To be grown successfully, robusta coffee demands a considerable labour input. In particular, harvesting coincides with harvesting of the main food crops and, as newly-opened land is recommended for the crop, this also involves a major labour input.

Intercropping is recommended and widely practised for the first 2–3 years of coffee production, Groundnuts, pigeon peas, cowpeas, green gram, and pineapples are the recommended intercrops, and the use of first-rains maize is also common.

Livestock

Sheep and goats are allowed to roam freely during the dry season, but when the crops are in the ground, they are kept tethered near the homestead, the tether being moved two or three times a day. Young boys are responsible for their care, as they are for the herding of cattle. Livestock are not integrated into a mixed farming economy but are kept primarily for investment, especially for bridewealth and to exchange for staple foodstuffs when these are in short supply. Sheep are used for their milk and they may be sold when cash is needed. Goats are rarely used as such; they appear to play a more important role in traditional and ritual activities.

Fishing is an important source of animal protein and rivers, streams, ponds, and swamps are exploited throughout the year mainly by men and boys. The greatest quantities are caught during the dry season and at the beginning and end of the rains when the fish are migrating.

Hunting provides another important source of animal protein. It is concentrated in the dry season and early part of the rains.

'Hunts in the traditional hereditary hunting blocks, into which most of the uncultivated land in the area is divided, are regarded as the most important events in the hunting calendar. A wide fire trace is cut and burned around the hunting block (usually a square mile or more in extent) in November to December and is hunted by burning on one day between January and March. The annual burning of such large tracts of land has serious consequences for soil degradation and erosion but measures to control it or to encourage earlier burning are unlikely to be successful while game provide such a readily available source of protein and while no other method allows easy access to and control of the bush surrounding cultivated areas' (Booker 1983, para 8.17).

Pests and diseases

Imported insecticides are well known to Yei farmers but are seldom available. Insect pests can cause severe losses both in the field and in storage. Birds, especially the Sudan dioch (*Quelea quelea*), can cause major losses of the more palatable sorghums earlier in the season. Animal pests are less serious in the more densely populated areas.

Cassava mosaic virus (CMV) is widespread but, potentially, control could be achieved by the distribution of disease-free planting material. Bulking up and evaluation of possible CMV-resistant varieties was beginning at Yei Research Station but this work was disrupted by the war. Although disease-free stocks of local varieties can be achieved by roguing out diseased plants, it is unlikely that smallholders will be prepared to do this on their own fields because of tuber losses (Booker 1983).

By 1986 the virus problem had reached epidemic proportions, particularly in the fields farmed by refugees who considered the cassava crop to be barely viable.

Land

As indicated earlier, the roadside pattern of settlement is typical of the Yei area and has been common since resettlement (for health and administrative purposes) was enforced by the Colonial Authorities in the 1930s. Tothill (1948) commented that the population was gradually moving back into the bush following the relaxation of controls. However, the development efforts of the post-Addis Ababa Accord era (1972–84) appear to have reinforced the tendency of the rural population to live on the roadsides. The consequent exhaustion of roadside soils has meant that the main cultivation areas are generally found some distance from the main homestead (*tukl*), inside the bush. The distance from the *tukl* varies according to factors such as the density of the population and the quality of soils.

Temporary houses may be built in the bush if the fields are too far for daily travelling. The expansion of cash-cropping, particularly that of coffee, has led to a further increase in these bush fields.

Fields are usually cropped for 3–5 years followed by bush fallow for 2–3 years, and cassava often replaces the fallow. Estimates from Landsat interpretation (Booker 1983) show that this recurrent form of farming covered 26.8 per cent of Yei District in 1981–2 while semi-permanent farming covered 2.4 per cent (mainly in the vicinity of Yei town). Natural forest, used mainly for hunting and gathering, covered 53.4 per cent.

In 1983 there was a major movement of Ugandan refugees into the area and by 1984 their numbers equalled those of the indigenous population which had been estimated at 150 000 in 1981. Immigration continued with the continuing problems in Uganda, and those refugees not in United Nations High Commission for Refugees (UNHCR) camps, were concentrated in a strip of land within ten miles of the border creating probably the most densely populated rural area in southern Sudan and an extended area of permanent cultivation.

Average farm size (excluding fallow and grazing areas) is estimated at 1.5 ha (3.5 feddans). The major determinants of farm size are family size, the availability of additional labour, and the decision to grow cash-crops. Coffee and cassava fields often cover an additional 1–2 ha but tobacco cultivation has little effect on farm size, the average plot size being around 0.1 ha.

Labour

For agricultural activities in this area there is little strict division of labour between the sexes, although clearing and seed-bed preparation and, increasingly, cash-crop production are the responsibility of men and boys. Weeding and harvesting and post-harvesting activities are generally women's work.

The increasing importance and considerable cash returns from coffee growing in the Aloma Plateau has meant that hired labour is now of some significance in the area. The availability of cheap Ugandan labour from refugee settlements is also encouraging the cultivation of much larger areas of both subsistence and cash-crops. Men, for the most part, are hired by the day or month for all types of work, but particularly for clearing and initial digging. They are paid in cash or with food.

For the majority of farmers, however, traditional work parties still provide the bulk of non-family labour, alleviating bottle-necks, particularly at clearing, planting, and weeding, during the busiest months of April to August. There are two types of work party, the small, largely kindred groups, involving reciprocal labour, where farmers work in each others fields without remuneration until all are finished and, secondly, those organized by wealthier farmers who require extra labour for a particular task—generally preparing fields, weeding, and harvesting. A work party usually numbers between 15 and 20 and lasts from 1 to 3 days with remuneration in beer, food, and local tobacco.

Apart from the Zande area and the Aloma Plateau south of Yei, two other areas, both on the east bank of the Nile, are similar in climate and soil type to the Green Belt. These are the southern and western foothills of the Acholi Hills, inhabited by Acholi; and the western foothills of the Imatongs, inhabited by Lutaka. Here, a much smaller variety of crops is grown and fruit trees are practically absent. Coffee growing is less popular here and in the Zande areas, due mainly to more limited development efforts. Potatoes are grown on a small scale in the Aloma, Acholi, and Imatong areas. Further expansion appears to be inhibited by storage and marketing problems.

The Ironstone Plateau and Central Hills

The Ironstone Plateau area is characterized by soils of variable depth over ironstone gravels or indurated ironstone sheets. Slopes are generally less than 2° with a distinct 'catena' of soils from upper slope (shallow soils) to lower slope (deeper soils). The soils are mostly of low fertility and the vegetation is predominantly woodland savanna or tree/bush savanna. This discussion concentrates on: (a) the area around Wau, which is the Bahr el Ghazal regional capital, an important commercial centre, and the only railhead in southern Sudan, and (b) the area around Mundri in the Central Hills at the southern extreme of the Ironstone Plateau.

Much of the Ironstone Plateau is sparsely populated. The most important group (apart from the Dinka who inhabit the northern fringes) is the Jur (or Jo Luo), but numerous other minor tribes are to be found. All the non-Dinka speaking peoples show a considerable homogeneity (Booker 1984*a*). The Moru-Madi predominate in the Central Hills south of Mundri, with the Bongo-Mittu-Baka mainly found to the north of that town. Water is obtained from perennial streams or hand-dug wells which may be dug to a depth of 20 m or more. The proximity of a dry season water supply is the dominant factor limiting settlement, for most of the Ironstone Plateau is effectively waterless in the dry season and, therefore, uninhabited. Around Wau, however, the land is quite densely settled and approaching permanent cultivation. Climatically, this area falls into zone M3 (wet monsoon, medium-length wet season, 5–6 humid months, 4–5 dry months, annual rainfall 950–1300 mm).

In this area the widespread infestation by the tsetse (*Glossina morsitans*) virtually precludes the keeping of livestock, which is of significance for farming systems.

Sorghum is the major staple throughout the Ironstone Plateau, although there is some variation in the importance of different varieties. The most popular long-season varieties are known as Mabior and Mbarkan in Wau. Cassava is probably second in importance as a provider of calories although the number of varieties identified are far fewer than those available in the Green Belt. In the Yei area, for example, the rooting stock have been introduced mainly from the south west via the Zande area (Tothill 1948, p. 892). The sweet varieties are preferred as they are easier to prepare, but bitter varieties form an important cash-crop when dried and ground into flour.

Cassava remains a minor crop among those people who have access to cattle on the fringes of the Ironstone Plateau, for example, the Jur north and south-east of Wau. Here cattle replace cassava as an insurance against famine and a source of cash income, but the Jur do not keep cattle themselves.

The Jur provide the Dinka with iron goods, wild animals, hides for shields, and grain. In exchange, the Dinka give them smallstock and cattle, herd cattle for them, bring cattle to manure their fields during the dry season, and sometimes work for them as casual agricultural labourers. The Jur thus own cattle but do not look after them; they prefer to work as cultivators (Booker 1984*a*, para 2.03).

Sesame and groundnuts are important throughout the area. The black-seeded sesame is known as Jur-simsim in the Wau area and it is most popular in the Jur area. Elsewhere white-seeded sesame is more common (Booker 1984*b*, para 10.13). Bulrush millet is an important crop in the transition zone to the north of Wau but is rarely grown in Ironstone country proper. Finger millet has the opposite distribution and around Halima is sown in July and August, often as the first crop on newly cleared land in association with the sorghum cv. Bende, and sesame. Various beans and cowpeas are grown, usually in association with the main sorghum crop. Other minor crops, including pumpkin, maize, sweet potato, okra, yams, tomatoes, and cucumber, are usually planted around the *tukl*.

The main crop associations identified in a smallholder survey (Booker 1984*b*) were:

(a) Mabior (long-season sorghum), legume with sesame intersown at the first weeding;

(b) Bende (medium-season sorghum), sesame and/or finger millet;

(c) groundnuts and cassava.

Cropping calendar

Before the harvest of Mabior, the main sorghum crop, and following the peak of planting and weeding in the current year's crop, suitable new land is identified by the presence of good grasses such as *Hyparrhenia* spp. and tree species such as *Anogeissus* spp. which indicate deep soils. A fire line is cleared and weeded and trees may be burned around the plot for identification. Following the main sorghum harvest, from January onwards, grasses, shrubs, and trees are cut down and left to dry. Most burning is left as late as possible (March/April) so that both old and new grass growth can be destroyed and the need for primary cultivation is avoided. The new land is frequently adjacent to existing fields, particularly where the blocks of land are more consolidated. On currently cropped land, firelines are renewed during the September–October period to avoid the danger of bush fires spreading through the standing crops.

Land preparation prior to planting is minimal; the short-handled *maloda* (used by non-Jur tribes) and the long-handled *kwer* (used by the Jur), only scrape the surface to remove excessive weed growth.

The majority of crops are sown between May and July, but the exact timing depends on the distribution of rainfall, the degree of weed growth, and the availability of labour.

Outside the Jur area the priority for planting is given to the sorghum cv. Mabior/bean/sesame association, normally planted from the last 2 weeks of April to the end of May. In the Jur area planting is 1–3 weeks later and priority is given to the short-season sorghums and other *tukl* crops.

Sorghum and beans are dibbled into land which has usually been cropped before. A range of sorghum varieties can be mixed and sown with the beans and Mabior seed. Two to three weeks after germination, sesame is broadcast amongst the weed growth and incorporated into the soil when the weeds are cut down.

Groundnuts are dibbled as soon as the sorghum association plantings are nearing completion. Then, after 2–3 weeks cassava sticks are planted amongst the developing groundnut crop at the first weeding. As in the Green Belt, this association is usually planted on land that is at the end of the cropping cycle. The ground cover of groundnuts reduces the weed problem which can restrict the early growth of cassava severely when it is grown in a pure stand.

Small areas of white-seeded sesame varieties grown in pure stands, and of sweet potatoes grown on ridges, are planted during the May–July period.

The medium-duration sorghum cv. Bende is planted after the establishment and first weeding of the main food crops; it is usually broadcast as a mixture with sesame and finger millet in July/August. As in other areas, this mixture is sown either into the weed growth which is then cut down, cultivating the seed into the soil at the same time, or it is sown on new land which takes longer to prepare than previously cropped land. In the Moru area a late-planted field is called 'Nyarango' after the main sorghum cultivar in the crop association. This is very similar to the above and consists of sorghum cv. Nyarango, finger millet, beans, and sometimes sesame and bulrush millet (Catford 1951). The cut weeds provide the seeds with a mulch protection against birds which are a major problem at this time of year.

Outside the Jur areas, the first crops to be harvested are cassava and cassava leaves (June/July) followed by green maize, cowpea and cowpea leaves. In the absence of cassava in their cropping system, the Jur depend on short-season sorghums as their first crops of the new season. They are available from early August and are consumed immediately due to their poor storage qualities. Sesame ripens sequentially, as described for the Green Belt and allows some flexibility in harvesting according to need and labour availability. Short-season groundnuts are harvested in September along with other early crops, but the main groundnut harvest is in October. Finger millet is usually harvested in November, but when planted before July it is subject to bird damage so it is grown only on a limited scale.

The harvesting of the various sorghum varieties is staggered over several months and culminates in the main harvest of Mabior starting at the end of December.

Weeding is usually carried out twice during the growing season, the Jur using the long handled *kwer* while other tribes prefer the short-handled *maloda*. Shallow tilling or scraping, broadcasting seeds just before weeding and the inclusion of ground cover plants in crop mixtures, all serve to reduce labour requirements for weeding.

Weed infestation, particularly by *Striga* spp. is a major reason for abandoning a plot and the time of planting for the major crops takes into account the bird menace at particular times in the growing season. Apart from this very little appears to be done to avoid crop losses. Wild animals, particularly baboons and wild pigs, are considered by farmers to be the most serious cause of crop failure in the remote west and south of Wau District. The departure of many young men and boys in search of jobs or for schooling has reduced regular trapping and now there is a steady encroachment of wild animals into the area.

Cash-cropping is not important in this area. Local tobacco is the only true cash-crop and this is sold in relatively small quantities. Sorghum, groundnuts, cassava, and okra are the main subsistence crops sold.

Livestock calendar

Around 60 per cent of the Jur are thought to own cattle, about five per family compared with 20–30 per family among the Dinka, with slightly more owning goats (79 per cent) and slightly less owning sheep (46 per cent) (Booker 1984*b*). Elsewhere in the Ironstone Plateau the presence of tsetse largely precludes the keeping of livestock. In the dry season, because of the scarcity of grazing, the Jur arrange for their cattle to be herded by

Dinka friends in the *toich* areas, such as the Nyang and Thiet areas of Tonj District. Relationships are built up gradually as trade contracts between individual families. The Dinka are entitled to the milk produced and, eventually, when animals are withdrawn for sale, for example, they are entitled to one or two cows, depending on the number of progeny. Both Dinka and Jur regard cattle as an insurance against crop failure and vital to survival. Cattle also play a significant role in social life, providing the means for maintaining social ties and obligations essential for a cohesive society.

Fishing and hunting

This is a seasonal activity with fish being caught when the water level in streams and rivers is low and fish are relatively concentrated. Fish are also caught at the beginning and end of the dry season when they move up and down the main rivers of the Pongo, Gette, Jur, Sue, and Bussere. The most common methods of fishing involve the use of nets and poisons. There is a seasonal migration of people away from the Bazia–Wau road to the Sue River partly because of the fishing opportunities during the dry season and partly due to the water shortages experienced near the road during the dry season.

As in the Green Belt hunting is an important traditional dry-season male activity providing an important source of animal protein. Honey gathering is also important, providing a dietary supplement and an off-farm income.

Land

A Landsat study of Wau District (Booker 1984*a*) showed that some 14.2 per cent of the total area is under shifting cultivation ($R<33$) (Ruthenberg 1980), compared with 6.1 per cent under recurrent cultivation and only 2.7 per cent under permanent or semi-permanent cultivation. The latter is largely in the vicinity of Wau town (Booker 1984*a*).

Shifting cultivation is common in the Jur tribal area. Each homestead unit has between 3 and 5 blocks of land which may be as far as 30 km apart, sometimes in different village neighbourhoods and rarely close to roads. This scatter is probably due to the need to exploit those fertile parts of the catena which are widely separated between interfleuves. The blocks are generally circular in shape with paths radiating from the centrally located *tukl*. Thus they bear some similarities to the *katiri* system in the Mundri area and as described by Eyre *et al.* (1953). The Moru *katiri* is a roughly circular area of around 200–400 m radius which is cleared by a clan of six to eight families. It is selected from virgin land in June and then burned over. The central point is a termite mound or tree around which the homesteads are clustered. From these radiate cultivation segments, which are allocated by a clan elder and whose size varies with the number of adult women in each family. The segments are divided by straight pathways and the whole plot is protected from wild animals by a stockade around the circumference. The *katiri* is cultivated for two to three seasons and then left to fallow for five years or more. On re-opening, cultivation segments are reallocated to families. Two or three *katiris* at different stages in the cycle will be cultivated at the same time by each clan. Among the Jur of Wau District, cultivation varies from 3 to 7 years after which the *tukl* and block are moved. Fallowing varies from 10 to 15 years depending on soil type.

Semi-permanent settlements are to be found among the Belanda Bviri south-west of Wau town. The main homestead is found near the road but is moved every 5 to 10 years due to the decline in fertility. The main cultivation blocks, rectangular in shape, are found away from the road and are moved every 3 to 5 years.

Permanent settlements along roads are found in the Bai, Ndogo, and Belanda Bor areas west and south of Wau town. The *tukl* and *tukl* plots are located on the roads but the main fields are still some 1.5–5 km into the bush. One block, usually circular in shape, is cultivated at a time, for a period of about 3 years, but the fallow periods between cultivation are less and the intensity of land use in excess of $R = 60$ (Booker 1984*b*). The whole family moves to a temporary homestead at this block for the cultivation period from April to December, but returns to the main homestead every few days for ceremonies, provisions, and water.

A limited number of permanent part-time farmer settlements belonging to government officials or traders are found in the densely settled areas around Wau.

These latter settlement patterns date from the end of the disturbances in 1972 and allow farmers to benefit from access to markets, schools, dispensaries, and borehole water established by the Sudanese Government. The concentration of population along the roads may be one reason for shorter fallow periods in some areas. In the Khorgana area which has high fertility but a sparse population, the problems of damage by wild animals to distant crops seems to have led to intensified cropping.

Estimates of farm size (Booker 1984*b*) indicate smaller relative farm areas (6 feddans) among the Jur, probably due to the importance of cattle to this group. There is some movement of young Jur men to the Bongo/Balanda Bviri area south and west of Wau, looking for work to earn money to purchase cattle for future dowries. This and larger family sizes may be the reason for the larger farms, averaging about 12 feddans, in this area.

Labour

Division of labour within the family is similar to that found in the Green Belt. Men are responsible for land clearance, *tukl* pole cutting and honey collection, women for the harvesting, threshing, and storage of food crops, as well as for all domestic chores including fetching water and collecting firewood. Children, usually boys, are responsible for tending livestock. The sorghum-based system of the Jur gives rise to much more marked peaks of labour use than are found among tribes who also grow cassava in their cropping system.

Although the cash economy is not as developed in this area as it is around Yei, those areas where surplus crops are sold are also more likely to employ labour for cash. The Khorgana area in the extreme west of Wau District and the Balanda Bviri area mentioned above are cases in point. Among the Jur and in the remoter south (Belanda-Bor) area, work parties remunerated in beer and food are more common. As elsewhere in southern Sudan, this enables farmers to increase their area cultivated and to complete tasks more quickly—particularly important where large areas of sorghum must be planted at the beginning of the rains.

The Colluvial Plain

Along the fringe of the Ironstone Plateau, bordering the Flood Region is a belt of sandy colluvial deposits varying in width between 1 and 15 km. In the north it is cultivated by Jur, Bongo, some Dinka, especially in the Lakes District, and other tribes, and to the east by Moru, Mandari, and Nyangwara (Tothill 1948, p. 903). Soils are of low fertility but are deeper than on the Ironstone Plateau. The vegetation is predominantly deciduous woodland transitional to *Acacia* woodland savanna. Annual rainfall varies from 900 mm in the north to 1200 mm in the south. The average rainy season around Rumbek lasts around 177 days from late April to early October.

Unlike the bordering Ironstone Plateau and Flood Region areas, water is readily available on the Colluvial Plain through shallow, hand-dug wells or perennial streams. As the land is not subject to flooding and is largely free of tsetse, the area provides an important site for wet season cattle camps. Around the town of Rumbek, the belt is considerably wider than to the north-west and south-east and is densely settled by Dinka cultivators.

These Dinka are of two main groups: the Dinka Agar who occupy the area to the north and east of Rumbek; and the Dinka Gok, centring on Cueibet to the west of Rumbek. Tribal conflicts and disputes are common and arise particularly in relation to competition over grazing and water for livestock. The more serious conflicts can lead to considerable local disruption.

The area has an extremely poor infrastructure even by southern Sudan standards and this makes imports, most significantly of sorghum, costly and irregular.

Cultivation plays a much more important role in the livelihood of the

Dinka in the area than for those Nilotics living in the Flood Region itself. Cultivable land is much less restricted on the Colluvial Plain, allowing a wider variety of crops to be grown and providing a greater diversity of diet. Although arable and livestock systems are, as yet, rarely physically integrated in any way, the latter plays an important role in providing a means of investment and saving, and a reserve which can be utilized in the all too frequent times of famine. Long-season sorghum is the basic staple throughout the area.

Cropping calendar

Where new land is to be cultivated, bush clearing is carried out in the dry season. Before the rains begin, all grass is burned except for *harig* fields (see below).

Cultivation begins in April when the rains are thought to have become reliable and even the heavier soils are easily workable. The first crops are planted in the limited fertile area around the *tukl* for example, short-season (3-month) sorghum cv. Rapjang, maize, pumpkin, and okra.

Short-season groundnuts (*terekeka*) are the first crop planted in the main fields and the first crop to mature. They play an important role in relieving hunger. The groundnut plantings are followed by the main mixtures of long-season sorghum cv. Kec, bulrush millet, sesame, and cowpeas. Kec is the main staple, it takes 6–7 months to mature but is the highest yielding and is preferred for its taste. It is said to tolerate mid-season drought and flood, and its successful harvest often determines the family's prospect of a year without hunger (Duncan 1978).

Apart from groundnuts and some *dura* which are usually planted using a dibbling stick, most seeds are broadcast on to the cleared land which is then turned by the cultivator on his knees using a flat hoe. The level of weed kill is poor and cultivation of further land is curtailed earlier than would otherwise be the case in order to begin the weeding. Weeding is usually done once during the growing season, with the peak period in July.

Late plantings, as elsewhere, are usually on newly-opened land, allowing enough time for the more labour-intensive preparation involved. Long-term bush fallows are rare because of the density of population. Perhaps, for this reason, a form of *harig* cultivation is commonly used for late planted crops. It involves firelining the area concerned in November, leaving it unburned through the dry season until the new growth of weeds and grass is underway, then burning off the vegetation and planting the crop—normally green gram plus Bambara groundnuts and late cowpeas. The benefits of the system are the reduced or zero labour requirement for weeding and the production of a friable seed bed, fertilized by the residual ash.

Disease and pest controls are rarely applied and shifting cultivation is not considered a means of controlling pests and disease. *Striga* is very widespread and a major factor depressing yields.

Crop harvesting begins in July, although the main early harvest is in August and September. Groundnuts are followed by maize and short-season sorghum and these early crops are usually consumed as they are harvested because stores of the previous year's crop are exhausted.

Long-season groundnuts (*agar*) have a dormancy period which permits some of the crop to be left in the ground at harvest, producing a light volunteer crop in July of the following year. The main crop is dug up by *maloda* in September and dried before being stored on a raised platform in an *atang*, a universal spherical container of woven grass or Borassus palm. Such containers are also used for storing sorghum which has been dried on the head (on raised platforms), threshed and winnowed. The various types of sorghum are harvested from August and harvesting culminates in January with the main harvest of Kec.

As on the Ironstone Plateau, tobacco is the only true cash crop. A dark, dwarf type is grown and it is made into pipe tobacco and snuff. It is grown mainly at an old cattle camp near permanent water and is planted in October, transplanted in December, and harvested in March.

Livestock calendar

The seasonal movement of cattle in search of grazing in the Rumbek area is particularly pronounced among the Yek and Parial tribes of the Dinka Agar to the east of Rumbek. It is less pronounced among the Dinka Gok. The main dry season grazing areas for the former are around Lake Nyubor to the north of the permanent settlements. These pastoralists usually begin to move their cattle in search of grazing and water in November or December, while around Cueibet and Malou Pec to the west, herds are not usually moved until well into the dry season.

The lesser *toiches* (flood plains) mainly along the rivers Bahr Na'am and Bahr Gel which run north on to the plain, tend to be used when the first rains have fallen in April/May and cattle have returned from the main dry season areas. A few villages which are near these rivers have grazing close by for longer periods than other villages.

With the rising flood, the cattle are brought to the high land, although the need to avoid tsetse-infested areas of bush restricts the land available. For the same reason, cattle only come to the homestead for short periods at the beginning and end of the wet season, if at all. About 80 per cent of wet season grazing is within 7 km of the permanent homestead, but some is between 14 and 42 km away (Booker 1984*c*). Unlike the Dinka of Gogrial and Jonglei the Agar, Apak, and Gok Dinka do not usually construct cattle-byres at the main compound (Duncan 1978) and cultivated fields rarely benefit from cattle manure.

Due to the limited wet season grazing and the low protein content of the grass, cattle tend to be in poor condition by the late wet season. The grazing areas are burned over at this time, beginning with the higher ground and following the line of the receding flood waters. The residual soil moisture supports a growth of new grass for the cattle as they move back to the dry season grazing areas. Between one-third and a half of the farmers feed their stock on crop residues and regrowth of *dura* between November and February (Duncan 1978). A few cattle may be kept at home throughout the dry season, usually because they are too sick to go to the *toich*, but also to provide milk for the family. On the whole, however, cropping and livestock systems fall far short of any systematic integration. Considerable distances may be covered during the dry season, ranging from about 4 km at Mattiang to 35 km at Karic and Timic (Booker 1984*c*). The cattle are kept in camps (*wut*) which consist of several family cattle-hearths (*gol*), the number depending on the available grazing. Duncan (1978) found average family herd size to be 9.5 cows. The cattle, but not the small calves, are released at about 9 a.m. to graze for about seven hours, then they are tethered to individual pegs for the night. Cows are milked twice a day by women or girls, once before release in the morning and again in the evening after tethering.

The most important use of cattle manure is as fuel. Dry season cattle camps are normally in areas with little or no available firewood and the use of dung to fuel fires for cooking, for warmth, and to discourage biting insects is very necessary. Residual ash is used for brushing cattle to keep off insects, for personal adornment, and for cleaning teeth. The dung used as fertilizer (mentioned by 40 per cent of householders in the Booker survey) is usually used on dry season tobacco, dry season maize and, sometimes, for wet season maize and vegetable production on the edge of the homestead compound.

Cattle diseases are of great concern to the pastoralists and the value of vaccination, antibiotics, and trypanocides is widely appreciated even though, for various reasons, they are not widely used (Duncan 1978).

With the possible exception of the aged, disabled, and expectant mothers, the whole family usually travels to the main cattle or tobacco camp, for periods of one to three months once the harvest is completed. An important activity during this time is catching fish for drying and later consumption or sale. The women and girls also make clarified butter for use in the coming rainy season, while the young men often go hunting—much of the meat being dried and kept for later use.

Land

Around Rumbek there are localized areas of considerable land pressure. The area immediately to the east of Rumbek is one such area and here the land is intensively farmed and of very low fertility. More land is potentially available to the south, but movement is inhibited by the presence of tsetse fly and lack of water. Where such movement has occurred (for example, from Pacong towards Mattiang) the dissociation between cattle and cropping systems tends to be greater.

The average size of farm cultivated has been variously estimated at 4.5 feddans (Duncan 1978) and 3.45 feddans (Booker 1987*c*). Land is commonly held in one block around the homestead, although declining fertility may lead to some households cultivating more fertile areas some miles away.

The main rotational practice appears to be the opening up of *harig* fields for late planting with green gram and associated crops. Fallow periods vary widely from one to 20 years, but there is a high incidence of 1-year fallows caused by labour shortages, and it is likely that planned fallows necessitated by loss of soil fertility are in excess of 5 years (Booker 1984*c*). Plots are generally cultivated for 6–7 years giving an *R* factor of around 50.

Labour

Division of labour by sex and age is similar to that found in other areas. The family supplies most of the labour used in the fields. Non-family labour paid with cash is rare, but communal working parties are again very important relieving major labour peaks at weeding and harvesting and also assisting in bush clearing and *tukl* construction.

The Dinka have two types of working party: *kaya*—a reciprocal work party where food or beer is not strictly necessary and; *kutkut* a work party where labour is hired for food and beer. Sorghum is the main food provided either as *assida* (porridge) or brewed as beer. Meat, mainly goat, is also given. This form of work party is associated with more prosperous households who have food left over from the previous season. Households with an inadequate grain supply join a work party to eat well. Thus the relative wealth of different households may be reinforced over time. This underlines the importance of the seasonal labour constraint on area cultivated and hence food production in the area.

Livestock are usually cared for by young male family members although girls may also be involved. Although some conflict is implied between crop husbandry and livestock, particularly during the main sorghum harvest when the herds are moved to the *toich*, this is largely mitigated by a network of relationships between several families which permits frequent sharing of herding duties.

Ox ploughs

In contrast to the flood plain area, ox ploughs are widely used on the Colluvial Plain. A small scale training project has been run in Rumbek since 1975 by the Ministry of Agriculture in collaboration with the Sudan Council of Churches (SCC), and an ox ploughing service was available in several major villages in the early 1980s. The uptake has been minimal, mainly because of the lack of management skills, shortage of water and feed and the tsetse problem. Despite this, the progress made was encouraging given the expected resistance by Dinka farmers to the use of oxen for draught power. One of the most promising areas for development would seem to be the establishment of contractors in each major village area (beginning with those villages with best access to year-round grazing and water) who would hire out handlers to interested farmers.

The Flood Region

To the north-east of the Ironstone Plateau, the underlying Basement Complex is completely masked by deep alluvium. The Sudd swamps form the centre of a large structural basin, a vast imperceptibly sloping region whose landscape is broken only by low, sandy ridges—the courses of numerous former rivers.

Rainfall and the overflowing of the Nile during the wet season give rise to various degrees of flooding with major implications for land use. Apart from the permanent water surfaces of the Nile and certain lakes, the Flood Region falls into four main areas (Barbour 1961).

1. Permanent swamps along the Nile which are perennially flooded.
2. The flood plain, or *toich*, inundated from the rivers for 4–6 months each year.
3. Intermediate land, largely flooded by poor drainage of rainfall.
4. Higher areas, within or beyond the flood plain which escape river flooding and have permeable soils, thus avoiding flooding by rainfall.

Susceptibility to flooding, as implied in areas (3) and (4), is not necessarily related to relative altitude and the terms intermediate and higher are hence somewhat misleading.

The main areas of cultivation are found on the ridges which are not subject to flooding. However, some tribes will also plant on the higher intermediate land as the floods recede, increasing the potential cropping area and providing some insurance against particularly poor rainfall. In the dry season the intermediate land and the receding margins of the *toich* are grazed by cattle herds and small stock.

Climatically, the Flood Region falls largely into the dry monsoon zone, with 3–4 humid months, 4–7 dry months, and an annual rainfall of 750–1000 mm.

The rains usually begin in late May and end in early October. However, annual rainfall varies considerably throughout the region and, 'as often as not, although the rainfall may be up to average, incidence of fall is spasmodic, distribution poor, and in consequence crops fail badly' (Tothill 1948, p. 185).

Soils are highly variable within the region both in terms of texture and fertility, but clays predominate and these can be very difficult to work, tending to contract and crack when dry and to swell and become impermeable when wet.

In the dry season even the largest river channels dry up or form a discontinuous series of water holes. Owing to the impermeable nature of the soils, this stagnant surface water is the only practical supply for the majority of the area's inhabitants.

As implied earlier, the population of the Flood Region is concentrated in those areas of land which remain free from flooding in the wet season. Starting from the north and east, Barbour (1961) lists these areas of concentration as:

1. The Shilluk and Dinka ridges which are found on either side of Lake No.
2. The eastern Nuer areas beside the Sobat near Nasir.
3. The island of high land near Waat.
4. The ridge running north from Bor almost as far as Fangak.
5. The ridge within Zeraf Island which lies close to the Zeraf Channel.
6. The highland of western Nuer District which lies between Bentiu and Adok.
7. The ridge on the north side of the Bahr el Ghazal between Ghabat el Arab (Wang Kai) and Riangnou.
8. The intermittent belt of high land stretching from the Aweil District of Bahr el Ghazal south-eastwards as far as Yirol through the swamps at the foot of the Ironstone Plateau.

The last area refers to the Colluvial Plain described in the preceding section. The relative importance of the primary occupations, cattle rearing, crop husbandry, and fishing vary according to the areas of grazing and cultivable land available. However, among the Nuer, Dinka, Shilluk, and Anuak, the types of crops grown, of cattle kept, and prevailing attitudes towards them are fairly consistent (Barbour 1961, p. 238). The following discussion refers mainly to the Twic Dinka living in the Jonglei area.

The unreliability of physical conditions means that the flood region is naturally more suited to pastoral production systems. Under present technological conditions, 'the steady yield of milk and the occasional surplus of meat that pastoral production ensures, will always provide a more reliable means of survival than crop husbandry' (MFEP 1983).

It is certainly clear that in terms of cash income, livestock are much more important to the Nilotics of the Flood Region than are crops. They are also a vital famine reserve and play an extremely important role in their social systems. However, at least two writers (Barbour 1961; Duncan 1978) have concluded that the greater part of these peoples' diet derives from crops grown and not their livestock.

Cropping calendar

In the Flood Region, the number of crops and of crop varieties is much more restricted than in other areas. Sorghum and maize are the crops grown most extensively, the two main varieties of maize having been introduced since the Addis Ababa Accord (El Sammani 1984). Lesser crops include sesame, groundnuts, beans (*Phaseolus* spp.), and tobacco. Both the maize and the sorghum varieties grown widely are resistant to water-logging.

In May, sowing begins with tobacco and maize and takes place around the *luak* where manure and ashes are used to increase fertility. Sorghum is sown in the main fields by the men and does not usually benefit from manuring. Resowing, often necessary where the young seedlings have been washed away, and weeding, begin soon after sowing. Weed infestation is very heavy and often two or more weedings are necessary in each cropping period (El Sammani 1984).

Harvesting of first season crops begins with maize in early August, followed by sorghum from mid-August to the end of September. Extended resowings give rise to a prolonged harvest period, most of the ripening crop being consumed as it becomes available because the previous year's harvest is usually used up.

In October, the land is cleared and resown with a second crop of sorghum. Resowing is limited this time, but weeding is crucial because of limited soil moisture. Harvesting is again extended but finishes in January. The success of this second crop depends heavily on adequate rainfall late in the season and the water retained in the soil. Yields are higher in the early-season crop. El Sammani (1984) gives sorghum yields of about 230 kg feddan^{-1} for the first crop compared with 150 kg feddan^{-1} for the second.

Ratoon crops of sorghum can be taken in a good rainy season (Tothill 1948) and, after harvest, the stalks are usually grazed by the few animals remaining at the homestead.

Most of the implements used are made locally apart from the hoe heads, which are often made by the Jur. The implements tend to be very inefficient, the hoes and digging sticks, for example, barely scratch the soil surface.

While flooding and drought are the main hazards to cropping in the area, pests and diseases also have a marked depressing effect on yields. The most prevalent in the Twic Dinka area are birds, grasshoppers, worms (possibly army worms), stem borers, aphids (*Aphis sorghi*), and sorghum head smut (El Sammani 1984).

Infestations of weeds, pests, and diseases are made worse because no form of rotation is practised, the system being virtually one of monocropping. Tothill (1948) describes the prevailing land use intensity as that of shifting cultivation with four to six years of cropping followed by twelve or more years of fallow. The Twic area, however, would appear to be under virtually permanent cultivation (El Sammani 1984). In the latter area, farm size was found to average 1.43 feddans, of which almost all was devoted to sorghum. Variations are thought to be due to the area of well drained land available, the availability of family labour, the degree of weed infestation, and the motivation of a farmer to cultivate.

Tothill (1948), describes the cattle-owning Nilotics as improvident and poor cultivators, with little interest in new crops such as cotton, sweet potatoes, and cassava, but their attitudes are hardly surprising given the problems of soils and climate that they face. Storage of any surplus is also difficult because of the high humidities prevailing throughout the year. Whatever the reasons, famines are common and large quantities of grain must be imported annually from central Sudan. These are paid for by the sale of cattle.

Livestock

The Nilotics' way of life is very much geared to the needs of their animals. Cattle are kept mainly as a source of food, primarily milk but also for meat. Their conspicuous use in social obligations etc. are thus of secondary importance. The importance of cattle in traditional ceremonies is discussed by Lienhardt (1961).

While crops are grown in the main areas of settlement, livestock follow a seasonal pattern of transhumance. In the dry season, they follow the receding waters of the Nile flood plain on to the traditional grazing areas of the *toich*. With the arrival of the rains, they retreat to the dryland areas which are free from flooding. A few animals, mainly cows with calves at foot, are kept in the village area throughout the dry season to provide milk for those people remaining.

The dryland habitats, woodland, grassland, and agricultural land, support between 89 and 95 per cent of the cattle in the Jonglei Canal area during the wet season while the wetland habitats, the *toich* and the swamp which is less important, support around 60 per cent in the dry season (MFEP 1983).

Only one bull is kept from any one dam and the low ratio of bulls, plus the practice of tethering overnight may lower herd fertility. There are two mating seasons. The most important is in November/December when the cattle go down to the *toich*. The other is at the beginning of the rains, from May to July, when fresh grass has appeared. The main calving period is unclear but falls mainly between July and January. Cows are milked twice a day at about 12 hour intervals by non-menstruating women, girls, or uninitiated boys.

A herd refers to a group of animals under the care and management of one man, although they may not necessarily belong to him or spend a full year in his care. Herders are frequently the oldest, unmarried sons. A network of relationships between several families permits the sharing of herding duties and reduces conflicting demands on labour, particularly during the growing season.

In a sample of 48 *gols* (family cattle hearths), it was found that 48.3 per cent had herd sizes between 41 and 50 (MFEP 1983). Because of the practice of sharing herding duties, cattle numbers per household are somewhat lower. Dividing the number of cattle recorded by the number of households, Duncan (1978) derived a figure of 9.5 head in Lakes Province in 1976. His herd size figure of 44 was similar to that obtained by the Mefit-Babtie team (MFEP 1983).

Cattle losses and offtake Cattle diseases are of major concern to owners and the importance of vaccination is widely appreciated if not so widely taken up. Cattle are isolated where possible when they fall sick. Diseases which seriously affect adult performance include trypanosome infection, respiratory problems, and bovine abortion, while fascioliasis, contagious bovine pleuropneumonia (CBP), and cysticercosis are important causes of death. There was a severe epidemic of rinderpest in the area in 1980–1.

In the group studied by Mefit-Babtie, 29 per cent of all herd losses were due to deaths. Death, marriage celebrations, and sales, in that order of magnitude accounted for 73.1 per cent of all losses. More than half of the deaths occurred between January and April, when grazing is poor and many beasts are lost after becoming stuck in the mud at the water's edge. These figures, along with figures gathered for weight gain, suggest that at least among the Twic Dinka cattle, the most limiting period on production is the dry season and not the wet season as was previously thought.

The percentages of male and female cattle in the *gol* sample referred to above, were found to be 21 per cent and 79 per cent, respectively. Similar proportions were found in all the areas studied by Mefit-Babtie. The very low proportion of males in the herd reflects excessive offtake of male stock in order to satisfy cultural, nutritional, and financial needs. Cattle sales typically peak between April and June. By this time the previous year's harvest is all but used up and so cattle must be sold to buy imported grain. More cattle have to be sold at this time to buy a given amount of grain because grain prices are at their highest while cattle prices are depressed by the large numbers being sold. Emigration, particularly that of young adult males, has recently exacerbated the longstanding short-fall in local harvests which necessitates excessive sales of cattle. The over-exploitation of castrates over 2 years of age has led to the sale of younger males and of females from the herd which, over time, is not sustainable.

Several conditions might improve sustainable offtake. These include:

1. Reduced mortality (particularly in the first two years of life) due to improved veterinary services.
2. Improved crop yields or improved availability of imported grain, thus reducing the need for high seasonal meat consumption and releasing male cattle of all ages for market.
3. Improved availability of consumer goods, which would stimulate, the desire to enter the market economy.
4. Improved terms of trade in relation to food prices, making the exchange of livestock for grain more attractive.
5. Improved communications which would serve to stimulate all of the preceding conditions (MFEP 1983).

Conclusions

As indicated at the beginning of this chapter, the inhabitants of southern Sudan face a daunting combination of physical and man-made constraints. In response to these, they have developed intricate systems of crop and livestock production which attempt to provide a reliable supply of food throughout the year and from one year to the next. In the Ironstone Plateau and Green Belt where cropping systems predominate, the smallholder minimizes the risk of total disaster by successive sowings of various sorghum cultivars, other grain crops, and cassava. In the Flood Plain, where natural conditions make cultivation even more precarious, livestock systems predominate, with complex systems of transhumance to ensure the best use of grazing, and of cultural interdependence to prevent individual and group disaster.

The level of technology required in all the systems discussed is not high, but is well understood. Until the infrastructure, institutional framework, and supporting agricultural services are improved, attempts to change existing strategies would be premature and would be wisely resisted by the majority of farmers.

References

Barbour, K. M. (1961). *The Republic of the Sudan. A regional geography*. University of London Press, London.

Booker Agriculture International Ltd. (1983). *Yei agricultural development district. Main report*. Government of the Sudan Regional Ministry of Agriculture and Natural Resources (RMANR), Juba.

Booker Agriculture International Ltd. (1984*a*). *Wau agricultural development district. Main report.* Government of the Sudan RMANR, Juba.
Booker Agriculture International Ltd. (1984*b*). *Wau agricultural development district. Smallholder survey.* Government of the Sudan, RMANR, Juba.
Booker Agriculture International Ltd. (1984*c*). *Results of the Rumbek smallholder survey.* Government of the Sudan RMANR, Juba.
Catford, J. R. (1951). Katiri cultivation of the Moru district of Equatoria. *Sudan Notes and Records*, **32**, 106–12.
de Schlippe, P. (1956). *Shifting cultivation in Africa. The Zande system of agriculture.* Routledge & Kegan Paul, London.
Duncan, A. (1978). *Results of a farm management survey carried out in Rumbek district, Lakes Province.* Government of the Sudan RMANR Planning Department, Juba.
El Sammani, M. O. (1984). *Jonglei canal: dynamics of planned change in the Twic area.* Graduate College Publications, Monograph 8, Graduate College, University of Khartoum.
Eyre, V. E. F., Ramsey, D. M., and Jewitt, T. N. (1953). *Agriculture, forests and soils of the Jur ironstone country of the Bahr el Ghazal Province, Sudan*, Ministry of Agriculture Bulletin, No. 9. Agricultural Publications Committee, Khartoum.
MFEP (Ministry of Finance and Economic Planning) (1983). *Development studies in the Jonglei Canal area: final report. Volume 10. Range ecology survey, livestock investigations and water supply.* Mefit-Babtie Srl., Glasgow.
Leinhardt, G. (1961). *Divinity and experience.* Oxford University Press.
Ruthenberg, H. (1980). *Farming systems in the tropics.* Oxford University Press.
SDIT (Southern Development Investigation Team). (1955). *Natural resources and development potential in the southern provinces of the Sudan. A preliminary report by the Southern Development Investigation Team, 1954, and Volume II Maps and diagrams.* Sudan Government, Khartoum.
Tothill, J. D. (ed.) (1948). *Agriculture in the Sudan.* Oxford University Press, London.
Twose, N. and Pogrund, B. (ed.) (1988). *War wounds: Sudanese people report on their war; development costs of conflict in southern Sudan.* Panos Publications Ltd, London.

Further reading

Howell, P., Lock, M., and Cobb, S. (ed.) (1988). *The Jonglei Canal: impact and opportunity.* Cambridge University Press, Cambridge.

18

Development programmes in non-irrigated rainland areas

H. R. J. DAVIES

Introduction

In the Sudan 80 per cent of the population depend on agriculture and related activities for their subsistence. Agriculture contributes some 40 per cent of GDP and accounts for some 90 per cent of the country's earnings from foreign export trade. The first comprehensive attempt at national planning, the *Ten Year Plan of Economics and Social Development, 1961/62–1970/71* (Ministry of Finance and Economics 1961) states 'As agriculture is the mainstay of the country's economy, considerable emphasis has been placed on its development'. Direct public sector investment in agriculture was to account for 41 per cent of the total, whilst investment in public health, roads, railways, and education would also have an influence on the agricultural sector. The same view is stated constantly in the planning documents. In the *Six Year Plan of Economic and Social Development 1977/78–1982/83* (MNP 1977) agriculture was allocated 32 per cent of total public sector investment (£S425m.). The private sector was expected to invest a further £S290m. giving an overall total of £S715m., but this target was beyond the indigenous sources of funding.

Changes in the significance of the various classes of cropland between 1960 and 1988 are given in Fig. 18.1. For 1987/8, total cropland was estimated at 16.1 million feddans, 87 per cent of which is rainland. In 1985, the year with the largest area reported under cultivation, only 23 million feddans was used, amounting to 3.5 per cent of the total land area of the Sudan. Even allowing for considerable under-reporting for southern Sudan and the desert areas in the north this, nevertheless, indicates considerable potential for further expansion.

Concern for the agricultural sector in the 1960s stemmed from:

1. Fears, expressed since the 1930s, about the Sudan's ability to feed itself.
2. The apparent lack of minerals of any significant commercial value.

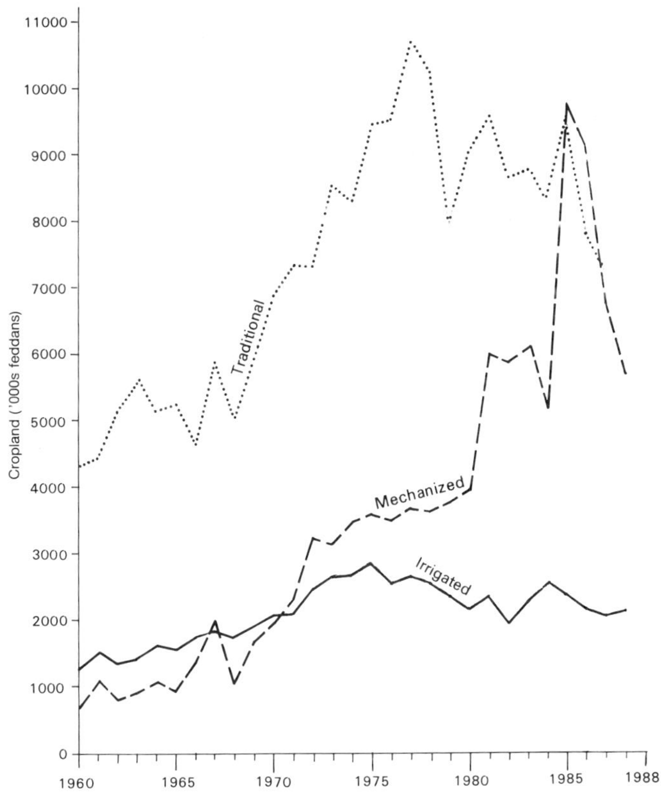

FIG. 18.1. Sudan: cropland, 1960–88.

(Today oil has been discovered but due to a variety of internal and external factors has not yet been brought into production.)

3. The need for a sound agricultural base for local industrial development.
4. Little prospect for alternative development through tourism or other tertiary activities.
5. Sudan's small but rapidly growing population (2.8 per cent p.a.: a doubling in 25 years). By 1983 the population had reached 20 million, double that at Independence.

These factors have remained relevant, but have been reinforced by further events which have made development of the rural sector even more urgent than before.

6. The drought years, 1968–1973 and the early 1980s, culminated in the famine of 1984/5 when the rains failed over large areas of the savanna zone. For many rainfall stations in the Sudan, 1984 was the driest year

on record. Government figures indicated that up to 8 million people (40 per cent of Sudan's population) were adversely affected by crop failures in 1984/5 and at that time Sudan was coping with up to 2 million refugees.

7. In 1955, 854 000 people (8.3 per cent of the population) were 'urban', but by 1983 the figure had increased to 4 154 000 (20.2 per cent of the population). This means an increase in demand on existing agricultural systems not only by numbers in the countryside but also by increased numbers of non-food producers living in the towns.
8. The rapidly increasing population was more than matched by increases in the number of domestic animals particularly along the southern edges of the Sahara. Here the human population increase was above the national average and in parts of Northern Kordofan the numbers of domestic animals may have increased ten-fold since the 1940s. Increased population and livestock have led to overgrazing and overcultivation and the resultant signs of desertification. Figure 18.2 shows that the risk of desertification is prominent throughout the whole of the sub-Saharan zone of the Sudan.
9. Attempts to develop manufacturing industry, which employs less than 5 per cent of the labour force, have not been very successful. For this reason, the burden on the agricultural sector to produce higher export earnings to pay for development has increased. The increasing potential market for food exports, especially meat, to Saudi Arabia and other Arab oil-rich states with rising living standards, has opened up new markets; the same states have also provided a potential source of capital for agricultural development.
10. The political disturbances in southern Sudan, up until 1972 and again in the 1980s, have excluded it from agricultural development although it is the area with the greatest potential in terms of rainfall reliability. Programmes for the south have been unfulfilled and most development has had to be confined to the drier north.

In some parts of the northern half of the country large-scale irrigation may help to solve some of these problems (see Chapter 19), but large areas even of this part of the country are without easy access to water for irrigation. Policy for such lands has been concerned with the mechanization of agriculture; modernization of traditional agriculture; provision of rural water supplies; and reorganization of the livestock industry with nomad settlement.

Mechanization of agriculture

This is something of a misnomer for, in practice, the only processes mechanized are ploughing and sowing; weeding and harvesting remain essentially

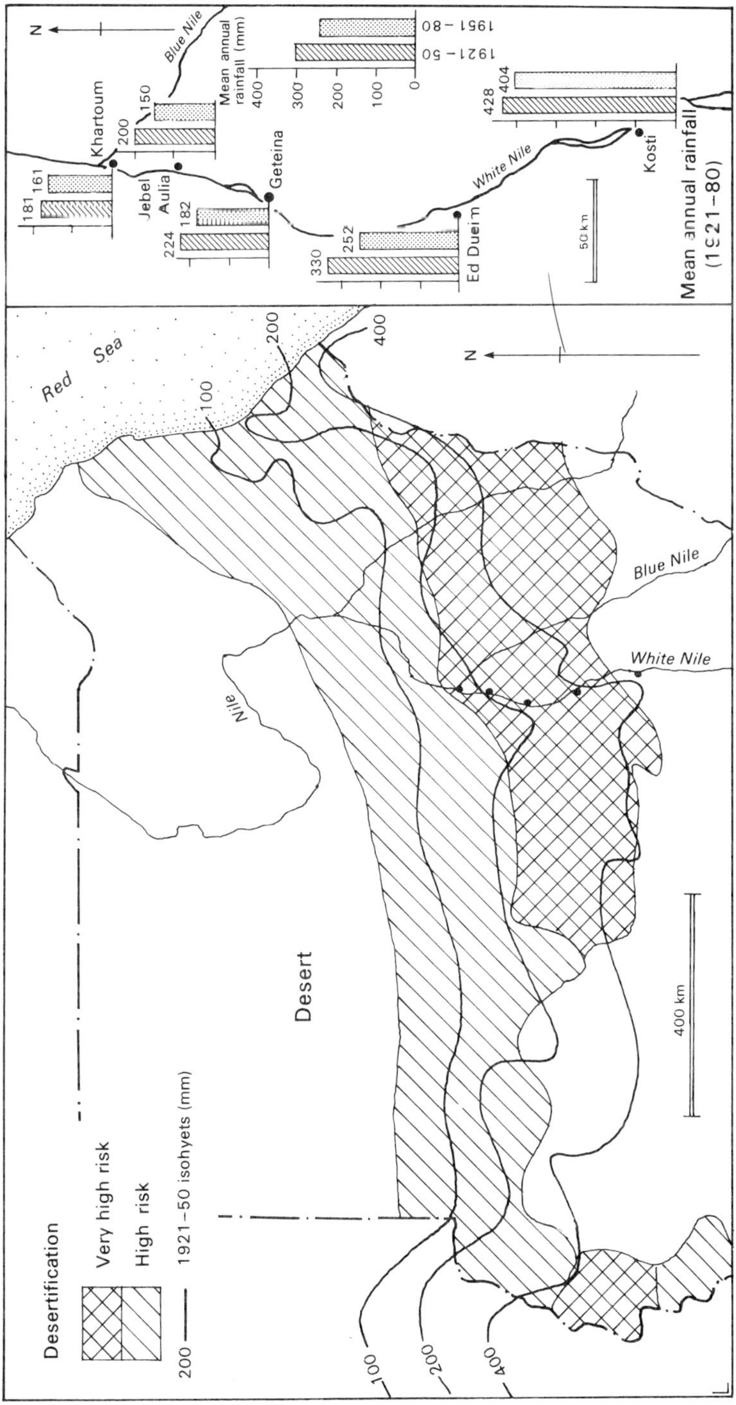

FIG. 18.2. Sudan: desertification.

hand operations. Combinable varieties of sorghum exist, but are not considered very palatable by Sudanese. Machine harvesting of sesame and cotton has not proved successful under Sudan rainland conditions.

The primary objective of mechanization was to expand sorghum production initially to make the Sudan self-sufficient but later to produce an exportable surplus to cover the costs of wheat imports. Wheat bread has become increasingly popular with urban dwellers and the better off social classes.

Mechanization was seen more as a means of increasing output per head, by using a smaller work force to cultivate a larger area, rather than as a way of increasing yields per feddan. It was also seen as a way of developing land suitable for agriculture, but left unused through lack of drinking water.

Mechanized cultivation began in 1945 on an area of 12 000 feddans on the clay plains at Ghadambaliya near Gedaref (Fig. 18.3). This site provided suitable land and was accessible to the Port Sudan–Sennar railway, which allowed the assembling of inputs and the marketing of the crop. In spite of fluctuations in fortune in the early years, success seemed assured by 1955. By that time state farms had proved unsuccessful and areas with better rainfall, to the south rather than the north of the 600 mm average annual isohyet (1921–50), were favoured for development (Davies 1964, p. 102).

In 1957 development between the Niles started at Dali and Mazmoum. By 1960 more than one million feddans were demarcated (870 000 feddans were already under cultivation) mainly in private enterprise farms of 1000–1500 feddans, under government supervision. Expansion of the system to new areas in western Sudan, an area suffering constant food supply crises began in 1968 in the Habila area of Southern Kordofan.

By 1981 the mechanized area topped 6 million feddans. The 1984/5 famine encouraged further expansion to over 9 million feddans, but due to poor results the figure for 1988 had fallen back to 5.7 million feddans (Fig. 18.1).

Mechanized rainland cultivation has become so popular among private entrepreneurs that it has been impossible for the government to ensure that it is kept under proper control. In 1985 it was estimated that 45 per cent of the area under mechanization was illegal in that it was outside the demarcated areas and, therefore, outside any semblance of government control (Earl 1985). In 1984 in the Gedaref area it was estimated that of 2 947 000 feddans under mechanized sorghum only 1 311 000 feddans (44 per cent) were in demarcated areas. Compared with 1983, this represented an increase of 41 per cent in the area of mechanized sorghum, of which 83 per cent was in undemarcated areas.

Originally, it was intended that the mechanization of agriculture would benefit cash crops, especially sesame and cotton, as well as food crops. In 1968, 34 650 feddans of American rainland cotton (2 per cent of the

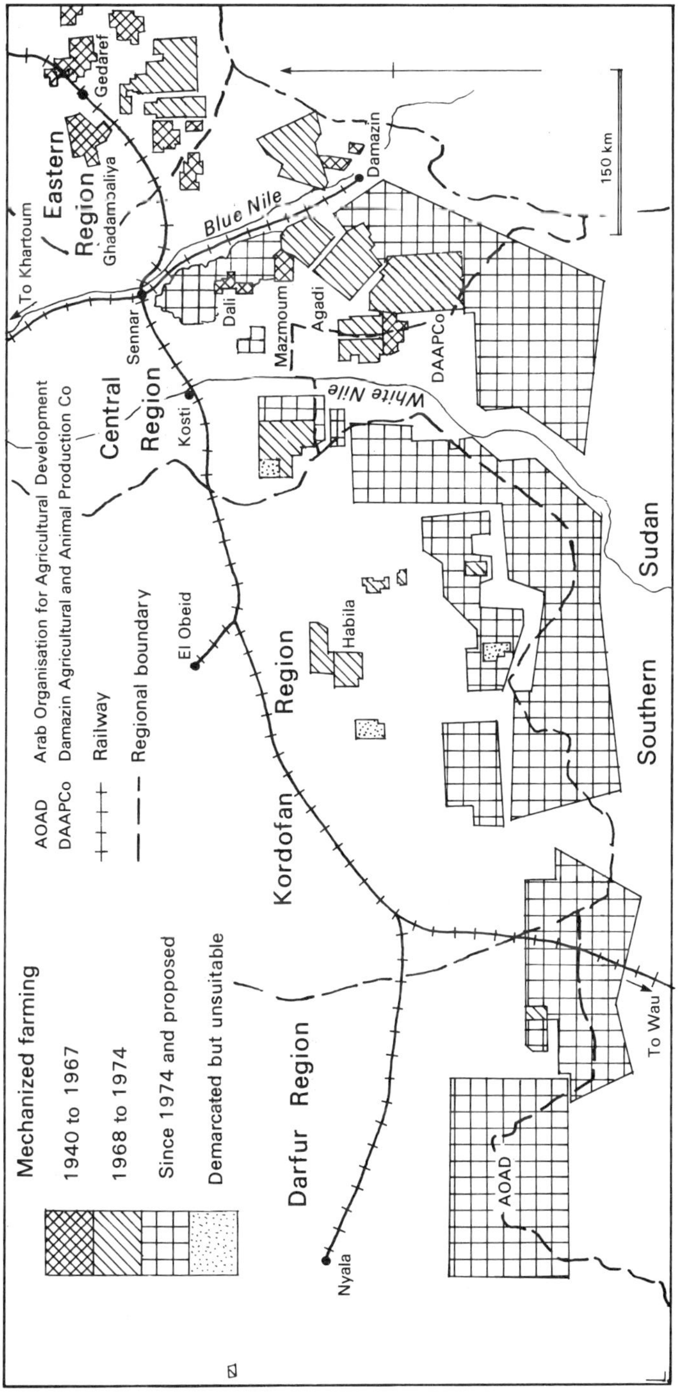

FIG. 18.3. Central Sudan: development of mechanized agriculture.

TABLE 18.1. Crops grown on mechanized farms

	1968/9 '000 feddans (%)	1987/8 '000 feddans (%)	1988/9 '000 feddans (%)
Sorghum	922 (72)	5615 (84)	4792 (84)
Bulrush millet	—	36 (<1)	19 (<1)
Sesame	328 (26)	1032 (15)	542 (10)
American cotton	35 (2)	—	—
Sunflower	—	47 (<1)	332 (6)

Source: Department of Agricultural Economics, MANR (various years).

total mechanized area) were cultivated. Of the remaining area, sorghum accounted for 72 per cent and sesame for 26 per cent. Now rainland cotton is an unrewarding enterprise and is not grown. Figures for 1987/8 and 1988/9 are given in Table 18.1 and indicate the appearance of a new oilseed, sunflower, among cash crops. Bulrush millet is the most important cereal staple after sorghum, especially on lighter soils with very variable rainfall in western Sudan. Before 1983, less than 5000 feddans of bulrush millet had been mechanized, but since the 1984 drought its popularity with mechanized farmers has increased.

It is questionable whether the effort put into sorghum production has proved successful. Since Independence until the 1984/5 famine, Sudan had a surplus of sorghum which contributed to the country's export trade. In 1982 it accounted for 22 per cent of exports by value, second only to cotton. However, the demand for sorghum in the Sudan has declined due to the increased popularity of wheat bread. The value of cereal exports (almost entirely sorghum) exceeded the value of cereal imports (mainly wheat) in four years in the 1960s, in 6 years in the 1970s and in the 1980s until 1984. In recent years the position has been made unclear by Food Aid. In 1987/8, the estimated total sorghum production for the Sudan was 1 379 000 tonnes, of which 62 per cent came from mechanized farms. This demonstrates clearly the current significance of mechanized sorghum production.

Although the primary aim of mechanization was to open up new lands for cultivation, it was also predicted that with good management yields should be improved. In rainland areas mechanized sorghum production gives much higher yields than traditional farming for the Sudan as a whole, but the figures for Southern Darfur and Southern Kordofan are much lower (Table 18.2). Government estimates show that yields of all food crops fell during the 1960s and early 1970s (SDEC 1976). There is also evidence of declining yields under mechanization in the Gedaref area over the same period (Shakkak 1977). More recent surveys have provided similar but less positive conclusions about mechanized agriculture (Affan

TABLE 18.2. Estimated sorghum yields: rainlands, 1986/7 (kg feddan^{-1})

	Mechanized	Traditional
Sudan rainland	307	171
Southern Kordofan	150	235
Southern Darfur	200	90
Southern Region	180	180

(Sudan irrigated 550 kg feddan^{-1} (1986/7); world sorghum 586 kg feddan^{-1} (1982)).
Source: Department of Agricultural Economics, MANR (various years).

1984, pp. 42–70). In practice, meaningful comparisons are difficult because traditional cultivation penetrates much nearer the desert margins. Construction of bunds across gentle slopes to concentrate sheet flow, especially on the clays, allows satisfactory cultivation by traditional means in areas unacceptable to the mechanized farmer. Furthermore, sorghum (and bulrush millet) have always been planted speculatively near the desert margin in the hope of good rains. Such factors tend to reduce overall yields for the traditional sector. By contrast, mechanized agriculture has opened up new cropland in wetter areas with more reliable rainfall. These areas are much more difficult for the traditional farmer to clear for cultivation and they have been avoided by northern Sudanese. Part of the complex relationship between rainfall and crop yield is illustrated for bulrush millet and sorghum grown in Northern Darfur and Northern Kordofan (Fig. 18.4). With falling yields per feddan and a population increasing at some 2.8 per cent per annum, larger areas have had to be cultivated in order to feed the population without recourse to even higher food imports.

Mechanized agriculture has been criticized severely in that it has led to much needless destruction of the vegetation and to a new form of bush fallowing, or land rotation, on a massive and disorganized scale, with land cultivated to exhaustion and then abandoned. After the first flush of new land fertility has been dissipated, yields per feddan have tended to decline gradually (Affan 1984, p. 46). A limited number of absentee lessees have dominated this rainland farming sector. Their interests have been in quick profits, not in the preservation of land productivity, sound ecological practice, or equitable development. By 1985, some 40 per cent of the cropped rainland had fallen into the hands of some 4000 investors (Abdel Ati 1986, p. 3).

Neither the Condominium Government nor the first Independence administrations in the Sudan were entirely happy with the development of large private enterprise rainland farms. When early attempts at state farming failed, a series of measures intended to enable peasants to become part of the mechanized agriculture movement followed, but these also failed. In 1968 the Government intervened directly again with the establishment of

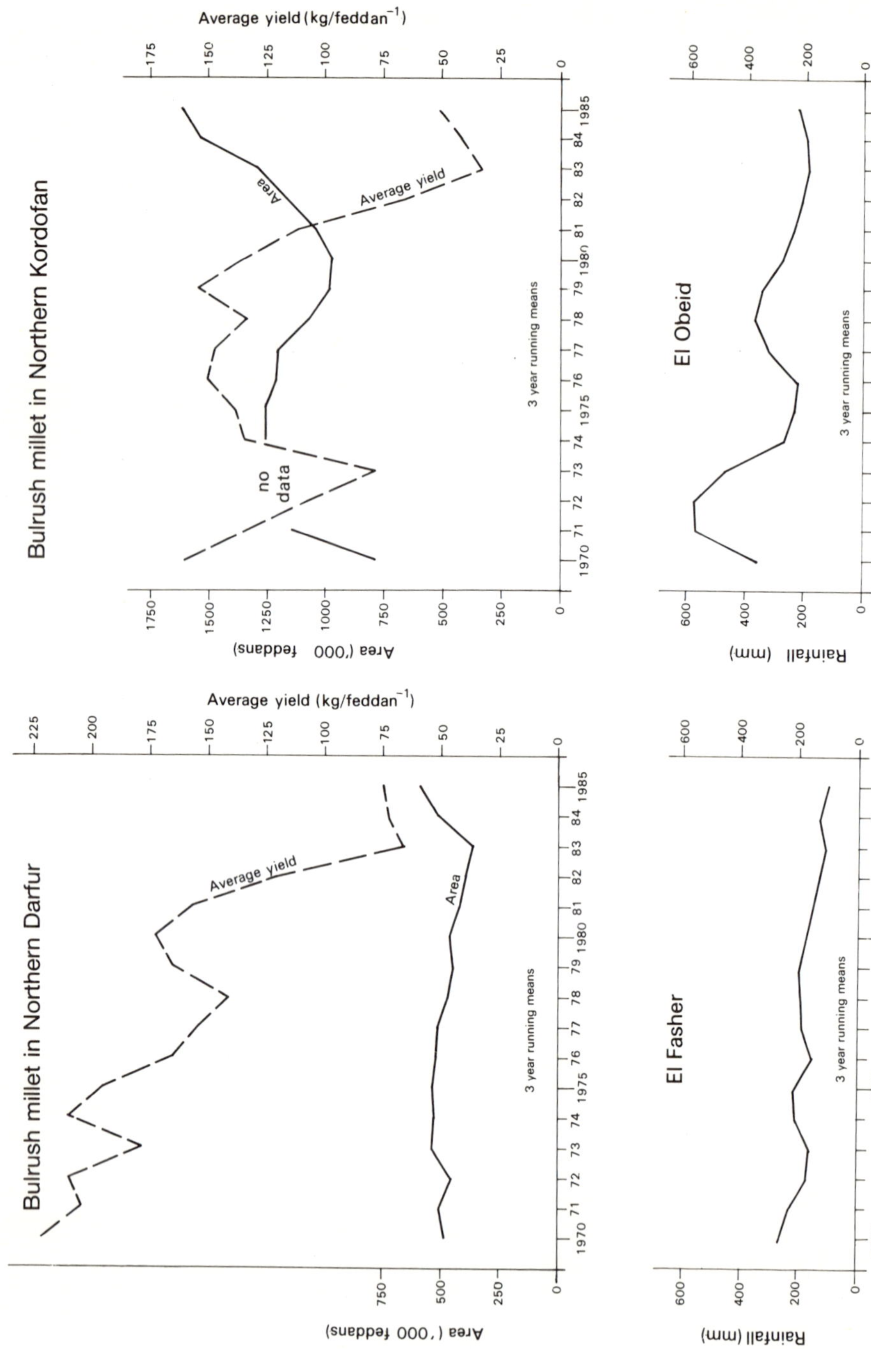
Bulrush millet in Northern Kordofan
Average yield (kg/feddan^{-1})
Area ('000 feddans)
Area
Average yield
no data
3 year running means
El Obeid
Rainfall (mm)
3 year running means
Bulrush millet in Northern Darfur
Average yield (kg/feddan^{-1})
Area ('000 feddans)
Average yield
Area
3 year running means
El Fasher
Rainfall (mm)
3 year running means

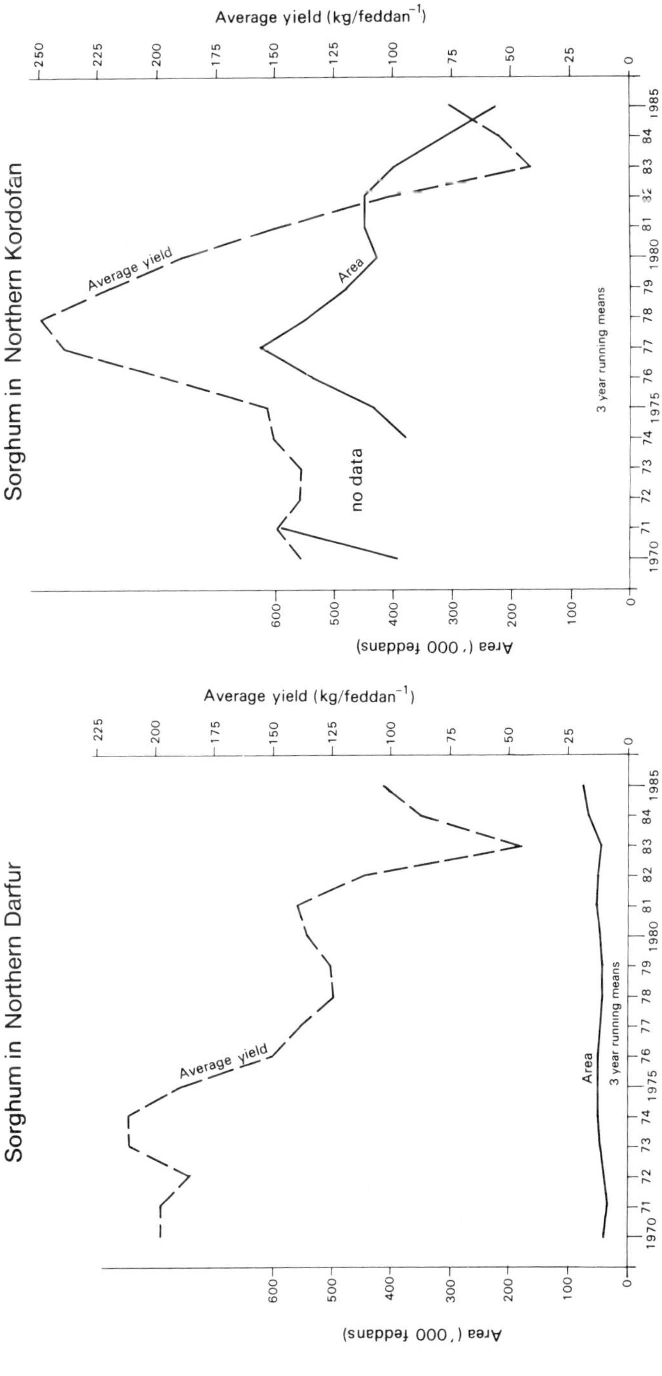

FIG. 18.4. The relationship between crop production and rainfall in Northern Darfur and Northern Kordofan.

the Mechanised Farming Corporation. By 1974 six important state farms were operating and these varied in size from 10 000 feddans for the smallest to 210 000 feddans for the Agadi state farm, financed with a loan from the Kuwait Fund for Arab Social and Economic Development. This demonstrates an increased awareness of the Sudan's development potential by oil-rich Arab States. Like the other state farms Agadi failed to be economically viable (Thimm 1979, p. 50). It was intended that sorghum production from state farms would build up a famine reserve. The poor results led to the 1981 decision to phase out state farms, a policy reiterated in 1983 (Earl 1985). Now, state farms operate on a much reduced scale for demonstration purposes, as pilot centres for innovations, or to explore the possibilities of opening up new areas. As a result private enterprise has had the field almost to itself and large-scale farming has proceeded virtually out of control since 1974.

From 1960, holdings were rented out on a short lease basis of 10 milliemes per feddan per year (equivalent to one penny per feddan) or £S10 per 1000-feddan holding (Davies 1964, p. 105). Since then the precise charges have changed, but have remained nominal. The arrangement was intended to attract wealthier urban dwellers, merchants, retired government officials, and others, to invest in agriculture. It proved successful but these investors regarded farming as a subsidiary money-making enterprise and they employed managers and agents to operate their farm holdings for them.

With this lack of commitment to the local agricultural area, rotations have been ignored and the rules requiring at least 25 per cent of land to be left fallow at any one time have not been enforced. The rate of soil exhaustion has, therefore, increased considerably. The ephemeral interest of investors and their managers has prevented a coherent pattern of new villages in the new cropping areas because the team working a farm do not live there all the time. They operate as 'suitcase farmers' who arrive in time to start the season and depart after harvest. If results are poor they move on next year to a new area.

During the 1970s, in an attempt to improve the position, leaseholders were allowed a second farm on the understanding that it was to facilitate rotation. The emergence of large land-holding companies was also facilitated by the new investment law of 1982, which encouraged the introduction of foreign capital into the rainland sector by offering good financial terms and a 25 year lease. By 1985, in Blue Nile Province alone, more than 2.5 million feddans had been allocated to 21 land companies in blocks exceeding 10 000 feddans. The two largest companies were allotted one million and 500 000 feddans, respectively, although, by that date, only 480 000 feddans had been cleared. Abdel Ati (1986, p. 16) estimated that in 1984 only 11 per cent of the area allocated to large land companies in the Sudan as a whole was cleared and under cultivation. Such land companies are often financed by Saudi Arabian or Gulf private capital and are managed by a European consortium (Kursany 1982). Abdel Ati (1986) suggested that some of these companies have little intention of farming normally but have obtained large holdings in order to obtain tax and other concessions with which to finance other marketing and financial activities in the Sudan. This would account for some apparent dilatoriness in commencing clearancc.

Apart from the attraction of foreign capital and expertise, it was believed that new large land companies could be supervised more easily and their activities regularized. This is a prerequisite for the emergence of a rational land use pattern in the Sudan—one which would reconcile the requirements of large-scale and traditional crop production, forest resource maintenance, and livestock grazing requirements.

Investigations into the profitability of mechanized farming carried out in the late 1950s and early 1960s showed that in years with a reasonable rainfall a large profit could be made from a well-organized 1000-feddan farm (Osman 1959; Davies 1964, p. 106). However, there could be difficulties. The 1959/60 season produced a bumper sorghum crop, labour shortages became severe at harvest time, and the crop could not be marketed easily because of an inadequate transport system in the Gedaref area. In that year many farmers claimed that they lost money because the plentiful crop depressed prices severely and harvesting costs were particularly high.

With the financial backing to get started in mechanized agriculture increased through a World Bank loan and through government incentives, some recent studies suggest that it is not so much the return on the crop that makes mechanized farming activities so profitable, but the various subsidies and credits available, financed largely by foreign loans (Ibnouf 1985; Affan 1984, p. 27; Abdel Ati 1986, p. 15). Data for the 1984/5 season in Southern Kordofan, produced by the MANR, suggest that costs for mechanized sorghum production are somewhat lower than those for traditional rainland cultivation. For sesame, although mechanized costs per feddan were lower, costs per harvested tonne were higher at £S664 compared with £S541 for traditional methods.

It can be concluded that the expansion of mechanized agriculture in the Sudan has made a major contribution to her self-sufficiency in sorghum, except in years of severe drought; in many years the value of her sorghum exports may have more than paid for her cereal imports. Private capital has been utilized in the agricultural sector to a much greater degree than before. Mechanized farming has also contributed modestly towards cash crop production, but through sesame and very recently sunflower, rather than cotton. The price paid for this success has been the destruction of vast areas of savanna woodland and the exhaustion of soils over large areas. It was hoped that the expansion of agriculture into new areas would lead to a more widespread distribution of the rural population and the development of a new modern system of permanent agricultural settlement to replace the old system of shifting cultivation, but this has not been achieved.

Modernization of traditional agriculture

So much effort in terms of finance and manpower has been put into the development of mechanized agriculture that it is hardly surprising so little has been done to modernize the traditional sector.

The apparently poor performance of the traditional sector, measured in terms of overall yields per feddan and its smaller and decreasing share of sorghum cultivation both in terms of area and production give an unfair picture of its importance to the Sudan economy and people. It is estimated that 70 per cent of those engaged in agriculture in the country as a whole depend upon traditional agriculture. The share of traditional rainland crops in area and in production is presented in Table 18.3.

The tendency for traditional cultivators to cultivate in drier areas with more variable rainfall and the effect of this on yields can be seen in Fig. 18.4. This desertward movement was more feasible in the wetter 1950s and 1960s and it was also made necessary by the rising rural population which more than doubled in Northern Kordofan between 1955 and 1983, exceeding the growth rate of the population of the Sudan as a whole. Farmers responded to the increasing population pressure, decreasing rainfall, and lower yields by bringing more feddans under cultivation (Fig. 18.4). Water saving through bunding and speculative crop cultivation assisted in pushing the desertward limit of cultivation beyond the 250 mm line with the inevitable disastrous effect on desert encroachment (Ibrahim 1978, p. 10).

Pesticides and fertilizers are rarely used in the traditional sector and, with unimproved seeds, yields tend to be low. The overall strategy of most traditional farmers has been minimization of input rather than maximization of output (Briggs 1984). This may overstate the case, but it remains true that peasants with small resources have to limit their inputs to a level which gives a return that they consider reasonable or adequate.

Individual traditional farmers tend to grow a greater variety of crops whilst mechanized farmers concentrate on one or two. In 1985/6, in the central zone of the Sudan traditional rainland farmers produced 17 per cent of the sorghum, 97 per cent of the millet, 66 per cent of the groundnuts and 3 per cent (1976: 7 per cent) of the cotton. Furthermore, traditional rainland cultivation of groundnuts has been shown (World Bank 1982) to be more cost effective than other systems of cultivation, although the vagaries of climate and variations in world market prices have a marked effect year by year.

Little effort has been made by the Sudanese Government to develop extension services, agricultural credit facilities, and other supportive

TABLE 18.3. The contribution of traditional rainland agriculture to crop production in the Sudan

	Sorghum	Millet	Sesame	Groundnuts	Cotton
Traditional rainland (% of Sudan total)					
1968/9					
Area ('000 feddans)	1439 (51)	1427 (99)	992 (75)	626 (87)	224 (20)
Production ('000 t)	379 (44)	266 (99)	113 (73)	115 (70)	25 (4)
1985/6					
Area ('000 feddans)	3215 (25)	3980 (96)	1340 (53)	805 (85)	58 (7)
Production ('000 t)	536 (16)	388 (95)	76 (57)	180 (66)	10 (2)
1986/7					
Area ('000 feddans)	2793 (24)	3590 (99)	1316 (50)	949 (80)	37 (4)
Production ('000 t)	478 (14)	306 (97)	99 (38)	213 (53)	7 (1)
Total rainland					
1968/9					
Area ('000 feddans)	2361	1427	1320	626	259
Production ('000 t)	741	266	154	115	29
Yield (kg feddan^{-1})	314	186	117	184	112
1985/6	11752	4106	2518	805	60
Area ('000 feddans)	2864	403	134	180	13
Production ('000 t)	244	98	53	224	217
Yield (kg feddan^{-1})					
1986/7					
Area ('000 feddans)	10983	3611	2612	949	39
Production ('000 t)	2993	312	264	213	10
Yield (kg feddan^{-1})	272	86	101	224	256
Sudan totals (rainland %)					
1968/9					
Area ('000 feddans)	2823 (84)	1436 (99)	1320 (100)	719 (87)	1118 (23)
Production ('000 t)	870 (85)	267 (99)	154 (100)	164 (70)	651 (4)
1985/6					
Area ('000 feddans)	12875 (91)	4128 (99)	2518 (100)	951 (85)	807 (7)
Production ('000 t)	3522 (81)	410 (98)	134 (100)	274 (66)	419 (3)
1986/7					
Area ('000 feddans)	11805 (93)	3626 (99)	2612 (100)	1182 (80)	861 (5)
Production ('000 t)	3445 (87)	317 (98)	264 (100)	399 (53)	556 (2)

Source: Department of Agricultural Economics, MANR (various years).

activities in the traditional rainland sector. In particular, the development of bulrush millet, a most significant crop on poorer sandy soils under desert margin conditions, and the main cereal crop after sorghum in western Sudan, has been sadly neglected.

All the Provinces in the central part of the country except for Northern Darfur have some kind of Extension Service (Saeed *et al.* 1986), but most of the effort has been on the irrigated areas and on mechanized rainland agriculture. The main rainland research station (Kenana Research Station) and its sub-stations at Tozi, Agadi, and Simsim have also been concerned primarily with mechanized farming, and their work has been hindered by shortages of trained staff, lack of finance, long travelling distances, and lack of transport. Failure to provide appropriate advice and a lack of understanding between officials and cultivators have also been noted at times.

Unlike the mechanized farmers, traditional farmers in the west of Sudan have failed to penetrate southwards and utilize the better land there. This is because traditional methods of land clearance and of land cultivation are much less suitable for the denser bush found further south. These areas are also much more remote from the main marketing routes and the problems encountered with unmade roads during the rainy season become more serious as the rainy season lengthens. Until recently, the indigenous people of areas such as the Nuba Mountains have been hostile to encroachment from the north.

The traditional farmer cultivates the land near his house fairly intensively applying to it a limited amount of animal dung and domestic rubbish for fertilization. Formerly, further away he adopted a system of bush fallowing whereby the land was cultivated for 3 to 5 years then rested. Further away still, the land was burned periodically and then a crop is taken (*harig* system). Today, increased population and poorer rains which reduce yields have led to the replacement of the *harig* system by a more intensive cultivation, to a shortening of fallows on other land, and a tendency to bring more land under cultivation every year. The effects are wind-blown erosion of the cultivated land, less land for grazing, shortages of wood for fuel, house construction, and other purposes, and, ultimately, land degradation and desertification. This problem was evident more than forty years ago in parts of Blue Nile and Northern Kordofan Provinces (Sudan Government 1944). Now that the existing strategies to cope with the vagaries of the rainy seasons are no longer viable, the situation has reached crisis point in many areas.

The Sudanese Government's main response to this situation has been its various attempts to associate peasant cultivators with the development of mechanized agriculture. After three years of state farming a participating cultivator scheme involving local people was initiated in 1948 in the Gedaref area. The scheme was based on a profit sharing arrangement with Government, but it failed and was formally abandoned in 1953. Reasons for failure included the lack of tradition for co-operation between farmers, the lack of a local system of profit sharing which could be used as a basis for the scheme, the small profit margin offered compared with the Gezira Scheme, and the ease with which sorghum could be sold illegally on the local market. Despite its failure, the object of the enterprise was laudable. It was envisaged that farmers would get used to machine operations through the Scheme for a few years, by which time they would have built up sufficient capital to enable a group of farmers to set up together on a mechanized farm of their own (Davies 1964, p. 104).

In 1959 a further attempt was made to associate small cultivators with mechanization. The development of co-operatives was planned for the Dali-Mazmoum area between the White and Blue Niles, and by the early 1960s there were 70 co-operative farms, each of 952 feddans, in this area. However, shortcomings in management and a good deal of mistrust between managers and members led to problems and the farms disappeared quite quickly.

In 1967, the Nuba Mountains Cotton Corporation was reconstituted in order to develop co-operative farming. It was established originally in 1924 to facilitate cotton production by providing seed, ginning, and marketing services to local Nuba farmers in the area. Its new function was to organize and encourage the grouping of farmers into units large enough to use a tractor effectively. Two million feddans were eventually allocated to the new Nuba Mountains Agricultural Production Corporation and by 1979 there were 70 schemes organized on a crop sharing basis and covering 93 000 feddans (Babiker *et al.* 1985, p. 49). Cotton cultivation was a condition of tenancy and each farmer was allocated 15 feddans of which 5 feddans were for cotton, and the other 10 feddans were divided equally between sorghum and fallow. The sorghum crop was entirely for the tenants whilst the profits from cotton were to be shared with the Corporation. In general, farmers cleared the land and prepared it for cultivation while the Corporation provided cotton seed, machinery, and advice. In recent years, however, the scheme has been largely abandoned, leaving only 30 000 feddans cropped in 1982.

In 1968, a decision was made to develop mechanized agriculture in the Habila area of Southern Kordofan. By 1979 there were 350 000 feddans under 'official' mechanization schemes, with a whole variety of private enterprise, government, and co-operative farms (Babiker *et al.* 1985, p. 50). There is little evidence to suggest that these mechanized schemes have fared any better than those in the Mazmoum area (Affan 1984, p. 25).

It can be concluded that the various attempts to associate the local peasant farmer with mechanized agriculture have had little success. In the Nuba Mountains area management problems have been compounded by distrust between the local Nuba and northerners trying to introduce new ideas. Furthermore, bureaucracy has tried to introduce new ideas and techniques before finding out what the local people really want (Affan 1984, p. 97). The divergence of viewpoint about priorities for development is a major difficulty which has been apparent in other areas in the Sudan (Davies 1986*a*).

Rural water supplies

Most of the rural population in the rainland areas of central Sudan are of nomadic, or semi-nomadic origin even if they are settled today. This tradition is expressed in the combination of animal rearing and crop growing and many farmers see themselves as pastoralist turned cultivator. Typically, such people have grown sorghum as a catch crop even in very marginal situations. It is this traditional practice together with the rising population which has led to the greater movement towards desert areas and, in turn, to desert encroachment.

Fears of local overgrazing and overcultivation in western Sudan, especially in Northern Kordofan were expressed as early as the 1930s (Sudan Government 1944). It was recommended, therefore, that after the Second World War a programme of rural water supply improvement should be implemented. A connection between the shortage of drinking water on the rainlands and the over-use of land around existing supplies was assumed and in 1946 a Rural Water Supplies and Soil Conservation Board was established within the Ministry of Agriculture. For the first fifteen years of its existence, this organization was concerned with excavating *hafirs* (surface water reservoirs) using modern earth moving machinery on Sudan's heavy clay soils. It was assumed that if more rural water supply points were created, people could move to new areas, reducing the pressure of man and animals around existing water points. Such a programme would also enable people to remain near their cultivated land. Without water points, many areas could only be occupied during the rainy season, or immediately after it, whilst ephemeral rain pools lasted. As the season progressed, people could make only the occasional foray to their farms and, after harvest, such areas were completely abandoned (Davies 1962).

It was argued that the new water provision policy would enable those concerned mainly with cropping to settle down and would also be of value to nomadic pastoralists. Adequate rural water supplies and a settled population were seen as prerequisites for the Sudan's development and for the provision of services. Furthermore, an expansion of rural water supplies would be essential for mechanized agriculture to develop on the hitherto waterless clay plains around Gedaref. The Ministry of Agriculture and officials of the Rural Water Supply and Soil Conservation Board charged with the rural water supplies programme saw little long-term future for the traditional farmer and believed that mechanized farming was the only conceivable way ahead (Jefferson 1956). In the first season (1947/8) 26 *hafirs* were dug. Lake Smith, the first and largest, was excavated at Ghadambaliya for the Mechanical Crop Production Scheme programme. During the first five seasons 138 *hafirs* were excavated in the Kordofan Region, the area most in need of improved rural water supplies, in order to allow population dispersal, to facilitate nomadic movement, and to develop an overland cattle route for marketing in Omdurman and for the meat canning factory at Kosti.

The policy of increasing rural water supplies even in fragile ecological zones was continued more intensively through the 1960s and early 1970s by

TABLE 18.4. New water point provision in the northern Sudan savanna zone

Province/region	Bores 1966–85	Wells 1966–80	Hafirs 1966–84	Reservoirs 1966–84
Northern Sudan				
Total water points	5655[a]	6045	278	17
West: desert margins (%)	26	19	32	40
Northern Kordofan (%)	19	15	16	11
Northern Darfur (%)	7	4	16	29 (Jebel Marra)
West: southern savanna (%)	17	25	24	18
Southern Kordofan (%)	6	19	24	12 (Nuba Mountains)
Southern Darfur (%)	11	6	0	6
Eastern region (%)	9	9	11	36 (Red Sea Hills)
Central region (%)	34	41	32	6
White Nile Province (%)	—	—	15	—
Blue Nile Province (%)	—	—	17	—
Khartoum and rest of North (%)	14	6	1	0
Total	100%	100%	100%	100%

[a] Not all successful.
Source: Unpublished data, Sudan National Water Administration (various years).

the National Water Administration. In recent years there have been more boreholes and wells dug and fewer *hafirs* excavated (Table 18.4). Of the 5570 bores drilled between 1966 and 1985 more than half were sunk between 1966 and 1972, and 900 of them were in Northern Kordofan. This change of policy has enabled the sandy *qoz* country of Kordofan to be opened up and the amount of drinking water available was no longer related to the local rainfall. The existence of bores and wells makes the establishment of permanent settlement easier but it removes the control element on the amount of cropping and the number of animals that an area can support.

Politically it became necessary to provide considerably more water points than were ecologically advisable and the result can be seen in East Kordofan District (Fig. 18.5). The creation of overlapping zones of influence from so many water points and the failure to adopt new systems of land use have led to serious degradation. The situation in east Kordofan and in the Butana, between the Blue Nile and Atbara Rivers, has been made worse by the different views on development expressed by administrators who saw better rural water provision as a way of relieving pressure on land, whereas the administered saw it as an opportunity to increase cropped areas and rear more livestock (Awad *et al.* 1985; Davies 1986*a*, p. 117).

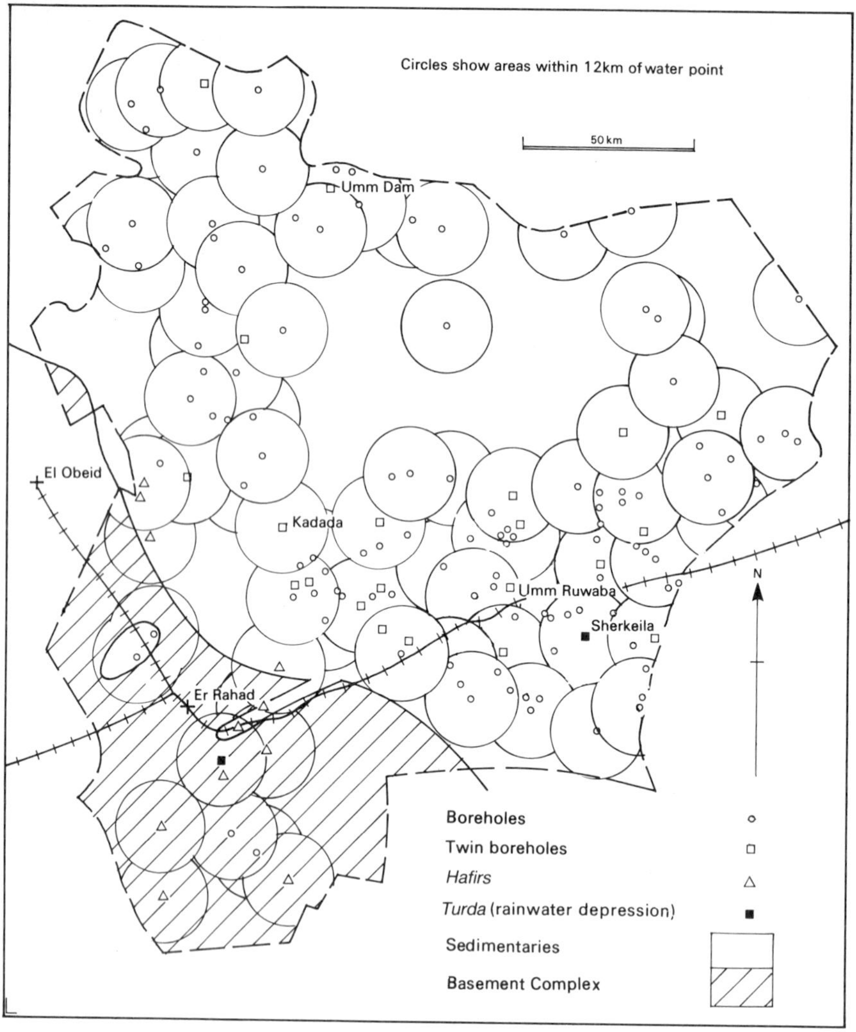

FIG. 18.5. Water supplies in East Kordofan District in 1979.

Livestock and nomad resettlement

In the Sudan, while there is an ever increasing market for meat to supply the growing towns and other nearby countries such as Saudi Arabia and the Gulf States, many Sudanese still see animals in social terms rather than in an economic context. The more animals you have, the greater your social status. Therefore, both sides consider an increase in livestock numbers to be advantageous. The feeling of a nomad interviewed by the author in 1986 demonstrated one problem associated with livestock rearing in the desert margins. He claimed to have had 100 camels before the drought of 1984 and had lost 50. He added that he must try to have 200 before the next drought struck. Livestock numbers in the Sudan increased very rapidly from the 1940s until the 1984 drought (Table 18.5a, b), so much so that the rangeland management problems which resulted now require a significant change in attitude to enable a satisfactory offtake of animals to occur. The Five Year Plan ending in 1976 and its successor, the Six Year Plan to 1982/3, both aimed to improve meat supplies by an annual rate of livestock increase of between 3.5 and 4.0 per cent. The nomad would approve of this policy and of the veterinary provision included, but his aim would not necessarily be to increase meat supplies.

The interaction of the various facets of rural policy to give results different from those anticipated can be demonstrated from experience in Northern Kordofan. As early as 1954 it was claimed that the area was 50 per cent overstocked (Harrison 1955), but comparative figures for livestock in the Dar Kababish area of the province (Table 18.5c) suggest a four-fold increase in livestock units over the 16 year period, 1953–69. This rapid increase was made possible by the provision of better rural water supplies, which instead of improving the situation as intended made it worse. If Harrison's 1955 estimate was correct the range was 600 per cent overstocked by 1969 and the situation has worsened since then. The significance of improved water supplies is demonstrated clearly by the five-fold increase in cattle. The annual offtake of animals is much too small and the ecological problem evident in this area is replicated in other parts of the central zone of Sudan (Table 18.5b). The effect of lower rainfall in recent

TABLE 18.5(a). Livestock in the Sudan[a], 1944–84

Year	Livestock[b] units (millions)	Increase in decade	
		Number	%
1944	6.10	—	—
1954	11.02	+4.92	+81
1964	14.35	+3.33	+30
1974	22.99	+8.64	+60
1976	23.09		
1981	34.30	+11.31 (1974–81)	+49
1984[c]	28.43	+5.44	+24
1981–4		−5.87	−17

Source: Department of Agricultural Economics, MANR (various years).
[a] All livestock figures for Sudan should be accepted cautiously.
[b] Livestock units based on the ratio: 1 cattle = 6 sheep = 6 goats = ⅔ camel.
[c] FAO estimate for September 1984: demonstrates the first effects of the drought.

TABLE 18.5(b). Livestock in the Sudan by region, 1976 and 1981 (millions)

	Cattle		Sheep		Goats		Camels		Livestock units		Increase in livestock units 1976–81 (%)
	1976	1981	1976	1981	1976	1981	1976	1981	1976	1981	
West: desert margins											
Northern Kordofan	0.94	1.19	2.47	2.76	1.68	1.98	0.85	0.97	2.76	3.27	+29
Northern Darfur	0.91	1.15	1.41	1.57	1.19	1.40	0.23	0.26	1.64	1.90	+16
West: savannas											
Southern Kordofan	1.47	7.86	0.83	0.93	0.90	0.82	0.00	0.00	1.76	8.15	+363
Southern Darfur	2.74	3.47	1.23	1.38	1.12	1.31	0.14	0.16	3.32	4.13	+24
Eastern region											
Kassala	0.64	0.81	1.59	1.78	0.91	1.09	0.57	0.65	1.82	2.15	+18
Red Sea	0.04	0.05	0.22	0.25	0.44	0.52	0.10	0.11	0.27	0.33	+22
Central region											
Gezira and White Nile Province	2.07	2.62	3.42	3.82	1.80	2.12	0.22	0.26	2.94	3.96	+35
Blue Nile Province	0.90	1.14	1.08	1.21	0.59	0.70	0.04	0.04	1.23	1.51	+23
Khartoum and northern Sudan	0.11	0.15	0.77	0.86	0.84	0.98	0.18	0.21	0.62	0.74	+19
Southern Sudan											
Upper Nile Province/Jonglei	2.83	3.59	1.22	1.36	0.84	0.98	0.00	0.00	3.18	3.98	+25
Bahr el Ghazal/Lakes/ Eastern Equatoria	2.73	3.45	1.97	2.20	1.15	1.35	0.03	0.03	3.28	4.08	+24
Western Equatoria	0.00	0.00	0.00	0.00	0.02	0.02	0.00	0.00	0.00	0.00	0
Sudan totals	15.38	25.48	16.21	18.12	11.48	13.27	2.36	2.69	22.82	34.20	+50

Source: Department of Agricultural Economics, MANR (various years).

TABLE 18.5(c). Livestock in the Dar Kababish area of Northern Kordofan, 1953–69

	1953 ('000s)	1969 ('000s)	Increase Number	%
Cattle	25	126	+101	+404
Sheep	141	643	+502	+356
Goats	52	65	+13	+25
Camels	63	256	+193	+306
Animal units	140	584	+444	+317

Source: Based on Khogali (1987).

years can be seen in the rapid increase in livestock in Southern Kordofan (Table 18.5b).

To reach a better balance between land, animals, and offtake the Sudanese Government has promoted two policies: to develop large ranches; and to settle nomads. Since 1977, the Government has recognized the possibility of attracting foreign capital, especially from the oil-rich Arab States, and the Six Year Plan (1977–83) lists £S240m. worth of proposed Arab Authority investments in the Sudan, more than half of which was directed towards the livestock sector. Proposals included a million feddan project for cattle production in Southern Darfur by a West German Company using Gulf capital as well as a ranching project of over one million feddans by a Luxemburg registered company using Arab capital. The success to date of such ranches is difficult to estimate but little real development seems to have arisen yet. One projected ranch with 500 000 feddans in Blue Nile Province has turned entirely to mechanized crop production because livestock rearing had too many problems (Earl 1985).

The rapid expansion of mechanized and traditional rainland farming and of irrigated farming has increased pressure on nomads' grazing lands and migration routes (Fig. 18.6; Fig. 18.7). The decreased rainfall in the north has curtailed journeys northwards into the desert and the migrations southwards which were undertaken only rarely after poor rains have become the norm. In Darfur and Kordofan Regions such new regular movements southwards towards the Bahr el Ghazal have increased conflict between sedentary cultivators and nomadic pastoralists (Khogali 1987). The troubled situation in the south has also led to raiding and fighting over grazing in Southern Darfur.

One solution to these problems might be nomadic settlement, and some has occurred spontaneously around newly-created water yards. Other nomads have given up livestock rearing at least temporarily as a result of the 1984/5 drought and they have drifted towards the towns. The Beja people of the Red Sea Hills were most severely affected and many have

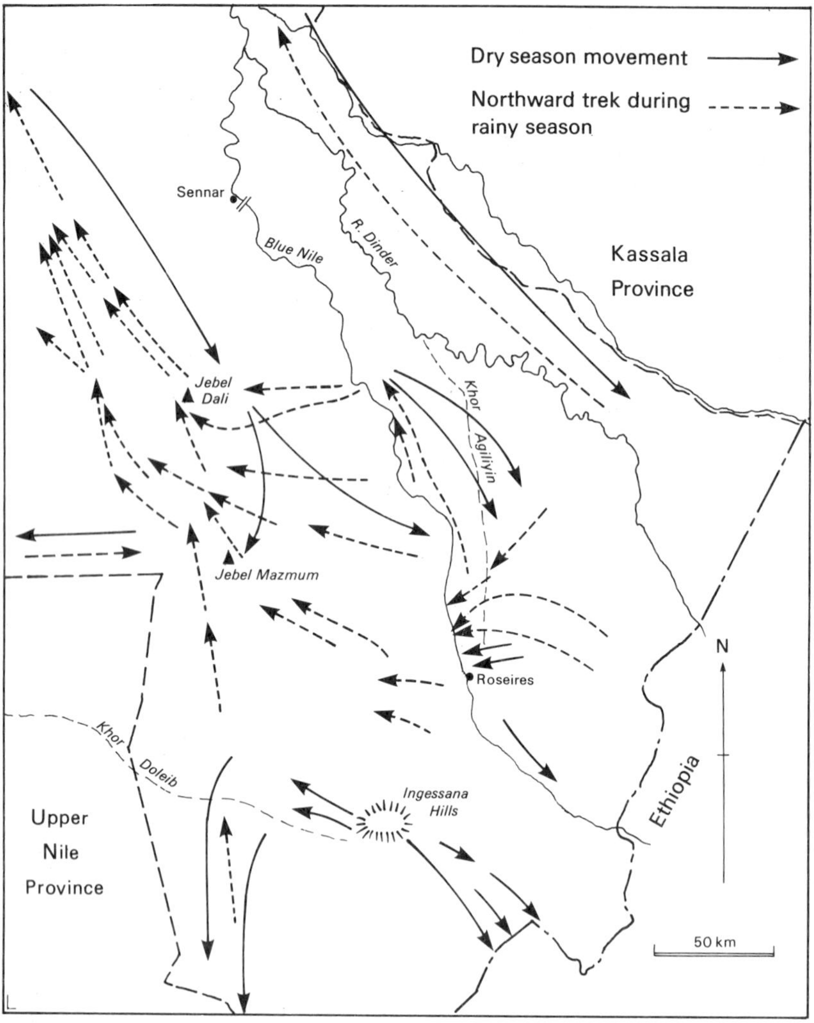

FIG. 18.6. Nomadism in Blue Nile Province, 1955.

drifted into the shanties around Port Sudan. Some, usually with considerable reluctance, have taken up rainland farming or irrigation tenancies.

Another possibility was to involve nomads with ranching, so that the family settled down and only a small number of men folk moved away for short periods to take advantage of the rainy season grazing. Most of the time grazing would take place over a series of paddocks grazed in rotation. Five small pilot projects of this type were set up in Darfur and Kordofan regions in 1968, but they have been labelled as failures by Thimm (1979, pp. 33–8). Khogali (1987) believes that this system could work and that the

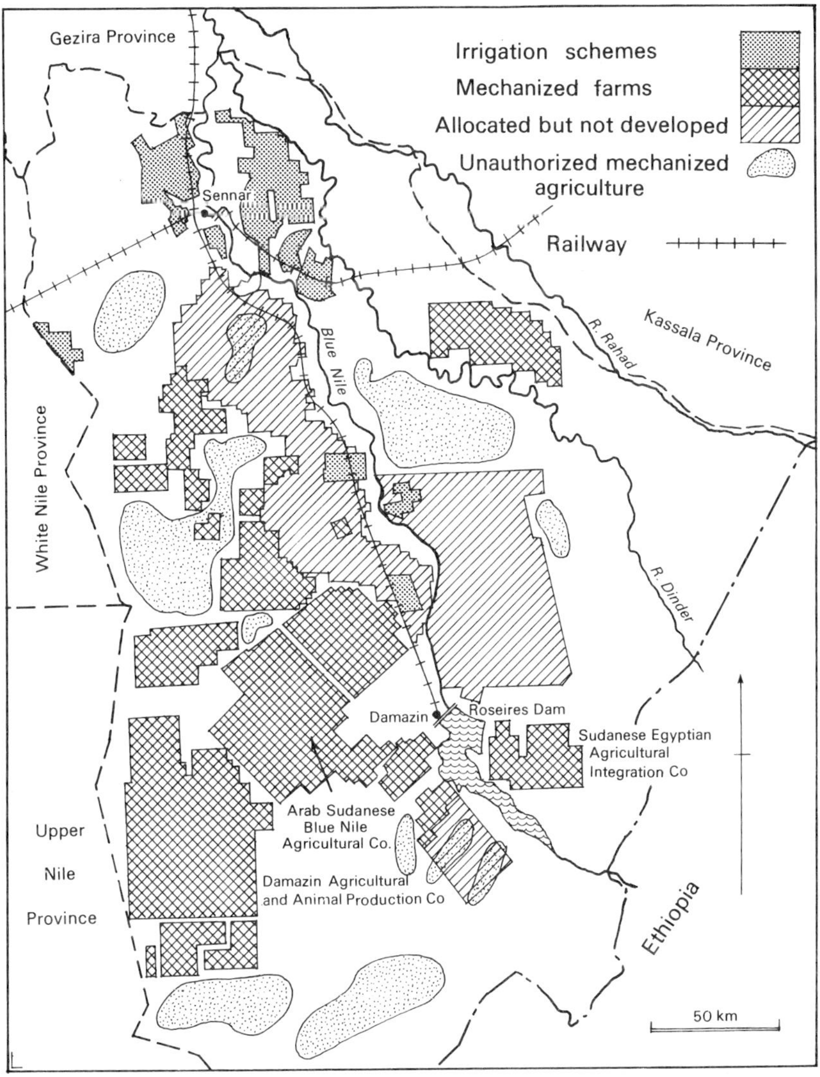

FIG. 18.7. Schemes in Blue Nile Province, 1986.

problem with the Gereih el Sarha project, the most important one, was not so much the system as its management and application. One outstanding conclusion was that the number of grazed animals would have to be reduced drastically. This kind of programme for nomadic settlement coupled with improved livestock has long been advocated by those who believe that conversion to settled cultivator is not the best way forward for most nomads (Davies 1966).

Clearly, the livestock situation could be changed with a policy of pasture improvement, increased availability of irrigated fodder and of processed crop residues, especially *ombaz* (oilseed cake derived from sesame and cotton seed). Some other ways in which irrigated agriculture and livestock rearing can be combined are currently in operation among the Kawahla of the White Nile (Khogali 1986). The livestock sector would benefit particularly from a coherent land use plan bringing together rainland farming, grazing, irrigated agriculture, and forestry. Livestock in relation to irrigation is discussed in Chapter 19.

The South

Most of the policy proposals for rainland development have been confined to the north of the Sudan. The one exception is the northern part of Upper Nile Province where the main programme has involved the extension of mechanized farming from 30 000 feddans in 1968 to 495 000 feddans in 1986, but reduced to 155 000 feddans in 1988.

Immediately after Independence in 1956 there were great hopes for agricultural development in the south. The *Ten Year Plan (1961–1971)* included the development of irrigation in Upper Nile Province and the development of rice production at Aweil (Chapter 19). Other proposed developments included coffee grown on government and private plantations and supplemented by outgrowers, and tea, developed on somewhat similar lines but on a smaller scale. Both these ventures were to be developed in Equatoria with a view to saving foreign exchange and assisting development in backward areas. A wide range of fruits including bananas, mangoes, pineapples, and various citrus fruits could be grown easily, and a canning factory was built with Soviet aid at Wau. A cigarette factory using southern-grown tobacco was to be established in Juba, and a scheme to grow kenaf and build a fibre factory for bags and sacks was envisaged at Tonj.

Already, various experiments in rural development had been attempted, notably the Zande Scheme and resettlement programme centred on Yambio, which provided services and schools and promoted cotton growing and the production of other crops in Western Equatoria. Work had also begun on a small industrial complex at Nzara based on cotton textiles, soap, and lumbering.

After the Addis Ababa Accord of 1972, hopes for development of the south rose again. Some of these were expressed in the 1977–83 Plan which envisaged the revival of plans to expand rice (Chapter 19), coffee, tobacco, and tea. By 1974, however, there were less than 1000 feddans under these crops in the south. Arab Authority funds amounting to £S515m. were earmarked for the expansion of tea, coffee, and rice, and further smaller sums were earmarked for tobacco and pineapple production. The Wau tomato, fruit, and vegetable canning factory which produced less than 500 tonnes in 1973 was to be revitalized as was the Zande Scheme. Other aspects of agricultural development included expansion of sorghum, maize, and cassava production.

In recent years continued unrest in the south has resulted in major setbacks to development while transport remains another problem. Although the Southern Region appears to be self-sufficient in sorghum in normal times, there are regular shortages in Juba, Malakal, and Wau. This is because areas in the north of the region, such as around Renk, market their excess crop to the north because of the deficiencies in the river transport service. The war in the south has simply reinforced such problems. Western Equatoria often produces crop surpluses while there are shortages in parts of Bahr el Ghazal and Upper Nile. At present, however, there is no satisfactory way of distributing one area's crop surplus to relieve another area's shortage. The main crops grown in southern Sudan are shown in Table 18.6 but the self-sufficiency of the Southern Region viewed as a whole is more apparent than real (Binayson and Dima 1986, p. 58).

There are large numbers of livestock in southern Sudan (Table 18.5b), but most belong to Dinka, Nuer, and Shilluk who keep them as a source of prestige and honour and are unwilling to sell. Livestock represent a potential

TABLE 18.6. Main crops in southern Sudan[a] ('000 feddans)

	1963 ('000 feddans)	1979 ('000 feddans)	Increase 1963–79 (%)
Sorghum	332	1022	+208
Bulrush millet	12	77	+542
Finger millet	40	93	+133
Maize	43	154	+258
Rice	—	23	—
Groundnuts	102	247	+142
Cassava	—	119	—
Beans	5	76	—
Sweet potatoes	—	72	—
Cotton (American)	29	30	—
Sesame	36	225	+525
Others	97 (includes rice and sweet potatoes)	362	—
Total	696	2500	+259

(1986 ('000 feddans): Sorghum 895; Bulrush millet 50; Sesame 66; Groundnuts 60; Cotton (American) 10).

[a] Crop figures for southern Sudan must be taken as rough estimates.

Source: Department of Agricultural Economics, MANR (various years).

economic asset of considerable importance which has so far defied rational development. The unwillingness of such tribes and of the nomads of the north to sell cattle led to the failure of the Kosti meat canning scheme in the 1950s and is one of the factors leading to proposals for large-scale ranching based upon foreign capital. Current offtake in southern Sudan is estimated at 1 per cent whereas potential offtake is said to be 10–15 per cent (Binayson and Dima 1986, p. 59).

Conclusions: desertification and the future

At various times during the twentieth century, especially during the runs of drier than average years, fears have been expressed about the expansion of the Sahara southwards. Relevant information about the drought of the 1940s was collected and analysed (Sudan Government 1944; Stebbing 1953), but concern turned to optimism in many quarters with good rains in the 1950s and early 1960s. Since then the problem has returned in a much more serious form.

Tangible evidence for desertification can be seen throughout the crisis zone delineated in Fig. 18.2. Visual observations in White Nile Province suggest that the vegetation boundaries have moved southwards by some 100 km since the 1950s (Alam El Din 1986). Evidence of advancing dunes have been reported from most areas especially in Northern Kordofan around Bara. The Mechanised Farming Corporation's review of activities (MFC 1984) concluded that 'Uncontrolled expansion of mechanized farming during the past years has devastated the forestry and pasture wealth of the country particularly in marginal rainfall areas.' Evidence of the decline in palatable species and *gizu* grazing as a result of overgrazing has been presented for many areas (Khogali 1986, p. 71). Evidence for increased dust in the atmosphere based upon the number of days with visibility of less than 100 metres (Ibrahim 1978, p. 17) has also been presented for various meteorological stations, notably for El Fasher. Other reports (SDEC 1976; Trilsbach 1983; Ibrahim 1978, p. 15) have demonstrated declining yields in the basic food crops (see also Fig. 18.4), and it should be noted that this evidence pre-dates the 1984 famine.

Two questions are relevant—whether the 1980s are a period of general rainfall decline (Hulme 1985), and whether rainfall is the only real problem? Evidence from the Gereih el Sarha project in Darfur suggests that if livestock numbers were reduced and rotational grazing controlled, then rejuvenation would be possible. Similarly, work in White Nile Province suggests that land degradation and desertification are more closely related to land misuse than to rainfall (Trilsbach 1986). Work in the Jebel Marra has shown that with protection the regeneration of woody savanna and forest is possible within the crisis zone (Ahmed 1983).

These studies, carried out before the 1984/5 famine, suggest that rainfall is not the key factor in the desertification process. Nevertheless, in a period of general rainfall decline, it seems likely that man and climate are both contributing factors.

Sudan has been aware of desertification since the mid-1970s when its Desert Encroachment Control and Rehabilitation Programme produced a major report (DECARP 1976). This document still forms the basis of Sudan's strategy on desert encroachment namely: to limit desert encroachment with dune fixation, shelter belts, fire line construction, and other vegetation protection activities, and to shift development southwards by creating mechanized farming schemes and ranches, and by providing a new, more southerly overland stock marketing route from Darfur to Omdurman.

The effects of the 1984/5 drought have brought renewed interest in the ideas presented originally by DECARP and the foreign interest and funds available have increased enormously. A plethora of projects put forward by donors have been approved (e.g. Euroconsult 1986) and it is to be hoped that the new activity will not result in failure by trying to do too much too quickly. Investigations by Davies (1968*a*, *b*) on the social acceptability of development to the people involved are relevant in this context.

Studies by Arifi (1978*a*) and Khalifa (1984) have emphasized the need for a more wholistic regional approach to development, while Affan (1984) and Davies (1986*a*, p. 120) maintain that the aims of the Sudanese Government and the aspirations of the rural population will only be reconciled by a less bureaucratic and more fundamental approach.

The future development of dryland agriculture in central Sudan requires a reduction of pressure by people and livestock in the more ecologically fragile areas and a reassessment of land use strategy. With increased numbers of people and livestock and the possibility of a semi-permanent decline in mean annual rainfall, the present low intensity systems of land use with their traditional drought damage limitation strategies are no longer viable. Developments in other parts of the Sudanese economy have a bearing on the development of dryland agriculture in the north. In the south no policy can be implemented until a political solution to the unrest has been reached.

References

Sources of statistical information are primarily the *Sudan yearbook of agricultural statistics*, *Current agricultural statistics*, and *Agricultural situation and outlook*, all produced by the Statistics Section of the Department of Agricultural Economics, within the Ministry of Agriculture, Food and Natural Resources (1974–88).

Abdel Ati, H. A. (1986). *The problems of the agricultural sector in Sudan: an account of policy gaps*. Department of Geography, University of Khartoum.

Affan, K. (1984). *Towards an appraisal of tractorization experience in the rainlands of Sudan*, Development Studies and Research Centre Monograph, No. 19. Development Studies and Research Centre, Khartoum.

Ahmed, A. A. (1983). Forest reserves and woodland savanna regeneration on the sub-Saharan massif of Jebel Marra, Democratic Republic of the Sudan. *Vegetatio*, **54**, 65–78.

Alam El Din, I. O. (1986). A note on vegetation. In *Rural development in White Nile Province, Sudan* (ed. H. R. J. Davies), pp. 43–4. United Nations University, Tokyo.

Arifi, S. A. El (1978). *Development strategy for the arid zone of western Sudan*. Department of Geography, University of Khartoum.

Awad, A. A. Al, Mohammed, Y. A., and Tayeb, S. A. El (1985). The impact of improved rural water supplies on the environment: the case of East Kordofan District. In *Natural resources and rural development in arid lands: case studies from Sudan* (ed. H. R. J. Davies), pp. 16–29. United Nations University, Tokyo.

Babiker, A. A., Musnad, H. A., and Shaddad, M. Z. (1985). Wood resources and their use in the Nuba Mountains. In *Natural resources and rural development in arid lands: case studies from Sudan* (ed. H. R. J. Davies), pp. 30–59. United Nations University, Tokyo.

Binayson, S. A-K. and Dima, S. J. (1986). Background and economy of southern Sudan. In *The agricultural sector of Sudan. Policy and system studies* (ed. A. B. Zahlan and W. Y. Magar), pp. 44–70. Ithaca Press, London.

Briggs, J. (1984). *Peasant farmers and crop choice: a study of agricultural decision making in central Sudan*, Occasional Paper, No. 13. Department of Geography, University of Glasgow.

Davies, H. R. J. (1962). Economic and social geography of the Sudan between the White and Blue Niles. Unpublished, B.Litt. thesis. University of Oxford.

Davies, H. R. J. (1964). An agricultural revolution in the African tropics: the development of mechanized agriculture on the clay plains of the Republic of Sudan. *Tijdschrift voor Economische en Sociale Geografie*, **55**, 101–8.

Davies, H. R. J. (1966). Nomadism in the Sudan. *Tijdschrift voor Economische en Sociale Geografie*, **57**, 193–202.

Davies, H. R. J. (1986*a*). The human factor in development: some lessons from rural Sudan? *Applied Geography*, **6**, 107–21.

Davies, H. R. J. (ed.) (1986*b*). *Rural development in White Nile Province, Sudan*. United Nations University, Tokyo.

DECARP (Desert Encroachment Control and Rehabilitation Programme) (1976). *Sudan's Desert Encroachment Control and Rehabilitation Programme*. Khartoum.

Earl, D. E. (1985). *Sudan: the economics of wood energy production on mechanized farms*. Energy Research Council/USAID Sudan Renewable Energy Project, Khartoum.

Euroconsult (1986). *Reassessment rehabilitation programme, Kordofan and Darfur: Final Report to the Ministry of Finance and Economic Planning, Republic of Sudan*. Euroconsult, Arnhem, The Netherlands.

Harrison, M. N. (1955). *A report on a grazing survey of the Sudan*. Ministry of Animal Resources, Khartoum.

Hulme, M. (1985). Rainfall variability and consequential hydrological changes in central Sudan. Unpublished Ph.D. thesis. University of Wales.

Ibnouf, M. A. O. (1985). An economic analysis of mechanized food production schemes in the central plains of the Sudan. Unpublished Ph.D. thesis. Michigan State University.

Ibrahim, F. N. (1978). *The problem of desertification in the Republic of the Sudan with special reference to Northern Darfur Province*, Science Research Council Monograph No. 8. Science Research Council, Khartoum.

Jefferson, J. H. K. (1956). *Soil conservation in the Sudan*. Khartoum, Ministry of Agriculture, Khartoum.

Khalifa, M. D. El (ed.) (1984). *Environmental impact of the southern stock route, Sudan*. Institute of Environmental Studies, University of Khartoum.

Khogali, M. M. (1986). The changing nature of nomadism in the northern White Nile region. In *Rural development in White Nile Province, Sudan* (ed. H. R. J. Davies), pp. 65–72. United Nations University, Tokyo.

Khogali, M. M. (1987). An assessment of Gereih al Sarha pilot project scheme for the settlement of the nomads and improvement of the livestock sector in the Sudan. *Geo Journal*, **14**, 53–62.

Kursany, I. (1982). *The dynamics and limits of capitalist development in Sudanese agriculture*, Development Studies and Research Centre Seminar Series, No. **17**. University of Khartoum.

MFC (Mechanised Farming Corporation) (1984). *Task force report on revised role of the Mechanized Farming Corporation*. MFC, Khartoum.

Ministry of Finance and Economics (1961). *The Ten Year Plan of Economic and Social Development, 1961/2–1970/1*. Ministry of Finance and Economics, Khartoum.

MNP (Ministry of National Planning) (1977). *The Six Year Plan of Economic and Social Development, 1977/8–1982/3* (2 vols). MNP, Khartoum.

Osman, A. A. (1959). Results of four case studies at Dali-Mazmoum Mechanical Crop Production Scheme season 1957/58. Unpublished results, Ministry of Agriculture, Khartoum.

Saeed, H. M., Ahmed, A. J. M., and Tahir, M. H. El (1986). Agricultural research and extension. In *The agricultural sector of Sudan. Policy and system studies* (ed. A. B. Zahlan and W. Y. Magar), pp. 381–407. Ithaca Press, London.

SDEC (Sudan's Desert Encroachment Control) (1976). *Summary presentation*. SDEC, Khartoum.

Shakkak, K. I. (1977). Mechanization of agriculture in the clay plains of the Sudan, with special reference to the Gedaref district. Unpublished M.A. thesis. University of Khartoum.

Stebbing, E. P. (1953). *The creeping desert in the Sudan and elsewhere in Africa*. Sudan Government, Khartoum.

Sudan Government (1944). *Report of the Soil Conservation Committee*. McCorquodale and Co. (Sudan) Ltd., Khartoum.

Thimm, H. U. (1979). *Development projects in the Sudan....* United Nations University, Tokyo.

Trilsbach, A. (1983). Desertification and rural change in central Sudan. Unpublished Ph.D. thesis. University of Wales.

Trilsbach, A. (1986). Agriculture west of the White Nile. In *Rural development in White Nile Province, Sudan* (ed. H. R. J. Davies), pp. 46–64. United Nations University, Tokyo.

World Bank (1982). *A study of cost of production and comparative advantages of crops in Sudan: a report to the Ministry of Finance and Economic Planning*. Khartoum.

Further reading

Abu Sin, M. H. (1986). Man's socio-economic response to drought in the White Nile. In *Rural development in White Nile Province, Sudan* (ed. H. R. J. Davies), pp. 96–107. United Nations University, Tokyo.

Adams, M. E. (1981). The Baggara problem: attempts at modern change in Southern Darfur and Southern Kordofan. In *Post-independence Sudan* (ed. Centre of African Studies), pp. 40–70. University of Edinburgh.

Arifi, S. A. El (1978*b*). *Local government and local participation in rural development in the Sudan*, Science Research Council Monograph, No. 10. Science Research Council, Khartoum.

Centre of African Studies (1981). *Post-independence Sudan*. Centre of African Studies, University of Edinburgh.

Davies, H. R. J. (ed.) (1985). *Natural resources and rural development in arid lands: case studies from Sudan*. United Nations University, Tokyo.

Lado, C. (1985). Environment, knowledge and change: a case study of peasant farming in Maridi District, southern Sudan. Unpublished Ph.D. thesis, University of Durham.

Nimeiri, S. (1977). *The five year plan (1970–75): some aspects of the plan and its performance*. Science Research Council Monograph, No. 1. Science Research Council, Khartoum.

SDIT (Southern Development Investigation Team) (1955). *Natural resources and development potential in the southern provinces of the Sudan: a preliminary report by the Southern Development Investigation Team, 1954, and Volume II, maps and diagrams*. Sudan Government, Khartoum.

Simpson, M. C. (1981). Large-scale mechanized rainfed farming developments in the Sudan. In *Post-independence Sudan* (ed. Centre of African Studies), pp. 197–202. Centre of African Studies, University of Edinburgh.

Trilsbach, A. and Hulme, M. (1984). Recent rainfall changes in central Sudan and their physical and human implications. *Transactions of the Institute of British Geographers*, NS **9**, 280–98.

United Nations Conference on Desertification (1977). *Desertification: its causes and consequences*. Proceedings of United Nations Conference, Nairobi, 1977. Pergamon, Oxford.

Zahlan, A. B. (ed.) (1984). *Agricultural bibliography of Sudan, 1974–1983*. Ithaca, London.

Zahlan, A. B., and Magar, W. Y. (ed.) (1986). *The agricultural sector of Sudan. Policy and system studies*. Ithaca Press, London.

19

Irrigation development programmes

H. R. J. DAVIES

Introduction

In 1956, the year of Independence, agricultural products accounted for 97 per cent of Sudan's foreign export earnings and three-quarters of the bank loans were for agricultural purposes (Ministry of Finance and Economics 1957). The most important export crop was cotton which provided 71 per cent of export earnings and an important source of budget revenue for the Sudanese Government through export taxes. More than 90 per cent of the cotton grown in the Sudan came from irrigated schemes.

It seemed logical for Sudan to look towards agriculture, especially irrigated agriculture, for its future development and to this end the first of the massive post-Independence extensions to the irrigated area was started. The Managil Extension with 800 000 feddans was destined to almost double the existing Gezira Scheme (Fig. 19.1).

The vagaries of the annual rainfall in the summer rainfall zone of central Sudan and the consequent fluctuations in yields from one year to the next prompted the Sudanese Government to plan the revolutionization of the whole of the agricultural system through the widespread extension of irrigation.

The Ten Year Plan, 1961–1971 (Ministry of Finance and Economics 1961) states unequivocally: 'The importance of irrigation water to the economy of Sudan can hardly be overemphasised;' and the Six Year Plan, 1977–83 (Ministry of National Planning 1977) comments: 'It is noticeable that both in ongoing and new projects irrigation has claimed the biggest share (of public sector investment in agriculture).'

The new Managil Extension and most of the further 1.1 million feddans of planned irrigation expansion between 1961 and 1971 were situated in the north and central areas of the country, which must rely almost wholly on irrigation water. Wetter areas with a much longer and more reliable rainy season lie to the south where irrigation is needed only as a supplement to

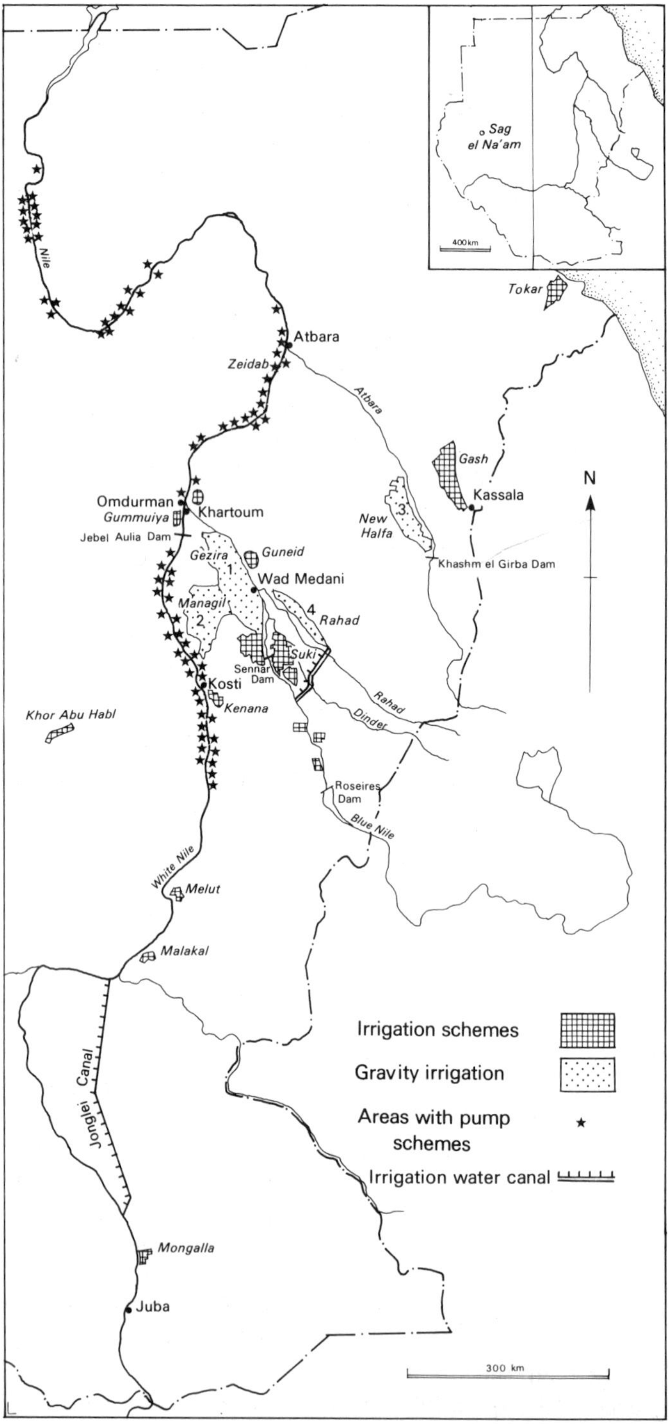

FIG. 19.1. Sudan irrigation schemes.

rainfall during the dry season. Therefore, the factors favouring irrigation expansion in the northern and central areas, were as follows:

1. Sudan's future development was seen to rest on its ability to export agricultural produce, especially irrigated cotton; the northern and central areas were much nearer to Port Sudan, Sudan's only major seaport, and they were already connected to it by rail. The climatic advantages of the south were outweighed by its relative remoteness.
2. The northern part of the country was more economically advanced and had a more developed infrastructure than the south. It was already producing important export commodities and, in the short-term at least, was a sounder investment.
3. The people of the northern areas of Sudan were more politically aware, better educated, and more sophisticated than those of the south. In 1956 there was only one secondary school in the south and very few students at the University College of Khartoum (formerly Gordon Memorial College and later the University of Khartoum) were from the south.
4. At Independence, only 27 per cent of the population lived in the three southern provinces of the Sudan. Southerners were quite different ethnically and culturally from the northerners, and the northerners dominated the Sudanese Government.
5. After Independence, southerners felt that the south was being treated as a 'colony'. This feeling bred resentment and led to civil war, rendering it almost impossible to implement development schemes in the south.

The Nile Waters Agreements and 'century storage'

Future prospects for the expansion of irrigation lay in developing the potential of the Nile system. However, the 1929 Nile Waters Agreement was a stumbling block. Under this treaty it was assumed that the average annual Nile flow at Aswan was 82 milliard m^3, of which 30 milliards reached the Mediterranean, 48 milliards were allocated to Egypt and only 4 milliards were allocated to the Sudan. The agreement also regulated how the Sudan could extract its share of water to fill the Sennar reservoir and operate its pump schemes. In essence, the object of the agreement was to protect Egyptian agriculture by safeguarding Egypt's irrigation water supplies and allowing the Nile silt to reach Egypt.

For the Managil Extension to be irrigated properly and the irrigated area to be expanded by a further 1.1 million feddans, as envisaged in the Ten Year Plan, the amount of water stored for use in the Sudan would have to exceed the quantities allowed under the agreement. Sudan rejected the

1929 Agreement on the grounds that she was not a true signatory under the Condominium. The Egyptians believed that the concept of 'century storage' would enable their area under irrigation to be increased and all seasonal irrigation to be converted to perennial irrigation, a concept which would also require revision of the 1929 Agreement.

'Century storage' implied a system of water flow regulation which, over the period of a century, would prevent all seasonal and annual variations in the flow of the Nile, so that virtually no water would flow into the Mediterranean. There were two views as to how this should be done. The more scientific approach involved the creation of storage reservoirs in various parts of the Nile Basin, from which water would be released as necessary for use in northern Sudan and Egypt. The main areas of storage would avoid as far as possible those areas in the Sahara with very high rates of evaporation (Hurst 1952). A simpler idea was the construction of a single high dam at Aswan in the middle of the Sahara Desert. This involved storage mainly within Egyptian territory. By the late 1950s, with the Sudanese decision to preserve full independence, the Egyptians moved rapidly towards the second more politically attractive method of achieving 'century storage'.

The new Nile Waters Agreement of 1959 (summarized in Keesing 1959) took into full account Sudan's increasing need for more irrigation water, the concept of 'century storage', and the Egyptian desire to build a high dam at Aswan. It was assumed that the average annual flow of the Nile at Aswan was 84 milliard m^3 and it was agreed that this notional figure should be shared between the two countries so that Sudan should be allowed 18.5 milliards of water against Egypt's 55.5 milliards (i.e. a ratio of 1:3 compared with 1:12 in 1929). The new agreement estimated losses of 13 milliards in the Sudd (i.e. 50 per cent of the water flowing into the Sudd) and losses of 10 milliards from evaporation and other factors. Provision was made in the agreement for investigation into a Jonglei diversion canal which would bypass a part of the Sudd and reduce evaporation losses considerably. The cost of work on such a project and any water savings would be shared equally between the two countries and it was predicted confidently that such a canal would ultimately increase the notional flow of the Nile by 10 per cent.

In view of the variations in the annual flow of the Nile, the length of time required to fill the new High Dam reservoir and the knowledge that Sudan would not need the whole of its water allocation immediately, an arrangement was made whereby Egypt could borrow up to 1.5 milliards a year until 1977 and then pay back the excess drawings over the following ten years.

Other parts of the agreement included compensation for 60–70 000 Sudanese nationals who would have to be resettled elsewhere in the Sudan once the reservoir was full. This resettlement was scheduled for completion by July 1963. As the agreement embodied the concept of 'century storage' it was agreed that it would remain in force for a hundred years.

Large irrigation schemes

It was assumed in the Sudan that the only really sound way to bring about development would be through large irrigation schemes. Pump schemes of various sizes would have some additional importance and to the peasant the *saqqiya* (or Persian wheel), the *shaduf* (of ancient Egypt) and *gerif* (riverbank) lands would remain vital. Satisfactory irrigation of the expanded Gezira required much greater storage than could be provided by the existing Sennar Dam, so a new dam at Damazin on the Blue Nile above Roseires was planned. This Roseires Dam would store enough water not only for the Gezira Extension but also for a proposed, but never undertaken, Kenana Extension of 1.2 million feddans.

The Atbara River remained the only large river in northern Sudan without water control facilities and the potential for a dam and irrigation scheme based on this river at Khashm el Girba had long been recognized. Furthermore, this would provide an area for the resettlement of Nubians displaced by the High Dam reservoir.

The 1959 Nile Waters Agreement enabled the Sudan to build both the proposed dams, which would be multipurpose and, like the Sennar Dam, would generate hydroelectricity. In the mid-1980s, almost 75 per cent of the electricity generated in the Sudan was hydroelectric power (Babiker *et al*. 1985).

After the 1959 Nile Waters Agreement the first priorities were the construction of the Roseires and Khashm el Girba Dams and the laying out of the Khashm el Girba Irrigation Scheme. Both dams were scheduled for completion during the Ten Year Plan period and the Khashm el Girba Scheme had to be in operation by 1964.

When the Kenana Extension was not developed, there was a considerable quantity of unused water still available from the Blue Nile. Irrigation feasibility studies were undertaken between 1961 and 1971 on the clay plains in the Rahad and Dinder Rivers area. However, the flow of both these rivers was unpredictable and the problem lay in supplying the area with sufficient water. The eventual solution was to pump water from the Blue Nile at Meina opposite Singa, a lift involving 7–15 m according to season, and thence transport it by gravity to the Rahad River, after crossing under the Dinder River by a siphon. Work began with Phase 1 (150 000 feddans) during the Five Year Plan, 1970–5 (Rahad Corporation 1977). Phase 2, which was completed during 1977–83, increased the scheme to 300 000 feddans.

Between 1955 and 1985, the number of very large irrigation schemes increased from one to four (Fig. 19.1) and the area under such schemes nearly trebled from one million to 2.8 million feddans. Other changes were also significant. Holdings in the old Gezira were of 40 feddans of which half was cultivated and the rest was left fallow at any one time; in the Managil Extension one-third was left fallow. Greater intensity was obtained in the Gezira by reducing the size of holdings, usually to 20 feddans, and by limiting the fallow area to one-quarter. In the Managil Extension fallow was soon eliminated altogether and at Khashm el Girba (now renamed New Halfa) and Rahad no provision for fallow was made.

Operation of large irrigation schemes

The general system of organization adopted for the new large irrigation schemes was similar to that already established in the old Gezira Scheme. The time set for the development of new schemes was limited and in the 1950s and 1960s the Gezira Scheme was still believed to be a success. The essential elements have been the provision and maintenance of dams and major canals by the Ministry of Irrigation with the management of the scheme in the hands of a government corporation: the Sudan Gezira Board (for Gezira/Managil); the New Halfa Agricultural Production Corporation; and the Rahad Corporation. Each scheme is divided up into tenancies and, with some exceptions, tenants agree to grow a fixed area of cotton and other crops as laid down by management. Cotton and wheat are marketed by the scheme management and profits after the sale of crops are divided three ways between the Sudanese Government, scheme management, and tenant. The tenant can dispose of other crops himself. Since 1981, the share arrangement has been changed from what was known as a 'joint account' system to an 'individual account' system, the latter relating the tenant's rewards more closely to his input.

The distribution of profits between the three parties has also varied over time, the original for the Gezira Scheme being Government 40 per cent, tenant 40 per cent and management 20 per cent, but the basic concepts have been in continuous use since 1925. Scheme organization is essentially on an authoritarian top/down basis which leaves little room for local needs and aspirations. There are tenants' organizations on each scheme and tenants are represented on the boards of management, but in practice they wield little real influence. In turn, the scheme management is charged with implementing government economic policy.

By the time the Gezira principles were being introduced to the New Halfa and Rahad Schemes, there was criticism of the Gezira system itself. The first of the main criticisms of the Gezira replication policy was the lack of freedom afforded to the tenants who, in practice, were little more than agricultural labourers. In practice, the sanctions are not applied efficiently, but the system as such leaves little room for peasant initiative and the aim of creating a responsible farming community on schemes is not attained under such a programme.

The second criticism was associated with the development of 'landlordism' (Arifi 1975), the tendency for tenants to find off-scheme jobs in towns or elsewhere, arranging for someone else to cultivate the tenancy on some sort of crop-sharing basis.

Education and other services are well provided on the large irrigation schemes but school leavers are not attracted to working within these schemes. Those with initiative and drive tend to move away, but often they were compelled to do so by the lack of new tenancies. One of the reasons for reducing the size of tenancies and for increasing the intensity of cultivation in the Gezira was to accommodate tenants' sons within the scheme.

When first established in 1925 the situation in the Gezira Scheme was very different from that found in the 1970s. Provided that skilled personnel were available, the scheme was technically sound, so expatriates were recruited to ensure success. Cotton (for cash), sorghum (for food) and hyacinth bean (*Lablab purpureus*) for fodder, all grew well and the use of fallow, which minimized the need for inputs, was well understood. There was a guaranteed market for long staple cotton in Lancashire and the surplus from other crops found a market either within the scheme or in neighbouring towns. The tripartite division of profits was also acceptable being based on traditional practices which were used on small, animal-operated irrigation schemes on the Blue Nile (Tothill 1948).

At the early stages of Sudan's development there appeared little option but to have an authoritarian top/down approach to development. The concept of a large-scale irrigation scheme was strange to the local society, but was justified because it guaranteed food supplies for the peasant cultivators' families and animals, as well as providing them with a cash income. The scheme's popularity was demonstrated by the large numbers of people who applied unsuccessfully to obtain tenancies.

Problems escalated in the Gezira because the system of management did not evolve as the situation changed. The tenants are no longer illiterate and they recognize the need to work together effectively on a large irrigation scheme. However, in spite of falling prices, cotton has remained the main cash-crop. Other crops have changed considerably over the years (Fig. 19.2), cultivation has become more intensive, and the need for increased inputs of fertilizer and labour has increased. Changes have occurred in one part of the system but not throughout, so that problems such as the supply and cost of inputs have created increasing strains on the system.

Although cotton provides a poor return to the cultivator, it is still Sudan's staple export. The Sudanese Government is genuinely concerned about the problems on large irrigation schemes, but fears that a suitable alternative to cotton may not be found and is very cautious in its approach to change. Such large sums of money are involved in foreign loans, necessary

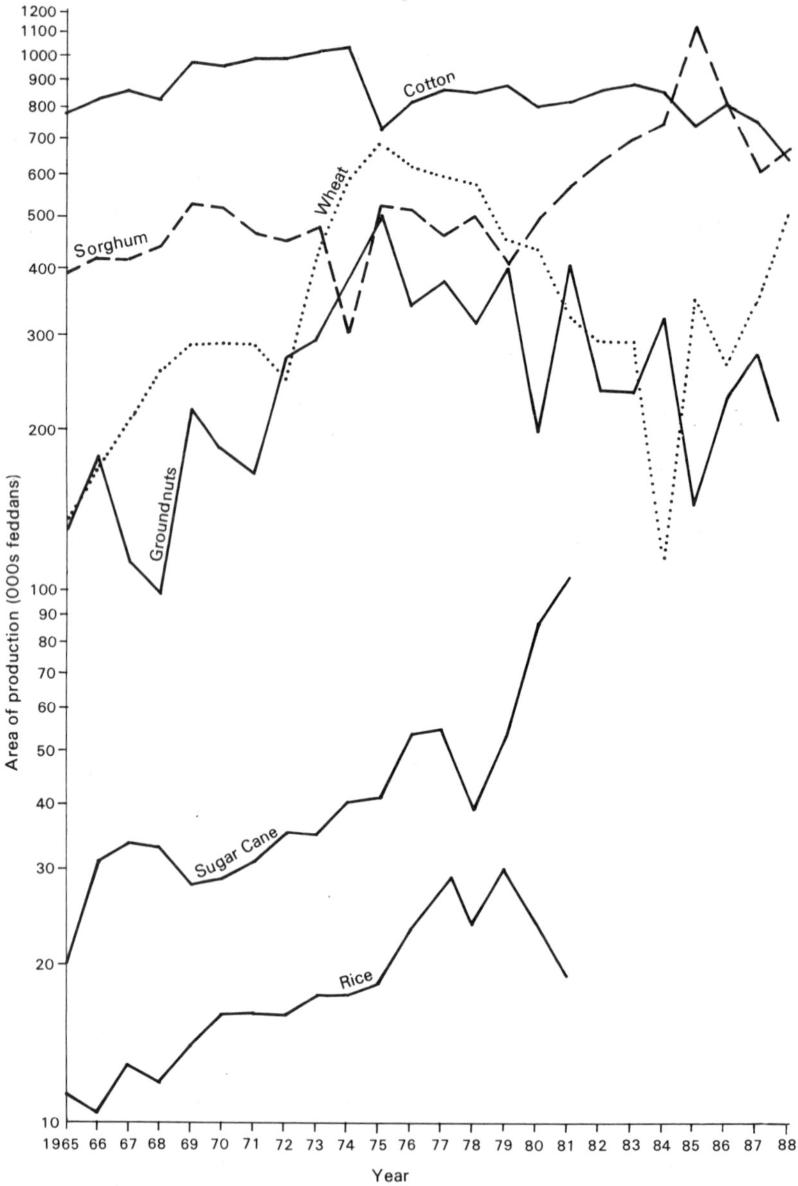

FIG. 19.2. Sudan: main irrigated crops, 1965–88.

imports, and government revenue that it is unrealistic to expect anything but a top/down approach to the management system. The result is an increasing conflict between management and participant, demonstrated clearly in the New Halfa Scheme discussed later.

Other factors have contributed to the shortcomings associated with large irrigation schemes. Rapid expansion of the irrigated area and increased intensity of land use on the schemes have coincided with prosperity in the oil-rich Middle East and economic difficulty in the Sudan. As a result, the increased demand for skilled management to maintain new schemes adequately has coincided with a brain drain of skilled manpower to Middle East countries which offer high salaries. For the educated classes, the attractions of the Khartoum conurbation are enormous. During the regular periodic shortages of essentials the capital is always supplied first. Here too is the centre of administration and influence. Although the large schemes are now linked together and to Khartoum and Port Sudan by a tarred all-weather road, other factors continue to operate against the welfare of rural development schemes at a time of skilled manpower shortage.

Importance of the large irrigation schemes to the production of Sudan's major crops can be seen from Table 19.1, but their performance has fallen short of expectations. The official targets may have been too optimistic, but they formed the basis upon which Sudan borrowed money and raised loans overseas. Clearly, the shortcomings of the large irrigation schemes have contributed to the creation of Sudan's estimated $US10 billion debt (Ballal 1987*a*).

Pump schemes and other forms of irrigation

In addition to the large state enterprises, there are over 2000 other irrigation schemes in the Sudan ranging from those of a few feddans, using ancient practices such as the *shaduf* and the animal operated *saqqiya* to the very large schemes using diesel or electric pumps. Increasingly, however, the *shaduf* and *saqqiya* are being replaced by small diesel and electric pumps. At present, in 1989 the largest single pump scheme is the 162 000-feddan Kenana Sugar Scheme on the White Nile.

In 1968/9, 21 per cent of the irrigated Egyptian cotton crop, about half the irrigated sorghum, and a wide range of fruit and vegetable crops were grown in private pump schemes. Before the New Halfa Scheme opened in 1964 private pump schemes were also responsible for growing the country's entire wheat crop.

The Sudanese Government had reservations about the development of large-scale, private irrigation schemes. These reservations, together with internal political factors, and the amount of indebtedness to the Agricultural Bank of the Sudan, caused in part by the falling world cotton prices, led to the Agrarian Reform Programme of 1970/1. This required all schemes using pumps with a diameter in excess of 6 inches to be nationalized (Magar 1986, p. 180), but in practice affected only White Nile and Blue Nile pump schemes; only one scheme was nationalized in the Northern

TABLE 19.1. Contribution of large schemes to production of main crops, 1985/6

	Large schemes[a]					Sudan			Six-year plan target		
	% of total irrigated area ('000 feddans)	% total irrigated production ('000 t)	% of Sudan total area ('000 feddans)	% of Sudan total production ('000 t)	Yield (kg feddan^{-1})	Total area ('000 feddans)	Total production ('000 t)	Yield (kg feddan^{-1})	Total area ('000 feddans)	Total production ('000 t)	Yield (kg feddan^{-1})
Sorghum	61	78	5	14	739	13155	3595	273	9100	1900	411
Wheat	75	69	75	69	511	360	199	553	890	311	750
Groundnuts	99	99	15	34	648	951	274	288	2900	839	541
Cotton											
Long staple	83	89	83	89	501	484	226	467	790	483	707
Medium staple	69	79	69	79	781	272	184	676	350	105	849
Short staple	0	0	0	0	0	60	13	217	95	19	200
Total cotton	78	85	72	82	542	816	423	518	1490	607	621

[a] Gezira/Managil; New Halfa; Rahad.
Source: based on Department of Agricultural Economics, MAFNR (1987); MNP (1977).

Region (Ali 1986). Since then, pump schemes have ceased to be important for cotton production.

A whole series of different organizational arrangements have been tried in an attempt to improve the efficiency and productivity of the larger pump schemes. Some arrangements have included an increase in the farmers' share of net profits and the setting aside of a part of the profits for local government and other social services. However, it seems that the profitability of many schemes has been attained only by subsidizing the water rate charges. The difference between these charges and the cost of providing water for government schemes on the White and Blue Niles was in excess of £S50m. over the three seasons, 1980–3 (Magar 1986, pp. 180–1).

Some relaxation of government control and some denationalization of pump schemes took place during the late 1970s. In 1979, 258 000 feddans were in privately-owned pump schemes of which 208 000 feddans were north of Khartoum. Nevertheless, this was only 6 per cent of the area under irrigation. Since 1984, plans have been prepared to denationalize many of the White Nile pump schemes with compensation to the former owners. However, the Sudanese Government will keep control of all the sugar schemes, apart from the Kenana Scheme, which is a joint venture between Government and foreign private capital. Co-operative pump schemes have, so far, proved particularly inefficient in northern Sudan (Suliman 1988).

Along the Nile in the Northern Region, basin irrigation is a traditional method of irrigation which utilizes the abandoned meanders of the river which are still liable to inundation during the flood season. The area flooded depends on the height of the Nile flood in any one season (Table 19.2). In this situation, pumps are used to minimize variations in the irrigable area from year to year, thus converting it from seasonal to perennial irrigation. Basin cultivation adapted in this way remains a vital part of the agricultural production along the Nile in the Northern Region. Much of the growth in the area under pump schemes in the Northern Region owes its origin to basin irrigation.

Gerif cultivation, which refers to land that can be cultivated after the

TABLE 19.2. Basin irrigation in the Northern Region

	Flooded area (feddans)		
	Shendi Reach	Dongola Reach	Total
1938, very good Nile	41 000	62 280	103 280
1941, poor Nile	6050	4680	10 730
Good average year	38 000	40 000	78 000

Source: Barbour (1961).

TABLE 19.3. *Gerif* cultivation on the main Nile in the Northern Region

	Year	
	1971/2	1972/3
Area (feddans)	25031	1246

Based upon returns for sorghum, bulrush millet, wheat, maize, haricot beans, faba beans, chickpeas, onions and barley.
Source: MAFNR (1974).

river levels begin to fall is another form of irrigation dependent upon the Nile flood. The role of pumps and other water raising devices for irrigating *gerif* land is the same as for basin irrigation, but there are no reliable figures. Annual variations in the area of *gerif* cultivation (Table 19.3) demonstrate the importance of the variability of Nile flow in the Northern Region.

Some areas away from the Niles are also irrigated. These vary from tiny garden and orchard plots using well water to rather larger enterprises based upon the rainy season flood of rivers such as the Gash and Baraka (Tokar Delta) supplied by water from Ethiopia, and Kor Abu Habl supplied by flow from the Nuba Mountains. Several other pilot irrigation projects have been developed of which the best known is the Sag el Na'am Scheme for irrigating 3000 feddans in North Darfur. Here, pumps are lifting ground water from a depth of 90 m, but the high irrigation costs are justified on social rather than economic grounds (Thimm 1979, p. 27).

The Gash, Baraka, and Khor Abu Habl Schemes provide widely different areas suitable for irrigation from year to year. The Khor Abu Habl is the smallest and has never been very successful. It began in 1947 with plans to irrigate up to 15000 feddans, a figure subsequently reduced to 5000 feddans. It was planned to produce short staple American cotton and as late as 1971/2 it had 2615 feddans under cotton. Since then, however, the scheme has been considered uneconomic and appears to have been slowly abandoned. Previously, yields compared favourably with rainland cotton grown in the Kordofan Region.

The waters of the Gash and the Baraka have been used to promote crop production and dry season grazing for generations, even though they display an alarming degree of variability from year to year. For the Gash, estimated total discharges in the flood season vary by a factor of ten, while variations on the Baraka are even greater. Data on the areas under cultivation are incomplete, but provide an indication of variations in crop area from year to year (Table 19.4). Variations in the Tokar Delta reflect variations in flood. Those for the Gash Delta do not correlate so readily, because of the rapid expansion of the alternatives of rainland sorghum

TABLE 19.4. Variations in the area under main crops in the Gash and Tokar Deltas (1975/6 to 1981/2) ('000 feddans)

	1975/6	1976/7	1977/8	1978/9	1979/80	1980/1	1981/2
Gash Delta							
Sorghum	50	16	32	24	10	20	37
Castor	46	11	17	34	23	27	9
Millet	—	—	—	—		5	5
Total	96	27	49	58	33	52	51
Tokar Delta							
Cotton	64	13	46	22	13	14	20
Sorghum	28	12	28	64	10	26	22
Millet	8	7	22	20	10	12	15
Total	100	32	96	106	33	52	57

Source: Yearbooks of Agricultural Statistics plus supplements (various years).

production in the Gedaref area and the development of the New Halfa Scheme. In the Gash Delta, the area under sorghum increased from 27 000 feddans in 1982/3 to 54 000 in 1983/4 (a poor year for rain) and to 65 000 feddans in 1984/5 (the year of rain failure). It remained at this level (62 000 feddans) in 1985/6, but dropped to 30 000 feddans in 1986 despite improved rainfall. This was not due to lower yields because sorghum yields at Gash compare well with those at New Halfa and on the Gedaref rainlands (Table 19.5).

The value of irrigation development

Economic assessment

The primary aim of irrigation development was to improve economic development by boosting exports and substituting imports. Over the 30

TABLE 19.5. Sorghum yields in eastern Sudan (kg feddan^{-1})

	1984/5	1985/6	1986/7
Gash Delta (flood)	400	440	600
New Halfa (gravity irrigation)	500	380	550
Gedaref MCPS[a] (rainland)	78	300	345

[a] Mechanical Crop Production Schemes
Source: based on Yearbooks of Agricultural Statistics plus supplements (various years).

year period from 1956 to 1986 the population of the Sudan approximately doubled to about 21 million. Over the same period cotton production, still the main source of export revenue, increased by 50 per cent (60 per cent under irrigation); sorghum production increased four-fold (five-fold under irrigation); and irrigated wheat production increased 11-fold.

In 1956, the groundnut crop was 65 000 t and virtually none of it was cultivated under irrigation; by 1986 the crop had increased more than four-fold and more than one-third was grown with irrigation. Sugar production was negligible in 1956; today it is an important crop under irrigation. In 1985, 518 000 t were produced.

By contrast, because of changes in eating habits, particularly among urban dwellers, wheat imports have increased nine-fold since 1956. Sugar imports amounted to 10 per cent of all imports by value in 1956 and by 1985 Sudan, for the second year running, had recorded a small surplus due to further developments of the Kenana Scheme.

Import substitution has had only mixed success. Imports of sorghum are no longer required in normal years, but the import of wheat has increased dramatically. The export drive based on cotton under irrigation has not kept pace with population growth. In 1956, agricultural exports accounted for over 94 per cent of trade by value; in 1983/4 they still accounted for 93 per cent. By now, trade was somewhat more diversified and less dependent upon cotton, whose relative importance had decreased by 21 per cent. In spite of efforts to improve Sudan's economy through irrigation, the proportion of exports from irrigated land was less in 1983/4 than in 1956 (Table 19.6), although this was partly due to the import substitution programme.

Irrigation was also seen as a way of increasing yields and reducing yield fluctuations from year to year. In fact, the trend was more towards increased use of inputs, with fertilizer imports up from 2 per cent by value in 1956 to about 10 per cent in 1983/4. Unfortunately, shortages of inputs, particularly fuel, fertilizers, and pesticides at critical times, have led to

TABLE 19.6. Sudan: exports by value 1956–84

Commodity	1956 %	1983/4 %	% Change
Cotton	71.0	50.0	−21.0
Oil seeds and cake (sesame and groundnuts)	10.5	16.9	+6.4
Livestock and livestock products	3.9	12.1	+8.2
Gum arabic	8.2	9.4	+1.2
Sorghum	0.6	4.6	+4.0
Others	5.8	7.0	+1.2
Imports as % of exports	69.0	180.0	
Proportion of exports from irrigated land	65.0	55.0	

Source: Ministry of Finance and Economics (1956); Europa Publications Ltd (1988).

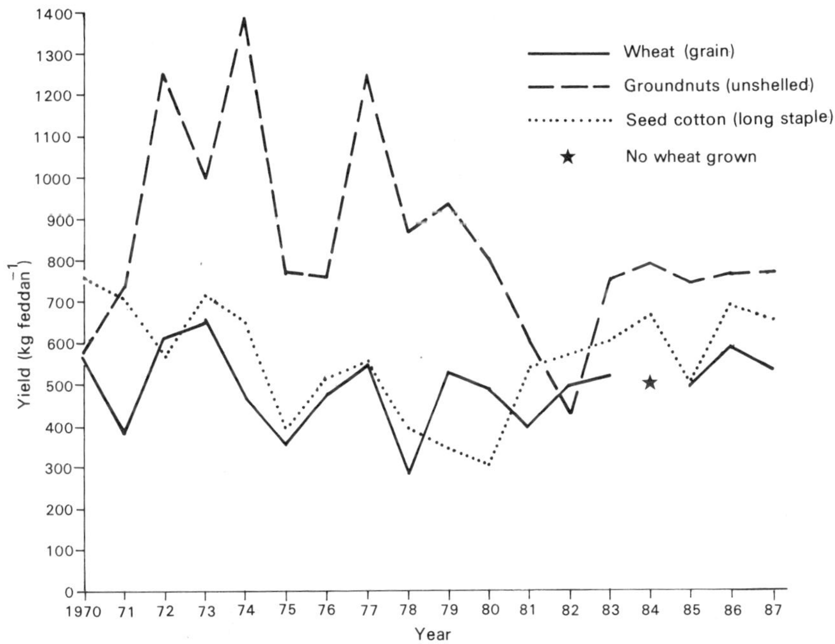

FIG. 19.3. Crop yields in the Gezira/Managil Scheme, 1970–87.

great fluctuations in yields (Fig. 19.3). One of the most disturbing features has been the tendency for yields to fall on new schemes such as New Halfa and Rahad. Over the 7 year period, 1979–86, average yields of seed cotton were 0.52 t feddan^{-1} for New Halfa compared with 0.84 t for Rahad, a much newer scheme.

Irrigation expansion has not been the answer to Sudan's economic problems. Imports, which amounted to 69 per cent of exports by value in 1956 had grown to 180 per cent of exports by 1983/4; over the same period foreign debts rose from $US16m. to $US6 billion (Ministry of Finance and Economics 1956). The current development package for the rehabilitation of the Gezira Scheme involves a total sum of US$264m. (Table 19.7) and reflects the deteriorated situation on many irrigation schemes.

Social assessment

There were social as well as economic reasons for developing irrigation.

Permanent settlement Irrigation was seen as a way of providing the rural peasant with security by ensuring him a regular food supply and source of cash income irrespective of the vagaries of the rainy season. Catch crops

TABLE 19.7. Sources of funding for the Gezira Scheme Rehabilitation Programme (total sum involved: US$254m.)

Sources	% of total
Kuwait	31
World Bank (through IDA)	30
Arab Fund	16
Japan	10
Saudi Arabia	7
United Kingdom	3
Italy	3
	100

Source: *Sudanow*, July 1987.

could still be grown on rainland near the scheme but they would be additional rather than part of the essential subsistence requirements. More permanent settlement and cash income were seen as a basis for providing improved health, education, and other services.

The 1955 census reported that 14 per cent of the population was nomadic, but this excluded many semi-nomadic movements (Davies 1988). In many quarters nomadism was seen as the antithesis of development and so both the New Halfa and Rahad Schemes included the settlement of the Butana nomads within their programmes. The 1983 census reported a reduction in nomadism to 11 per cent and, in Kassala Province containing both the New Halfa and Rahad Schemes, nomadism was reduced from 40 per cent to 12 per cent. However, the success of settlement programmes has been more apparent than real, for the number of nomads has actually increased substantially in the Sudan from 1 406 000 in 1955 to 2 192 000 in 1983. Even in Kassala Province much nomadism still exists in practice for many tenants put more effort into off-scheme livestock rearing than into their tenancies. On the Gummuiya Scheme, a 10 000-feddan pump scheme south of Omdurman, intended to produce fruit and vegetables, much of the land is used for producing irrigated fodder crops such as *bersim* and the short-season sorghum cv. Abu Sabaeen (Briggs 1986).

Management versus farmer It has been argued that on irrigated schemes farmers should specialize in cash-crops so as to obtain a cash income with which to purchase basic necessities. Not every farmer should need to grow sorghum and, as the crop can be grown on rainland outside the scheme, irrigated land should be utilized for more valuable crops. No provision was made for sorghum in the original programmes for the New Halfa and Rahad Schemes, but pressure from tenants changed this (Table 19.8).

TABLE 19.8. Area of some large schemes under main crops, 1984/5 and 1985/6

	Gezira/Managil		New Halfa[a]		Rahad[b]		Suki[c]	
	84/5	85/6	84/5	85/6	84/5	85/6	84/5	85/6
Total area ('000 feddans)	1113	1324	185	241	266	248	58	63
Notional size[d]	1813		401		300		87	
Cotton (per cent)	43	30	49	29	49	48	43	43
Sorghum (per cent)	38	44	33	58	26	36	36	55
Wheat (per cent)	0	18	0	12	0	0	0	0
Groundnuts (per cent)	19	8	18	1	25	16	21	2
Cropping intensity (per cent)	61	73	46	60	89	83	67	72

[a] Original crops: cotton, wheat, groundnuts, with sugar-cane on a Government plantation.
[b] Original crops: cotton, groundnuts.
[c] Pump scheme on Blue Nile.
[d] From Nile Waters Master Plan.
Source: based upon Yearbooks of Agricultural Statistics plus supplements (various years).

1986/7 saw a drop to 37 per cent of the cropped area under sorghum at New Halfa and 27 per cent at Rahad.

The New Halfa Scheme has not been as successful as anticipated (Davies 1986*a*; Thimm 1979, p. 22) and it highlights the frequent difference in perception of development between planner and farmer. Since the 1960s the Sudanese Government has aimed at improving the trade balance by boosting agricultural exports and substituting imports. The New Halfa Scheme should produce cotton for export, wheat and sugar, from a Government sugar plantation, to reduce imports, and groundnuts to fix nitrogen, thus reducing the need for imported fertilizers. This could provide a basis for the development of an industrial node based upon crop processing. Labour could be provided by resettled Nubians displaced by the Aswan High Dam and also local Shukriya nomads, who could be settled. In 1980 there were 22 367 tenants on the New Halfa Scheme of which 67 per cent were nomads (about half were Shukriya) and 29 per cent were Nubians (Abu Sin 1983). 330 000 feddans were to be cultivated by these tenants and a further 25 000 feddans were for a sugar plantation.

The nomads resented the development of the New Halfa Scheme on their ancestral land and they took up their share of tenancies to retain their rights, not because they approved of the Scheme. This is demonstrated in contemporary Shukriya folk songs (Abu Sin 1985). The nomads regard development as primarily the provision of a secure food supply for themselves and fodder for their animals; money is important but not their first priority. They are disillusioned with the scheme because of the exclusion of livestock, the original omission of sorghum from the crop schedule, and the authoritarian management.

The Nubians who are irrigation farmers and wheat-bread eaters should have been more favourably disposed towards the Scheme, for it was planned in order to persuade them to settle there. Like the nomads, they dislike the authoritarian management and they argue that with more freedom and more voice in the management of the scheme they could make it more successful. They maintain that many services and improvements to their standard of living promised by the Sudanese Government were never fulfilled. Both nomad and Nubian point out that their lack of confidence in the scheme management is justified by the shortcomings of the sugar plantation and factory. In Abu Sin's 1985 survey, 24 000 feddans of sugar cane were grown but only 15 000 feddans were harvested. Sugar production reached 76 000 t in 1974, but by 1982 it had fallen to 36 000 t. There appears to have been some improvement since then with 60 000 t produced in 1985. Tenants have put most of their effort into off-scheme activities such as livestock and trading. This is best demonstrated by the cropping intensity which was 46 per cent in 1984/5 but rose to 60 per cent in 1985/6, the best result for many years, due partly to the effects of the 1984/5 drought (Table 19.8).

On the Rahad Scheme, an attempt was made to integrate livestock into the programme, with an allocation of 38 000 feddans for fodder for dairy development. At a later stage 10-feddan plots were also created for irrigated grazing land. Nevertheless, local nomads feel that their needs are not met and they resent the Rahad Corporation's attempts to cull tenants livestock by 20 per cent.

The effects of irrigation development on the local environment

The construction of dams across major rivers in the Sudan have had significant effects both upstream and downstream. When the Sennar and Jebel Aulia Dams were built, the increased river level upstream not only flooded former agricultural land but also affected the traditional pattern of land use in areas where agriculture had always relied upon the annual rise and fall of the river. On the other hand, the maintenance of a given water level for a period of time, has made irrigation by pump and other water-lifting devices easier and more economic where suitable land is available. Fishing is also more feasible. The situation has been different for the Roseires and Khashm el Girba Dams because of the different local environmental conditions.

In the case of the Jebel Aulia Dam, money was set aside for alternative livelihood schemes. The original plans included provision for people affected on the upstream and downstream sides of the dam, and the main provision was to provide government-operated irrigated areas. These included the Abdel Magid area of the Gezira Scheme and various pump schemes along the White Nile. The Gummuiya Scheme which was planned as compensation for those downstream of the dam did not proceed at the time, although the project was resurrected in 1974.

Downstream of the Khashm el Girba Dam there have been many largely unexpected effects. 95 per cent of the flow of the Atbara River takes place between July and December and although the silt content does not match that of rivers such as the Indus, it is greater than that of any other major river in the Nile Basin (Abdel Ati 1979, p. 33). The dam caused silt to become trapped within the reservoir and this halved its storage capacity between 1964 and 1979. It was feared (Ministry of Irrigation 1979, p. 17) that soon it would not be able to store enough water to serve the New Halfa Scheme effectively. Downstream, the effects have also been catastrophic. The annual fertilization of the river banks and bed no longer takes place for the river has become an eroder of the banks rather than a depositor of silt. The old system of peasant cultivation after the annual flood had subsided, has been severely restricted and rendered less profitable and this has been exacerbated both by the reduction in flow of the river, now confined between August and October due to the requirements of the New Halfa Scheme, and also by a rise in the water saltation levels (Abdel Ati 1979, p. 34).

The reduced flood of the Atbara River has left less land available for dry season grazing by the nomads and fishing along the river has almost ceased. The dam has prevented large quantities of uprooted trees and other timber reaching the lower Atbara and this has created a marked shortage of wood for fuel and construction. This shortage has, in turn, led to increased degradation of the rainland vegetation.

Abdel Ati (1979, pp. 35–7) reports the case of El Besli village which lost all its *gerif* land and many of its date palms through bank erosion. Formerly the village had a large number of dom trees (*Hyphaene thebiaca*) and a mat-making industry based on the fibred leaves of these trees. The loss of river-borne timbers led to the felling of many dom trees and to the virtual extinction of the mat-making industry.

Some of the people affected adversely by the Khashm el Girba Dam have been found tenancies and other work associated with the New Halfa Scheme. Small pumps have been installed to tap groundwater along the Atbara and 60 per cent of cultivators now rely on these (Abdel Ati 1979, p. 60). Of the 18 pumps operating in the late 1970s, 13 were privately owned, 4 were owned by co-operatives and one was government owned. The inevitable social and economic changes brought about are considerable and the problems created in the lower Atbara were recognized by the Sudanese Government by the late 1970s.

Current proposals envisage the development of irrigation based on ground water on the lower Atbara (10 000 feddans) and an expansion of the New Halfa Scheme by a further 500 000 feddans requiring the building of the Rumela Dam (Ministry of Irrigation 1979, p. 17). The present economic situation and the less than effective operation of existing large scale irrigation schemes in the Sudan have put these solutions a long way into the future.

The widespread development of irrigation along the White Nile has upset the traditional pattern of pastoral activity in the area. The almost continuous frontage of pump schemes has seriously inhibited river access for nomads who return to the river in the dry season to take advantage of the river bank and island grazing (Khogali 1986). This has caused frequent disputes between settlers and nomads and has led to increased overgrazing in the areas still accessible. River fluke infection has also become an increasing problem.

The White Nile pump schemes attract settlers from the west and these people also increase the pressure on the rainlands for farming, firewood, and construction timber (Trilsbach 1983, 1986). Together with the Khartoum area they provide a market for meat, cheese, and other livestock products, so there is every incentive for the pastoralist to maintain his animal numbers at a high level. Land degradation is the inevitable result.

The effects of irrigation development on the national situation in Sudan

Irrigation developments along Sudan's rivers have required very large capital investments. Estimates for the cost of the Rahad Scheme in 1977, subsequently found to be inadequate, amounted to considerably more than the country's annual export earnings. Irrigation schemes have also required a great deal of indirect investment in infrastructure; the only coherent tarred road network serves the Gezira, White Nile, and Khartoum Province areas and connects these with Port Sudan via the Rahad, New Halfa, and Gash Delta Schemes.

Investment in irrigation along the Niles and Atbara Rivers has taken place at the expense of other parts of the country and has been subsidized by the less developed areas, particularly the west. It has been argued that although Darfur and Kordofan provided nearly 50 per cent of Sudan's exports over the period 1984–6, they received only 5 per cent of the production inputs (Ballal 1987*b*).

Irrigation in the southern Sudan

In the south there are three main irrigation schemes along the Nile at Mongalla (40 000 feddans) and Melut (56 000 feddans), both for sugar production, and at Malakal (10 000 feddans) for rice production (Ministry of Irrigation 1979, pp. 13–14). Hardly any rice has been produced from the Malakal Scheme since the mid-1960s, and there has been little evidence of sugar production from either Mongalla or Melut. The Melut Pump Scheme, first scheduled for development in the Ten Year Plan, was to act as a demonstration project to the local Dinka and Shilluk to show the advantages of irrigation during the dry season and of cash crop production. The original scheme was for a three-course rotation of cotton–rice–fallow, but it became a sugar plantation after a decision in the early 1970s to concentrate rice production in the Gezira. By contrast, flood irrigation of rice seems to have been more successful at Aweil in Bahr el Ghazal Province. Although the area has fluctuated dramatically during the 1970s from 0 to 20 000 feddans, 10 000 feddans were reported under cultivation in 1981.

There are a number of plans to extend irrigation in the south (Binayson and Dima 1986, pp. 56–7). In addition to Mongalla, areas south of Malakal, earmarked for irrigation in the Nile Waters Master Plan (Ministry of Irrigation 1979, p. 13), totalled 250 000 feddans of which 50 000 were related to the Jonglei Canal project. Although the idea of reducing water losses from the Sudd area had been discussed throughout this century it was not until after the Second World War that any serious attempt was made to review the implications and the likely local impact of any canalization scheme (Sudan Government 1954). If plans for future expansion of irrigation in the Sudan (Table 19.9) take place, these will far exceed the Sudan's allocation under the 1959 Nile Waters Agreement. If development had gone ahead as planned, both the Sudan and Egypt would have used their allocations by about 1990. The Jonglei Canal is designed to save about 4 milliard m^3 of water in its first stage and ultimately up to 8 milliards (Ministry of Irrigation 1979, p. 13). These savings are to be shared equally between the Sudan and Egypt.

Southern Sudanese see the Jonglei Canal as a project designed to benefit the north not the south and, in spite of considerable efforts, the effects of the scheme on the Sudd region and its population of up to one million inhabitants cannot be fully anticipated. The protagonists point out the

TABLE 19.9. Programme for expansion of irrigation to AD 2000

Irrigated area ('000 feddans)	1979	2000	Increase	(%)
Nile upstream of Malakal	40	290	250	(625)
White Nile	499	905	406	(81)
Blue Nile	2697	3633	936	(35)
Main Nile	311	593	282	(91)
Atbara	401	971	570	(142)
Total area for Sudan	3948	6392	2444	(62)

Source: Ministry of Irrigation (1979).

advantages of improved navigation, prospects for irrigation development, and control over the wide annual variations in the flooded area, currently estimated at between 8000 and 19 000 km^2 (Democratic Republic of the Sudan 1980). Those opposed to the scheme point out the various imponderables and see the only definite result as a disruption of life in the south in favour of the benefits accrued by the north. Excavation of the 360 km canal commenced in June 1978, but in 1989 was suspended due to the war.

Closely allied to irrigation development is the construction of water flow control dams for generating electricity. These have been built at Sennar, Roseires, and Khashm el Girba, and most remaining potential HEP sites lie in southern Sudan. The attitude of the Sudanese Government in Khartoum, which is resented greatly by Southerners, is illustrated in the *Nile Waters Master Plan Summary* (Ministry of Irrigation 1979, p. 13): '. . . the distance from the main load centres around Khartoum is great . . . . implementation of major hydroelectric schemes in the southern region is not appropriate'.

Conclusions

The expansion of irrigation has failed to bring about the anticipated improvement in the Sudanese economy. Increases in the world prices of agricultural products have not kept pace with those in manufacturing due, in part, to rapid technological developments in crop production in other parts of the world.

Unlike some African countries, such as Nigeria, Sudan has been unable to turn to oil for its salvation because of internal political strife. Furthermore, the high oil prices of the 1970s have exacerbated her problems as have the increases in income of some of her oil-rich Arab neighbours which have led to a 'brain drain'.

With hindsight Sudan's indebtedness to foreign countries for development borrowing has not proved worthwhile. Too much money was channelled into consumption rather than development and developments were not always sufficiently well considered.

In some ways Sudan has been running to stand still for she is just as dependent upon the welfare of her agricultural sector today as she was thirty years ago. Sudan seems to have tried too hard to develop irrigation and too many new schemes without the necessary financial and manpower backing to ensure success.

Sudan's problems are not merely those relating to the availability of inputs, skills in marketing, and shortcomings in management. The top/down organization with insufficient concern for the needs and aspirations of those people directly involved in or affected by new schemes has created distrust between government and management on the one hand and participants on the other. While the Sudanese Government has been concerned with what it sees as national priorities, its authoritarian policy has encouraged cultivators with stifled initiative not farmers with initiative. The result has been to lower the social status of the irrigation scheme tenant.

Perhaps Sudan should have avoided the development of so many large and complex schemes, but it is questionable whether there was any real alternative way of bringing irrigation water to such vast areas of the country. Where agricultural production is vital to the national economy it is easier for the authorities to invest in large authoritarian-run schemes in order to ensure that crop changes take place in the national interest.

Sudan's modern schemes have each been viewed in isolation and little real thought has been given to the ways these modern schemes should interact with existing land users, be they camel nomads, cattle nomads, indigenous rainland or riverbank cultivators. Greater success in economic and social terms might have been achieved with some kind of parallel development of irrigation, rainland farming, and livestock rearing so that each was complementary not in competition.

TABLE 19.10. Estimated water losses in 1978 from some irrigation schemes in the Sudan

	% water loss
White Nile Schemes	54
New Halfa	38
Gezira/Managil	29
Average for all irrigation schemes	30

Source: Abdel Ati (1986).

With intensive development of irrigation, investment has been concentrated in the eastern and central parts of the country where irrigation water is available and Port Sudan is accessible. This has resulted in a lopsided development of the economy of the country. More attention also needs to be paid to methods of irrigation, for water might be saved if it was used more efficiently. Some estimates of water losses are given in Table 19.10.

There seems little immediate prospect of further irrigation expansion for Sudan is already using nearly all her share of the Nile waters and the Jonglei Canal cannot be completed until major political problems in the Sudan are resolved. The admitted shortcomings of many existing schemes and the large foreign debt arising from these developments demands a period of consolidation and rethinking.

References

Sources of statistical information are primarily the *Sudan yearbook of agricultural statistics, Current agricultural statistics*, and *Agricultural situation and outlook*, all produced by the Statistics Section of the Department of Agricultural Economics, within the Ministry of Agriculture, Food and Natural Resources (1974–88).

Abdel Ati, H. A. (1979). The impact of Khashm el Girba dam on the western lower Atbara area, Sudan. Unpublished M.Sc.Econ. thesis, University of Wales.

Abdel Ati, H. A. (1986). *The problems of the agricultural sector in Sudan: an account of policy gaps*. University of Khartoum, Department of Geography.

Abu Sin, M. E. (1983). *Livestock economy and attitude of tenants: Rahad and Khashm el Girba, a comparative study*. Rahad Corporation, Khartoum.

Abu Sin, M. E. (1985). Planners and participants perception of development in the semi-arid lands of the Sudan: a case study of Khashm el Girba Scheme. In *Natural resources and rural development in arid lands* (ed. H. R. J. Davies), pp. 66–80. United Nations University, Tokyo.

Ali, A. H. A. (1986). The irrigated sub-sector of agriculture in the Northern Region. In *The agricultural sector of Sudan. Policy and systems studies* (ed. A. B. Zahlan and W. Y. Magar), pp. 161–78. Ithaca Press, London.

Arifi, S. A. El (1975). Landlordism among small farmers: the case of the Gezira tenants in the Sudan. *Sudan Journal of Economic and Social Studies*, **1**, 10–14.

Babiker, A. A., Musnad, H. A., and Shaddad, M. Z. (1985). Wood resources and their use in the Nuba Mountains. In *Natural resources and rural development in arid lands* (ed. H. R. J. Davies), pp. 30–59. United Nations University, Tokyo.

Ballal, A. I. (1987*a*). Sudan debt crisis: the deep end. *Sudanow* **12** (5), 8–11.

Ballal, A. I. (1987*b*). Suffering from neglect: development in Darfur, *Sudanow*, **12** (6–7), 23–6.

Barbour, K. M. (1961). *The Republic of the Sudan*. Athlone, London.

Binayson, S. A.-K. and Dima, S. J. (1986). Background and economy of southern Sudan. In *The agricultural sector of Sudan*. Policy and systems studies (ed. A. B. Zahland and W. Y. Magar), pp. 44–70. Ithaca Press, London.

Briggs, J. (1986). The human and physical environments come together: a case study of the Gummuiya scheme. In *Rural development in White Nile Province* (ed. H. R. J. Davies), pp. 114–36. United Nations University, Tokyo.

Davies, H. R. J. (1986*a*). The human factor in development: some lessons from rural Sudan. *Applied Geography*, **6**, 107–21.

Davies, H. R. J. (1988). Population change in the Sudan since Independence. *Geography*, **63**, 249–55.

Europa Publications Ltd. (1988). *The Europa world yearbook*. Vol. 2, p. 2481.

Hurst, H. E. (1952). *The Nile*. Constable, London.

Keesing, (1959). Nile Waters Agreement. *Keesing's Contemporary Archives*, **12**, p. 17126.

Khogali, M. M. (1986). The changing nature of nomadism in the northern White Nile Region. In *Rural development in White Nile Province, Sudan* (ed. H. R. J. Davies), pp. 65–72. United Nations University, Tokyo.

Magar, W. Y. (1986). The White Nile pump schemes. In *The agricultural sector of*

Sudan: policy and systems studies (ed. A. B. Zahlan and W. Y. Magar), pp. 179–91. Ithaca Press, London.

Ministry of Finance and Economics (1956–60). *Economic Survey*, Khartoum (Annual publication).

Ministry of Finance and Economics (1961). *The Ten Year Plan of Economic and Social Development, 1961/2–1970/1*. Ministry of Finance and Economics, Khartoum.

Ministry of Irrigation (1979). *Nile Waters Master Plan Summary*, Khartoum.

MNP (Ministry of National Planning) (1977). *The Six Year Plan of Economic and Social Development 1977/78–1982/83* (2 Vols), Khartoum.

Rahad Corporation (1977). *Rahad irrigation project*. Rahad Corporation, Khartoum.

Sudan, Democratic Republic of (1980). *Jonglei Canal*. Executive Organ of the National Council for the development of the Jonglei Canal Area, Khartoum.

Sudan Government (1954). *The Equatorial Nile Project: report of the Jonglei Investigation Team* (5 Vols). Khartoum.

Sudanow (1987). The Gezira Scheme: making a comeback, *Sudanow*, **12** (6–7) 18–91.

Suliman, S. F. A. (1988). A comparative study of three forms of organization of irrigated agricultural production in northern Sudan: a case study of the Shendi area. Unpublished Ph.D. thesis, University of Wales.

Thimm, H-U. (1979). *Development projects in the Sudan. . . .* United Nations University, Tokyo.

Tothill, J. D. (ed.) (1948). *Agriculture in the Sudan*. Oxford University Press, London.

Trilsbach, A. (1983). Desertification and rural changes in central Sudan. Unpublished Ph.D. thesis. University of Wales, Swansea.

Trilsbach, A. (1986). Agriculture west of the White Nile. In *Rural development in White Nile Province, Sudan* (ed. H. R. J. Davies), pp. 49–64. United Nations University, Tokyo.

Further reading

Abdel Mageed, Y. (1981). Integrated river basin development: a challenge to the Nile Basin countries. *Sudan Notes and Records*, **62**, 69–81.

Barnett, A. S. (1977). *The Gezira Scheme: an illusion of development*. Cass, London.

Briggs, J. A. (1981). Problems of agricultural planning: observations from the Jamu'iya development scheme in the central Sudan. *Sudan Notes and Records*, **62**, 98–108.

Davies, H. R. J. (ed.) (1985). *Natural resources and rural development in arid lands: case studies from Sudan*. United Nations University, Tokyo.

Davies, H. R. J. (ed.) (1986*b*). *Rural development in White Nile Province, Sudan*. United Nations University, Tokyo.

Gaitskell, A. (1959). *Gezira: a story of development in the Sudan*. Faber, London.

SDIT (Southern Development Investigation Team (1955). *Natural resources and development potential in the southern provinces of the Sudan. A preliminary report by the Southern Development Investigation Team, 1954 and Volume II, Maps and diagrams*. Sudan Government, London.

Sørbø, G. M. (1977). *How to survive development: the story of New Halfa*. Science Research Council Monograph, No. 6. Khartoum.

Zahlan, A. B. (ed.) (1984). *Agricultural bibliography of the Sudan, 1974–1983*. Ithaca Press, London.

Zahlan, A. B. and Magar, W. Y. (ed.) (1986). *The agricultural sector of Sudan: policy and systems studies*. Ithaca Press, London.

20

Agricultural marketing and pricing: a synopsis of current problems

P. OESTERDIEKHOFF

Agricultural marketing as an element of development policy

The subject of agricultural marketing entered quite late into the discussion on planning and evaluation of development processes and projects. Although this deficiency had already been pointed out in the 1960s (Abbott 1967, p. 393, 1968, pp. 87–92; Jones 1972, pp. 18–19), the empirical research on agricultural markets is still excessively deficient—both in relation to other topics and in view of the political significance of the subject. It may well be that economic research was not prepared to turn towards agricultural marketing as long as the 'classical' tenet—that production is what matters (Abbott 1967, p. 393)—was pre-eminent.

This attitude was reflected clearly by the poor allocation of development funds for marketing projects, although governments in many countries frequently interfered in agricultural markets, particularly output markets. In particular, exportable goods were often subject to market policies and in many countries these were supported by marketing boards which were inherited from colonial authorities. The Sudanese Government intervened primarily in the marketing of export commodities, e.g. by the declaration of marketing rules, minimum prices, the collection of taxes, and through the establishment of parastatal monopolies. The Government exercised little influence on markets serving exclusively domestic needs, until deficiencies in food provision caused public responses.

In the early 1980s agricultural marketing and price policies in Africa became a prominent issue in development thinking. It was mainly the Berg Report (World Bank 1981) which put the organization of marketing, and particularly the structure of prices, at the core of policy considerations. In principle, it is held that the growth of agricultural production generally, and of export crops in particular, is limited by obstacles emanating from the sphere of marketing. The long-lasting neglect of all aspects of agricultural marketing in favour of production has been reversed now. Huge attention is paid to marketing and price policies, with production as a dependent factor.

This formula has also been applied to the economic situation in the Sudan (World Bank 1983). In fact, empirical results suggest that marketing and price policies have impaired production and supply significantly. It can be demonstrated that they exerted a strong influence in cases where available production capacities were not fully utilized or where there were adverse changes to the output structure in relation to structural adjustment policy. In the past, reference has been made to 'price disincentives' as a consequence of the heavy tax burden on producers of export goods, and of high trade margins (Beshai 1976; Oesterdiekhoff 1979, 1983). Now, as the export capacity of agriculture has become eroded progressively, the negative results are considered serious in official circles.

Rain-fed agriculture has received major attention recently because its import demand is low and its export potential is regarded as favourable, thus allowing high net exports in the event of increased output. The Sudanese Government has paid much attention to this aspect in recent policy planning and a special study group on marketing, pricing, and incentives, which is to contribute to a new strategy for rain-fed agriculture (Sudan Government 1986), has been set up. Following the pattern of the World Bank report of 1983, the group was concerned mainly with those items which have a particular bearing on 'producer incentives', with particular attention to the tax burden, exchange rate, and trade margins.

The purpose of this chapter is to outline the subject of an intensifying debate on new development policies for the Sudan. At the beginning, a general survey of the origin and structure of agricultural markets will be given, followed by an analysis of product-specific properties of marketing systems, price determination, and revenue distribution. Finally, some options of market and price policies will be discussed.

The organization of agricultural markets

A short history of marketing

When the Sudan came under the rule of the Turko-Egyptian empire, foreign trade experienced remarkable new impulses. Traditionally, foreign trade was concentrated on Egypt and produced a constant export surplus as a result of the great foreign demand for slaves, ostrich feathers, and gum arabic. During the Turkiya a rapid increase in trade took place based on goods imported from Europe via Egypt. This was accompanied by an increased number and final dominance of traders and commercial companies from the Levant and Asia minor. Apart from dominating foreign companies, rural traders from the northern Sudan, known as *jallaba* started to establish themselves in trade. Northern Sudan had had a central function in Egyptian–Sudanese trade until the beginning of the twentieth century and, as a result, had gained a comparatively high level of urbanization.

After the abolition of the Mahdiya, trade prospects improved greatly with the Condominium Government. One of its first measures was to combat dangers on transit roads and improve the transport system. The gradual extension of the transport network (roads, railway, and shipping) brought an immediate and impressive increase in agricultural production. Besides state-organized plantations directed to export markets, small peasants' commodity production also gained ground, delivering supplies to both domestic and foreign markets.

Nevertheless, peripheral areas remained in a state of autarchy well into the twentieth century. They had only a small and irregular demand for imported goods like tea, sugar, and soap which were sold occasionally by migratory traders. By itself, the agricultural sector could hardly form marketing linkages in the absence of an appreciable degree of regional specialization. It was mainly the compulsion exerted by tax collection which forced peasants into the money economy. In addition, the allurement of prestige goods, which attracted the traditional elite and migrant workers, led to a continuous expansion of cash-cropping in the areas of traditional rain-fed agriculture. The rapid increase in agricultural exports was accompanied by a continuing expansion of domestic markets for food and other agricultural goods. Many new *suqs* (souks) for local trade emerged and old market-places increased in importance. Besides the existing trading centres such as El Obeid and Omdurman, new centres of interregional trade, such as Kosti, Kassala, Gedaref, Malakal, Wau, and El Fasher, flourished (El Fatih Shaaeldin 1984, p. 39; El Haj Bilal Omer 1985, p. 80).

Although the *jallaba* traders contributed to the dissemination of monetary and market relations and their economic and political position was greatly enhanced, they represented only the lower social levels within the trade sector before the Sudan achieved political independence. Most of the mercantile capital was of foreign, chiefly British, origin and it dominated foreign trade as well as transport, insurance, banking, and the infant industrial manufacturing (El Fatih Shaaeldin 1984, pp. 32–3). Big merchants took advantage of their financial and organizational superiority by making a double profit; (1) by the sale of imported goods which were distributed by a ramified network of rural traders who were financially dependent on wholesale merchants and (2) through the collection of agricultural products for export and by employing dependent traders, mostly *jallaba*, as agents and commissioners.

The independent state intervened in several ways in favour of the national trade capital: Sudanese traders were allocated import licenses; their profit margins in foreign trade were enlarged by the system of duty-free, 'nil-value' imports; agro-commercial activities (including pump schemes and mechanized farming) were strongly supported; and only small tax rates were demanded from private traders on domestic markets. The Sudanization, while gaining ground progressively at the village level, also paved the way for national trade capital in wholesale commerce and foreign trade (Mahmoud 1984, pp. 38–41). Previously, these areas had been dominated by Europeans, Indians, Egyptians, and other foreign nationals. The nationalization of foreign companies at the beginning of the 1970s, which played havoc with foreign positions in commerce and banking, also indicated the growing political and economic significance of the Sudan's trading classes.

Institutions and stages of agricultural marketing

The nationalization of trade and commerce did not change the structure of agricultural marketing. The hierarchy of market participants still exists, headed by wholesale merchants engaged in foreign trade. These merchants command large amounts of capital and operate on many markets all at once, assisted by agents and rural traders who are often financially dependent. At the bottom of the commercial hierarchy are the rural traders (shop keepers, bush traders) who depend on the financial resources of wholesale merchants granted as suppliers credits.

Nowadays, the village shopkeepers are the central commercial institution of villages. They sell the every-day consumer goods, buy most of the peasants' cash crops and are the favoured source of informal credits to peasants, particularly before harvest when peasants urgently need money to finance domestic supplies.

Small peasants sell their crop to village shopkeepers for several reasons: it enables them to sell small quantities at a time, depending on the financial needs of the households; it saves transport expenses and the time required to visit markets; and it provides them with a means of obtaining money quickly by liquidating parts of their stock if they are faced with financial bottlenecks. The delivery of crops to the shopkeeper in repayment for credit is often fixed by an informal contract. These *sheil* credits enable peasants to cope with the difficult situation at the end of the cultivation period (Reeves and Frankenberger 1981, p. 37).

Although the place of *sheil* in the marketing process has often been described (Shaw 1966; Wilmington 1955; Mohamed Hashem Awad 1967; El Haj Bilal Omer 1985), there is no sound evidence of its role in the trade sector. Ahmed Humeida Ahmed Ali (1984, p. 11) has quoted the results of a study by the Agricultural Bank of Sudan (ABS) in which 50 per cent of the peasants in the Umm Ruwaba District were involved in *sheil*. In the western provinces in particular, the informal and adaptable character of *sheil* suits local requirements, but the importance of *sheil* is probably underestimated in empirical studies for the people involved have no interest in revealing the procedure. There is strong reason to assume that some sort of secret *sheil* exists in many transactions between peasants and shopkeepers (Keddeman 1978, p. 74).

Since informal credits are a trade contract as well as a financial agreement, the process of marketing is biased in various respects:

1. *Sheil* contracts generally imply a considerable reduction of producer prices. Results of surveys covering several villages and areas, indicate about a 50 per cent reduction in the market price (after harvest) granted to the peasants delivering crops in repayment of credits (Ahmed Humeida Ahmed Ali 1984, p. 11; Keddeman 1978, p. 74; Oesterdiekhoff 1979, p. 31 and p. 45). However, *sheil* prices differ between individual peasants depending on their creditworthiness, the highly indebted peasants receiving lower prices.

2. Peasants' scope for decision-making is limited severely by their obligation to service *sheil* credits. This also reduces their ability to make use of seasonal price spreads and to minimize risks of crop damages and food shortages. Furthermore, *sheil* peasants do not have a chance to enter into arbitrage transactions in order to benefit from regional price discrepancies.

3. Under *sheil* conditions the interaction of the peasant with the trader, who acts as an oligopolist on the financial market and as an oligopsonist on the commodity market at the same time, can change to a monopoly/monopsony relationship. At this extreme stage of completely mortgaged harvest which prevents peasants raising other credits, they operate on a forward market which is controlled by the trader/creditor. The latter can reap a double monopoly rent because of his dominant position at both markets which are interlinked by *sheil* transactions (Beshai 1976, p. 267).

4. The Government's minimum price regulations for the most important cash crops have virtually no influence on the actual producer prices as long as these are subject to hidden credit agreements. Pricing policy interventions by the Government intended to influence production and supply, will prove ineffective when confronted with *sheil*.

The position of the trader/moneylender who sometimes disposes of the transport and processing facilities, is based on his integration into the marketing system which provides the commodities, outlets, and funds required for his village business activities. To safeguard his commercial activities, the trader has to associate, directly or via an agent, with a wholesale merchant who supplies consumer goods, buys crops, and gives credits. His relationship with the village trader is often similar to that of the village trader with the peasants.

The wholesaler deals with both manufactured commodities and agricultural products and disposes of a large number of business relations to keep up the level of transactions. At this marketing stage, too, a combined oligopoly/oligopsony coincides with an atomistic structure of supply (Ahmed Hamza Khalifa 1976, p. 213). This configuration is stabilized by a high concentration of capital, storage, and transport facilities allowing the wholesale merchants to play a multifunctional role in the marketing system. The power of disposition over marketing functions deserves the highest attention in view of conditions which are characterized by notorious deficiencies of services. According to Reeves and Frankenberger (1981, p. 39) the oligopsony of trade companies and wholesalers on the auction market of Abu Haraz (Kordofan) is secured by their exclusive disposal of storage facilities and by the provision of payments in advance to their agents, who are also shop owners in Abu Haraz, so that they can buy crops in large quantities.

Even within the wholesale trade market, power is highly concentrated as illustrated by the market structure of Abu Haraz (Reeves and Frankenberger 1981, p. 38). Gum arabic, groundnuts, and *karkadé* from about twenty-five neighbouring villages are sold to shopkeepers in Abu Haraz who act as agents for fifteen trade companies and wholesalers of El Obeid. The number of buyers does not mean competitive conditions: one company alone bought up 63 per cent of the groundnut supply, the total supply of sesame and *karkadé* and half of the gum arabic offers in November 1980 (for market power compare also Louis Berger International 1983, pp. 24–5).

The parastatal companies, being a result of the nationalization of foreign trade, represent an alliance of government and private trade, as shown by the mixed capital share as well as by their market behaviour. The private sector holds a 70 per cent share of the Gum Arabic Company's capital and a 42 per cent share of the Sudan Oilseeds Company's capital. The board of directors of both companies is dominated by private businessmen and the boards are allowed to interfere in auction markets only when minimum prices are undercut or in case the markets are not cleared. As they are bound to operate as price takers at the official floor prices, they do not actually curtail the market power of private wholesalers. As a result of the private traders' political pressure, the export monopoly of the Sudan Oilseeds Company was repealed in 1980. Several private exporters as well as a number of buyers at the regional markets stimulated the demand for groundnuts and sesame which increased prices at auction and other markets. In 1986, however, the monopoly of the Sudan Oilseeds Company was re-established because there was evidence that demand stimulation produced domestic prices considerably above world market prices. In the 1986/7 season the Sudan Oilseeds Company announced a guaranteed fixed price for groundnuts at which the Company was obliged to buy up any quantities offered.

A survey of market channels and institutions would not be complete without considering the parallel markets. In fact, only a small proportion of the marketed agricultural surplus is sold at official auction markets (see below). Although evasion from the auction procedure is only allowed when transactions take place outside the range of the next official market, buyers and sellers have a common interest in avoiding tax payments levied by the rural councils. It is difficult for peasants to reach the auction markets due to lack of transport facilities, but auctioning is not attractive to peasants anyway, for they are likely to be exposed to highly concentrated demand power and price agreements among traders.

Smuggling is an important element of the parallel markets. It applies particularly to livestock and grain, as high prices can be obtained in the neighbouring deficit countries, Ethiopia, Chad, and the Central African Republic. Using data for 1978, it was estimated that the quantity of smuggled livestock was about seventeen times as high as the volume of official exports (El Fatih Shaaeldin and Siddig Umbada 1984, p. 25). Smuggling of grain from the Kassala/Gedaref area and from the coastal region of the Red Sea is also considerable (Antoun 1986, p. 15; FAO 1982, p. 15).

Analysis of individual markets

The marketing conditions of the most important cash crops are outlined in the following sections. Results of market processes and their impact on sectoral terms of trade are also considered where data are available. An analysis of the distribution of export revenue, as revealed by price structures and price policies, throws some light not only on the interaction of producers and traders, but also on the aims and strategies of the parastatal marketing bodies and the Sudanese Government.

Groundnuts

Supply and demand The volume of groundnuts harvested has shown a slight downward trend since the mid-1970s and a violent drop in 1982/3 due to the drought. The average output amounted to 830 620 t in the preceding five years while in 1982/3 only 501 400 t were harvested. The decrease continued in 1984/5 and 1985/6 with 378 000 and 274 000 t harvested, respectively; for the 1986/7 season, 350 000 t are expected (World Bank 1983, Vol. II, p. 117; MANR 1986*b*, p. 26).

Groundnuts are cultivated by irrigation, mainly in the Gezira and Rahad Schemes, but also in the traditional rain-fed areas which produce about two-thirds of the output. While the reduction in the cultivated area on irrigated schemes has been due to management decisions, the rain-fed cultivation has been impaired by bad climatic conditions. In 1984/5 in Northern Kordofan, once the centre of rain-fed cultivation, yield per feddan was only 12 per cent of the average for 1979–84. Productivity rose during the following year, but the cultivated area has decreased to less than

20 per cent of the five years' average. The centre of production has shifted to Southern Darfur which is less susceptible to drought and now furnishes more than half of the total groundnut area and output (MANR 1986*b*, p. 26).

Whereas the export share of the total output of shelled nuts reached a low of 1.6 per cent in 1982/3, direct domestic consumption and industrial processing have demanded continuously growing portions of the output. The sales of groundnut oil have also shifted from foreign to domestic markets. This trend can be traced back to 1979 for groundnuts and groundnut oil and, therefore, cannot be attributed simply to output deficiences due to the drought. In 1973–8, the domestic consumption of shelled nuts averaged 64 per cent of the available output, but by 1982/3 more than 98 per cent was required for the home market. It was mainly the demand of the edible oil industry which rose from 32 to 61 per cent of total output, which contributed to this increase of domestic use (Oesterdiekhoff 1988, p. 28).

Marketing channels and market structures Marketing of groundnuts from irrigated schemes has several special features compared to the marketing in rain-fed cultivation. In the Gezira Scheme, the most important producer of irrigated groundnuts, but also in the Rahad and El Suki Schemes, groundnuts were introduced quite late. Organized markets were not established, so the tenants sell their crop to small village traders and to wholesalers from Khartoum. A significant quantity of the harvest was marketed by *sheil* at the beginning of the 1970s (Farah Hassan Adam and Apaya 1973, p. 111; Ahmed Humeida Ahmed Ali and Zohair Mubarak Abdalla 1976, pp. 27–30).

The marketing co-operatives established for irrigated cultivation in the 1960s, were weakened through a credit embargo by the Sudan Gezira Board at the beginning of the 1970s and were finally excluded from the market through the boycott of the private trade. During their heyday, they had been able to sell more than half of the Gezira's groundnut harvest. At the beginning of the 1980s, the co-operative sector received fresh support from the ABS which started to play a more active part in several sectors of agricultural marketing. Currently, the bank grants loans to tenants amounting to 70 per cent of the value of the harvest expected. Fifty per cent of the loan is for harvest expenses (mainly labour costs), the remainder for transport and other post-harvest services. The marketing union, comprising fifty-seven co-operatives, and the ABS organize collective marketing through agents in Khartoum and the proceeds from sales are distributed to co-operative members after the repayment of debts (Mohamed Farah Gibriel 1984, pp. 10–11).

In traditional rain-fed cultivation, marketing is characterized by a hierarchy of markets and institutions. To obtain the urgently needed cash to pay back informal credits, peasants are usually compelled to sell during the seasonal price-low at harvest time. Investigations by Oesterdiekhoff

(1979, p. 34) and more recently World Bank (1983, Vol. III, p. 56) have shown that *sheil* prices amount to about 50 per cent of the market prices.

Although the auction markets are supposed to act as regional assembly points for recording, taxing, and selling the supply of groundnuts from a respective district, only part of the total harvest is offered at auction. At En Nahud, the largest auction market for groundnuts until the drought in 1983, it was estimated that the volume turned over comprised about 50 per cent of the marketed harvest, but another estimate suggested that little more than 35 per cent of the Sudan's total output was offered at auction (Mohamed Farah Gibriel 1984, p. 12). Nevertheless, the auction markets found at El Obeid, En Nahud, Er Rahad, and Umm Ruwaba in Kordofan, Nyala in Darfur, and Tendelti, Kosti, and Ed Dueim in Blue Nile Province are important as price leaders for local pricing in satellite markets.

The market at En Nahud (Louis Berger International 1983, pp. 22–5) is characteristic of other official markets, which are used for the marketing of sesame as well as groundnuts. The auction market is frequented by village traders, transport entrepreneurs, intermediate buyers (*dubbai*), and well-off peasants, who deliver about 20 per cent of the supply of groundnuts. Peasants visiting the market primarily to sell their own harvest, have limited time available for the lengthy auction procedures or protracted sales negotiations. So, compelled by the comparatively high monetary and opportunity costs of staying away from their holdings, they tend to sell after a short period of time, even if prices are unfavourable.

The *dubbai* take advantage of this situation by buying up the peasants' supplies outside the official market and offering them to wholesalers on the auction market. The *dubbai* are able to aggregate small quantities of groundnuts into lots of about 10 kantar each day, meeting about half the needs of large buyers (exporters, industrial processors) for quick transactions. However, they do not command sufficient capital for a period long enough to participate in speculative dealings. On the demand side, there are about twenty large wholesalers, a small group of oil-mill owners, and representatives of the Sudan Oilseeds Company. Oil-mill owners have insufficient working capital to build up a large stock of raw materials at the seasonal price-low. They are forced to undertake continuous purchases in order to make full use of production capacities and they are hit most by the seasonal price fluctuations.

Before being admitted to the auction procedure, private merchants have to prove possession of a certain amount of capital, have to provide a profit tax certification, a trade licence, and the guarantee of a bank or of an important person of commercial life. Only eleven of the twenty traders at En Nahud could meet these requirements in 1982/3. These traders had sufficient capital and storage facilities to build up stocks if prices were expected to rise or, alternatively, to shift stocks to the market of El Obeid. They would fill up the warehouses at the beginning of the season immediately after harvest when prices were still relatively low and sell later to oil-mill owners and agents of the Oilseeds Company. The latter becomes active in December when it has information on the international harvest and expected export prices. As auction market prices are strongly influenced by the export demand and, therefore, recover from their post-harvest low in December, speculators can profit greatly from the price increases during the season which may amount to 30 per cent and more within two months.

After the re-establishment of the export monopoly in 1986, the Sudan Oilseeds Company covered the auction markets with a widespread net of agents and additional transport facilities in order to reduce dependency on the large suppliers. In contrast to the formerly valid rule to intervene only in case the minimum prices were undercut, in 1986/7 the Company guaranteed a fixed price to all sellers which amounted to £S55 kantar^{-1} for the large auction markets (En Nahud, El Obeid, and Nyala). Since 1984, the small supply and the relatively high domestic demand have put domestic prices above export prices and the Sudan has lost several export markets. In order to regain lost positions, the Company was compelled to intervene in domestic markets and to put restrictions on the scope of action and the margins of wholesalers. By the introduction of a new price scheme, the market power of the traders was limited because the fixed price is coupled with the Company's buying guarantee. Furthermore, it reflects the actual market situation better when compared with the former floor prices.

In spite of these innovations, the characteristic dilemma of agricultural price policy remains unsolved. Producers must be granted higher price incentives in order to supply the edible oil industry and the export trade with sufficient quantities of groundnuts; at the same time the domestic price level must be stabilized in order to remain or to become competitive on foreign markets. This implies that the Sudan Oilseeds Company has to reconsider the aim of a maximum of trade profits and export revenues and to assume responsibilities for the long-term development of oilseeds cultivation and marketing.

Prices and price structures The secular trend of nominal auction prices shows only a slow increase during the 1970s (for price trends analogous figures from MANR 1985, MANR 1986*a*, World Bank 1983, Vol 3., and Sudan Government 1986 have been used). Between 1974 and 1978, even depressive tendencies loomed large reflecting the establishment of the Sudan Oilseeds Company. The abolition of the monopoly and the existence of two competing export companies in 1980/1 caused considerable buoyancy of market prices. While the auction prices at El Obeid rose from £S5.48 to £S9.99 kantar^{-1} within one year, prices at En Nahud almost tripled from £S4.31 to £S12.61. By the 1985/6 season the prices of both markets had increased almost ten-fold to £S54.00 and £S41.25, respectively, compared to the 1979/80 level.

From 1984/5 these rapid price increases made domestic prices exceed the level of export prices. In 1984/5, the average auction market price of groundnuts at El Obeid was 22 per cent above the export price equivalent of £S70.5 kantar^{-1}. Now, the Sudan Oilseeds Company, reintroduced as an export monopoly in 1986, operates with a fixed price at a level close to the previous year's auction prices. The movement of actual market prices which were far above the minimum prices during the period 1980 to 1985 has made the minimum prices an insignificant price bottom.

The Government receives a substantial share of the export revenue from groundnuts, for the Oilseeds Company pays 65 per cent of the gross profits as business profit tax, 5 per cent of the sales proceeds as development tax and 2 per cent as stabilization fees; the Government's share of net profits amounts to 58 per cent. The Government's increasing share of export proceeds is demonstrated by the continuously widening gap between export prices and auction market prices. With decreasing marketing expenses (as a portion of the marketed value), the trade margin had increased considerably in the mid-1970s, compared to previous years (Oesterdiekhoff 1979, p. 39). From the start of the Company's activities the auction market prices tended to remain behind the increase in export prices. It was not until the monopoly was abolished and there were drastic shortages in supply, that the domestic price level increased and finally exceeded the level of export prices. The period of monopolistic export marketing is clearly reflected in the real prices (auction prices deflated by the consumption price index) at that time; between 1974 and 1980 they decreased from £S1.73 to £S1.18 kantar^{-1} at El Obeid, and from £S1.55 to £S0.93 kantar^{-1} at En Nahud (Oesterdiekhoff 1988, p. 39).

A Government report has blamed the price policies of the Oilseeds Company for the decreasing export performance since these policies were not conducive for increased production (Sudan Government 1986, p. 27). In fact, production had been decreasing long before the drought precipitated a further downward trend. Strategies aimed at increasing output levels in order to increase the share of exports without impairing the growing domestic demand, should not be based on the expectation that climatic conditions will enable the production performance of the 1970s to be recovered. Besides measures to promote agricultural productivity, the marketing system needs to be overhauled fundamentally in order to ensure that producers actually come in the reach of price incentives, export revenues are distributed more equally and steady market conditions minimize the risks of cash-cropping for small peasants.

Sesame

Supply and demand Sesame is traditionally an important source of edible oil. It is still grown in many areas primarily for home consumption, being used both for oil processing and as a grain substitute. In mechanized farming sesame is second in importance to *dura*; its high profitability contributed considerably to the rapid expansion of mechanized farming at the beginning of the 1970s.

The long-term pattern of output and yields is similar to that of groundnuts. A marked increase at the end of the 1960s and during the first half of the 1970s was followed by a stagnation and slight recession which ran into a drastic set-back in terms of output and yields during the first half of the 1980s. The significance of traditional sesame cultivation has been continuously reduced during this set-back though, with the exception of 1984/5, it has always accounted for more than half of the total sesame output. During the 1970s the traditional sector accounted for more than 70 per cent of the total sesame output; during the period 1979–84 it accounted for 66 per cent, followed by a significant decrease during the drought to 46 per cent, and a partial recovery to 58 per cent in 1985/6 (MANR 1986*b*, p. 29).

The regional centres of smallholder sesame cultivation are located in Northern Kordofan, Southern Darfur, and the Blue Nile Province, while mechanized rain-fed farming is concentrated on Gedaref and Damazin. All these cultivation areas have suffered decreases in yield. The mechanized schemes which recorded an average yield of 126 kg $feddan^{-1}$ in 1979–84 recorded only 49 kg in 1985/6, and during the same period, the yield of traditional sesame cultivation declined from 95 to 57 kg $feddan^{-1}$ (MANR 1986*b*, p. 29).

During the 1970s domestic consumption demanded an increasing share of the total output of sesame which eventually amounted to more than 80 per cent. This increase was due to the expansion of oil processing facilities. During 1974/8 about 41 per cent of the mean total output of sesame was allocated to industrial processing, compared with 54 per cent during 1978–82 (World Bank 1983, p. 122). The increasing domestic demand for oilseeds was a consequence of the subsidized expansion of industrial processing facilities leading to overcapacities far in excess of available raw materials.

Marketing channels and market structures The marketing system for sesame is essentially identical to that of groundnuts. The traders are the same, they are handling sesame as well as groundnuts, both oilseeds are turned over on the same auction markets, the same customers buy oilseeds for industrial processing, and the Sudan Oilseeds Company exports sesame as well as groundnuts.

Traditional cultivators of sesame are facing the following specific marketing conditions:

1. Storage is possible only for a relatively short period as sesame is prone to insect and parasite damage resulting in considerable weight loss. The sale of small quantities of crop spread over a long period is

generally favoured by the peasants, but this is limited by storage conditions as well as from *sheil* obligations and financial bottlenecks.

2. In monetary terms the per feddan yields of sesame are higher and there is less risk compared with sorghum because sesame is grown in areas of higher rainfall. That is why sesame is of special interest to the traders and moneylenders in *sheil* business.
3. Due to its higher specific value and hence reduced relative transport costs, sesame is a favoured mechandise.

The largest auction markets are Gedaref in Kassala Province, El Obeid and Umm Ruwaba in Kordofan, and Singa, Kosti, and Tendelti in Blue Nile Province. Only a part of the sesame production is sold via auction markets. As with groundnuts, there is a parallel market which has an even higher volume of turnover. For the years 1965–7, the share of sesame output delivered to auction markets was estimated at 45 per cent of the total, whereas at the beginning of the 1970s it was valued at only 15–20 per cent (Beshai 1976, p. 196). Official production and market statistics show figures of 17.7, 3.1, and 11.1 per cent for the three years, 1978–81 (MNP 1980/1, p. 14; World Bank 1983, Vol. II, p. 122). In Kordofan, in 1981/2, only 6200 t, i.e. 7 per cent of the total crop of 88 000 t, were put up for auction at El Obeid and Umm Ruwaba. Besides this, a small portion went to village auction markets, but the largest part of the crop was sent immediately to the edible oil industry (Louis Berger International 1983, p. 33).

Following a period of oligopolistic competition between 1980 and 1986, the export of sesame came under the monopoly of the Sudan Oilseeds Company in 1986/7. As with groundnuts, the Sudan Oilseeds Company has announced a fixed price which, for the larger markets, is set at £S61 kantar^{-1} for the season 1986/7.

Prices and price structures After an intermediate phase of stagnating price development between 1974 coinciding with the Sudan Oilseeds Company's monopoly, and 1978, rapid price increases occurred indicating general inflationary tendencies and a growing competition between export and the increasing domestic demand in a situation of stagnating and even decreasing supply. After the monopoly was removed, the factors forcing prices up came into play and the level of nominal prices on auction markets increased almost 5-fold within four years until 1984/5. In Gedaref, prices increased from £S62.4 to £S73.0 kantar^{-1} between 1984/5 and 1986/7; at El Obeid the increase was from £S64.8 to £S83.0. The enforcement of a fixed price of £S61.0 in 1986/7 reduced the actual market prices to the 1984/5 level implying that, with an annual inflation rate of about 31.7 per cent (World Bank 1987, p. 202), real prices were more than halved.

Real auction market prices indicate more clearly than the nominal prices the depressive tendencies of the monopolistic period from 1974/5 until 1979/80. In Gedaref and El Obeid, the downward pressure on real prices was greater than in Umm Ruwaba so they occasionally amounted to less than half of the 1974/5 level. The fact that the minimum prices in El Obeid amounted to 49 and 45 per cent, respectively, of the export prices in the years 1976/7 and 1977/8, after they had been between 63 and 73 per cent before 1974/5 (Oesterdiekhoff 1979, p. 51), indicates that the real price losses are not due to decreasing export prices, but to deliberate price policy aimed at increasing the margin between domestic and export prices.

This intention becomes evident when the purchase prices, determined by the Company for purchases from merchants, are compared with export prices. During the 1976/7 season, the Company offered a purchase price exceeding that of the previous season by 1.2 per cent, although the export price had increased by 9.2 per cent. In 1977/8 it increased its price offer by 1.4 per cent, but could realize an export price increase of 22.3 per cent (Sidgi Awad Kaballo 1984, p. 38). It is, therefore, appropriate to mistrust the official declaration that the purchase prices and the minimum auction prices were determined in accordance with the international price development.

An econometric study based on market data from El Obeid for the period of 1942/3 to 1977/8 concludes that, in Kordofan, the small farm-holdings producing sesame are reacting positively to price incentives. However, their response depends more on the real than on the nominal price movement, so there is every reason to believe that peasants are not subject to money illusions (Sidgi Awad Kaballo 1984, pp. 17–22). Until recently, the Sudan Oilseeds Company has followed a price policy based on the assumption of low price elasticity of supply. It has influenced domestic prices adversely and extended the margin between domestic and export prices to such a degree that potential increases in supply are curtailed (Sidgi Awad Kaballo 1984, pp. 47–8).

With an adequate price policy it would be possible to increase both the profits of the Oilseeds Company and the income of the oilseeds cultivators. With a fixed price of £S61.0 kantar^{-1} for the 1986/7 season, which implies a reduction of the nominal and an even greater reduction of the real prices, there is reason to believe that a different course might be followed as long as there are no measures for strengthening the peasants' market position, such as the provision of improved credit and storage facilities, in order to counteract the negative effects of decreasing auction markets prices.

Gum arabic

Production and supply Gum arabic is probably the oldest export merchandise of the Sudan. It was exported initially to Egypt but also to Europe in the Middle Ages. On account of its high quality, it is the only export item for which the Sudan could achieve a quasi monopolistic position. However, the existence of a variety of substitutes prevents the Gum Arabic Company adopting a monopolistic export price policy.

The trees producing gum arabic grow on the sandy soils in a belt about 400 km wide stretching east–west across the Sudan. *Acacia senegal* (*hashab*) which produces the best quality white gum is found mainly in Kordofan while the inferior quality red gum produced by *A. sayal* (*tahl*) is collected mostly in Kassala, Blue Nile, and Upper Nile. Gum arabic is exported as unprocessed merchandise for which the Gum Arabic Company has the legal monopoly. It buys the gum from merchants who send the graded product to Port Sudan, or it intervenes directly in auction markets to prevent an undercutting of the minimum price. The quantities of gum arabic bought up by the Company have decreased over the years. In the 1960s, the mean annual supply of *hashab* and *tahl* to Port Sudan still amounted to 48 559 t, while in the 1970s it was only 36 241 t. The mean annual yield for the 6 year period, 1980–6, amounted to 33 610 t with minimum values of 14 048 and 20 422 t recorded in 1984/5 and 1985/6, respectively (figures based on records of the Gum Arabic Company, Khartoum).

This downward trend in production and supply has been the result of growing population pressure on arable and grazing land; overgrazing and the spread of mechanized farming leading to progressive soil erosion; cutting down stands of *Acacia* spp. to satisfy the growing demand for firewood and timber; competition for land and labour from other cash crops; and the southward migration from the gum belt to escape the drought.

Marketing channels and market structures At the primary market stage, gum marketing is identical to that of sesame and groundnuts. Most collectors transfer their products to village traders, lorry owners, and money lenders outside the auction market system. A quantitative assessment of the different marketing channels for gum arabic at the primary stage (GITEC 1980) indicated that almost 30 per cent of the total harvest was marketed through *sheil* contracts, another 40 per cent was bought up by village traders who were often acting as commissioners of large export merchants, and about 30 per cent was put to the auction by the collectors themselves (GITEC 1980, p. 36). In a study of agricultural market structures in Kordofan these results were, in general, confirmed, although the latter figure was estimated only at 20–30 per cent (Louis Berger International 1983, p. 47).

Gum arabic is particularly suitable for *sheil* contracts for it is completely commercialized and there are no risks with regard to its cultivation and harvesting. As peasants are normally unable to meet their debt obligations, traders can maintain a safe and cheap source of gum arabic by turning over their debts into the following season. Their profits were estimated at 70–100 per cent of invested capital during the market season of 4 to 5 months (GITEC 1980, p. 40). The prices realized by collectors who offered their goods at auction markets in many cases even lagged behind the minimum prices. As the gum is graded after sale, the traders have an opportunity to claim price cuts on account of alleged inferior quality (GITEC 1980, p. 42).

The most important auction markets are El Obeid, Umm Ruwaba, and En Nahud in Kordofan, Singa, Kosti, and Tendelti in the Blue Nile Province, Nyala and Ed Daein in Darfur, and Gedaref in the Kassala Province. The Ministry of Finance and National Economy publishes the minimum price before the gum tapping begins early in October; if actual prices fall short of the announced minimum price, the Company is obliged to purchase gum at the auction markets in order to stabilize prices. At about the same time the Company announces the Port Sudan delivery price, the price at which it buys the gum delivered by merchants to the Company's warehouses in Port Sudan. These prices are valid for one season, from the beginning of October until the end of September the following year.

After the auction markets, the gum is cleaned and graded. A large part is traded as cleaned *hashab* without further processing; the gum selected for uniform size and with a light colouring is classified as hand-picked selected, for which higher prices are obtained; small quantities are also traded as uncleaned (natural) and as breakage (dust).

Since 1969 the entire export business has been carried out by the Gum Arabic Company which holds a monopoly licence; before that the export of gum was shared between five large commercial firms. The Company's primary aims are:

(1) to stabilize Government revenues (in foreign exchange and domestic currency) from the export of gum arabic;

(2) to guarantee minimum prices on the auction markets;

(3) to support and promote production (e.g. reforestation, supply of drinking water to collectors).

The Company contributes very little investment in fixed capital to the market services (storage, processing, transport). This is out of proportion to its high gross margin (i.e. the difference between Port Sudan delivery and export prices) of 40–50 per cent of the export price and its profit rate of 20–25 per cent of the turnover (Louis Berger International 1983, p. 50).

Prices, taxes, and income distribution Since gum arabic is entirely for export, it cannot be turned over on uncontrolled markets to the same extent as groundnuts and sesame. The activities of the Gum Arabic Company have a greater impact on marketing than those of the Sudan Oilseeds Company which tends to be confronted with withdrawals of supplies and shifts to parallel markets in response to rigid interventions. Thus, the vertical price structure of gum is determined more directly by administrative price-fixing than is the case for sesame and groundnuts. The price-fixing takes a strict course from top to bottom, starting with the minimum export price from which, after deducting handling costs, export charges, and the profit margin of the company, the Port Sudan delivery price is calculated; the minimum auction price is determined after deducting transport costs, market fees, taxes, and the margins of the traders. This price is by no means the lower limit of the farm gate price, but a guarantee for the village traders who do their best to increase their profit margins by increasing the gap between the floor price and what is paid to the farmers (Blake 1983, p. 17).

Since 1979/80 there has been an increasing gap between export prices and domestic market prices, the former increasing much more than the latter. Before that time, export prices moved almost in accordance with domestic prices. While the minimum auction price has remained comparatively steady at about 70 per cent of the Port Sudan delivery price, the share of both these prices in the export prices has decreased continuously in the eighties. Apart from the boom in 1973/4, the Port Sudan delivery price had always been above 50 per cent of the minimum export price. In 1979/80, however, it dropped to 49 per cent, in 1980/81 to 44.5 per cent, in 1983/4 to 40.6 per cent, and in 1984/5 to 33.9 per cent (Oesterdiekhoff 1988, p. 64). At the same time the export prices in dollars remained stable or increased by only 10.3 per cent in 1984/5, while the domestic prices calculated in dollars dropped. The devaluation of the Sudanese currency and the stable dollar export prices enabled the Government to extend the share of public revenues in export earnings as measured in domestic currency (Blake 1983, p. 22). In this way, it managed to enlarge the trade margin of the Gum Arabic Company despite increasing prices on the domestic market and without forcing up the dollar export price. The increase in nominal prices on domestic markets was limited to the average inflation rate so the real producers' prices stagnated or even dropped.

This price policy has had a significant effect on the distribution of export revenues among collectors, traders, and the state. Before 1978/9 the producers' share was usually between 34 per cent and 38 per cent of the total export revenue; in the following years up to 1981/2 it dropped to 25 per cent (Blake 1983, p. 33; GITEC 1980, p. 42), and the downward trend continued until 1985/6 when collectors received 17.5 per cent of the export revenue (Oesterdiekhoff 1988, p. 66). Considerable increases in the domestic price of gum arabic accompanied the massive rises in the minimum export prices in 1985/6 and 1986/7. In 1986/7, minimum auction prices increased almost six-fold to £S300 kantar^{-1} compared with 1984/5. Government intended to give additional production incentives after the heavy output losses in 1984/5. Owing to its price and exchange policy, increases in the domestic price of gum arabic did not have a negative bearing on the share of export revenue reaped by the state and the trading sector. Eventually, however, the relatively stable dollar export price had to be given up. For the season 1985/6, the Ministry of Commerce increased the minimum export price for the 'hand-picked selected' quality from 1600 to $US2000 per tonne, and in 1986/7 the floor price was increased threefold to $US6000 per tonne (Sudanow 1986, No. 9/10, p. 18).

The outstandingly high export charges levied by the Sudanese Government on gum arabic range between 30 and 50 per cent of the total export proceeds. These can be traced back to the colonial era when the Government paid great attention to the gum trade. From an economic point of view, high taxes and fees on exports might be justified if the price elasticity of the home supplies and the international demand were such that they allowed a shift of tax to foreign customers. This is the case when domestic suppliers respond flexibly to price movements while the demand of importing countries is rather less elastic. In the past, such a market situation may have prevailed; nowadays with the numerous substitutes available it is questionable whether demand is inelastic. This supports the view that the Government's export revenue is a burden not to foreign customers but to the domestic suppliers. The fact that there is only little competition among traders, limited participation by the gum collectors in organized market procedures, highly unequal distribution of market power among traders and collectors, and lack of transportation and financial resources, suggests that a large part or even the total tax burden, is shifted to the collectors. As the feasibility to administer prices is considerably greater on gum markets than on sesame or groundnut markets, the Government could have adopted an appropriate minimum price policy to achieve a different distribution of tax. With the minimum prices lagging behind the market prices, however, the price policy has not prevented a backward shift of the tax burden; on the contrary, it was part of a strict market control which debited gum collectors for fiscal purposes.

Livestock

Breeding and supply In 1985, the Sudan was estimated to have 19.4 million cattle, 25.3 million sheep, 15.5 million goats, and 2.7 million camels. The annual growth rates in the 1970s and at the beginning of the 1980s amounted to 3.0, 5.7, 3.6, and 1.6 per cent for cattle, goats, sheep, and camels, respectively (AAAID 1982, pp. 9–10; Livestock and Meat Marketing Corporation, personal communication). In some areas, however, the drought resulted in a dramatic decline in animal numbers. Estimates for 1984, the worst year of the drought, put losses of cattle, sheep, goats, and camels by settled as well as migratory pastoralists in Kordofan at 80, 60, 20, and 10 per cent, respectively (El Sammani 1986).

While the western provinces are the most important surplus areas, the large terminal markets are located in the urban centres of the central region. Hence the interregional marketing channels run from west to east. The long distance between the places of production and consumption remains the dominant structural characteristic of livestock marketing in Sudan. During the rainy season most of these distances have to be covered on the hoof.

The Livestock and Meat Marketing Corporation, Khartoum (personal communication) has estimated the annual offtake at 9, 25, 35, and 1 per cent for cattle, sheep, goats, and camels, respectively; in 1984/5 the number of heads amounted to 2.22, 5.06, 5.45, and 0.12 million. Of the total offtake, the proportion for home consumption and that put on the market varies according to the kind of animal and regional particularities. However, only a small part of the market offtake is put on official markets, for at organized local markets people have to pay in cash, whereas on the parallel markets they are offered various kinds of informal credits. Furthermore, the slow auction procedure has discouraged sellers and buyers who prefer to make use of the services of unofficial brokers. About 80 per cent of the sheep and camels are marketed outside the official market of En Nahud and, in El Obeid, according to the estimates of market authorities, about 50 per cent of the sheep and 20 per cent of the cattle are disposed of on informal markets (Louis Berger International 1983, p. 73).

In northern Sudan in 1980, the share of the official markets amounts to about 17, 20, 5, and 24 per cent of total markets for cattle, sheep, goats, and camels, respectively (calculation based on figures of MNP 1980/1 and AAAID 1982, p. 10). Although available figures are likely to be inaccurate, it is a fact that the official markets on which transactions are controlled, registered, and taxed, represent only a very small portion of the total volume of livestock trade.

Stages of marketing Three market levels can be identified, two in the production areas and one in the most important area of final demand. Primary markets are found in small towns, villages, and at assembly points (e.g. wells); these are supplied directly from breeders or bush traders who visit nomadic camps in search of animals for sale. The secondary markets are located on the outskirts of larger towns in the production areas (El Obeid, En Nahud, Nyala, El Fasher). Most of their supply comes not from the breeders but from small traders and the bush traders who sell to the agents of wholesale merchants and to local butchers. A considerable proportion of the livestock arriving at these markets is finally exported. Cattle are marketed by auction in groups of four or five head; for the other animals the contracts are negotiated individually between buyers and sellers.

A typical phenomenon of the secondary markets is the large number of brokers (*damins*), who guarantee that the seller is the legal owner;

commissioners; agents of wholesale merchants (*wakeels*); and other groups of people specialized in certain services (Zohair Mubarak Abdalla 1974, pp. 31–4).

The terminal markets are located near the Three Towns and in Port Sudan. Omdurman is not only the transit centre and terminal market for a large proportion of the officially marketed livestock, it is also the financial and organizational pivot of Sudan's livestock marketing. The largest part of the working capital on the markets in the production areas, of the credits to butchers, and of the capital and know-how in the export business, are provided by the wholesale merchants of Omdurman (AAAID 1982, p. 89). It was estimated that in the mid-1970s the seven largest merchants controlled about half of the livestock trade in the Sudan. By means of a net of agents and dependent traders, they also dominate the secondary markets, while on the final demand markets they are facing an atomistic demand from the side of the butchers who depend upon their credits (Zohair Mubarak Abdalla 1974, p. 96, p. 100).

Usually, the herds arriving in Omdurman are sent to feed lots for 40 to 50 days in order to recover from the trek and to fatten. The feed lots are owned by merchants who count among the largest buyers on the secondary markets in the western provinces. The concentrated feed is provided by feed mills which are also situated around Omdurman and Khartoum. They are particularly favoured for investments by Sudanese and Arab agro-industrial companies (AAAID 1982, pp. 159–61).

Prices and price policy Until 1978 the nominal price movement was characterized by a continuous upward trend almost in line with the overall inflation rate. The very large number of enforced sales during the drought stopped the upward trend of prices, but since 1985/6, a counter effect has been noticeable as the reconstruction of herds after the drought has caused a shortage of supply and an increase in demand, particularly of cattle (Table 20.1). Price rises have been far more dramatic in the production areas than on final consumption markets.

The development of illegal exports has probably had an even greater

TABLE 20.1. Actual market prices of cattle and sheep (£S per head)

	Omdurman		El Obeid		Nyala		El Fasher	
	Cattle	Sheep	Cattle	Sheep	Cattle	Sheep	Cattle	Sheep
July 1984	623	140	292	51	254	39	289	74
July 1985	788	208	368	74	185	43	196	65
June 1986	1374	450	916	202	983	156	2235	184

Source: Statistical Records of the Livestock and Meat Marketing Corporation, Khartoum.

impact on the market situation. It is estimated that besides an unknown number of cattle, more than 600 000 sheep are smuggled out of the country annually. Considering the total annual offtake of 5–6 million sheep and a regular market supply of about 1 million sheep, smuggling results in a significant cut in domestic supplies and has contributed to the huge price increases of recent years.

In Khartoum where meat prices and bread prices act as a political signal, there were drastic increases in meat prices in 1986 due to these market disturbances. Ceiling prices for cattle and sheep meat ordered by the Sudanese Government, proved to be inefficient in the face of a growing supply deficit. Furthermore, the official price control is limited to the last stage of marketing, the butchers' shops which are themselves victims of the market irritations. With a sheep price of £S325–750 and an average weight per head of 12–14 kg, an administered price ceiling of £S10.5 per kg for meat is unrealistic; the same applies to cattle meat for which the retail price was limited to £S7.5 per kg, while in October 1986 cattle with an average weight of 190–210 kg, were traded at £S2250 per head (Sudanow 1986, No. 9/10, p. 30).

As the Ministry of Commerce could not intervene successfully at the retail level and was not prepared to take action at preceding market stages, it is currently trying to take measures with an overall impact on the meat market. Exports of livestock are banned, but this may affect only legal exports and result in an increase of livestock smuggling. Concomitantly, the Government has started to import meat from Australia with the intention of improving the supply situation and exerting pressure on prices. While this sounds rather paradoxical for a country with such a wealth of livestock, the official explanation given is that the increasing degree of monopolistic market domination by big merchants and the undesired price effects can only be controlled by imports of meat (Sudanow 1986, No. 9/10, p. 30). This is a helpless, though popular response to the market situation by a government which does not dare to manipulate directly.

Grain

Production and supply The most important grain crops are sorghum (*dura*), millet, and wheat, but sorghum with 60–70 per cent of the total harvest volume is the country's staple food crop. Millet and the cheapest kind of *dura* (*feterita*) are cultivated and consumed by the rural population; the poorer urban classes also buy the cheap *dura*, while people with medium and high incomes consume the better *dura* varieties (*safra* and *duber*). Wheat has gained in importance in recent years, tripling its share of overall grain consumption from 8.0 per cent in 1978/9 to 24.8 per cent in 1982/3; at the same time, the share of millet consumption decreased from 29.5 to 10.1 per cent, and the share of *dura* stabilized at about 65 per cent (FAO 1986, pp. 59–60).

Grain is cultivated in all of the three agricultural subsectors. The irrigation schemes, mainly the Gezira Scheme, supply most of the wheat output but contribute only about 10 per cent to the *dura* harvest. The mechanized rain-fed cultivation is by far the most important producer of *dura*, but its contribution to the output of millet is insignificant. Traditional rain-fed agriculture supplies more than 95 per cent of the millet and 20–30 per cent of the *dura*.

Although the traditional sector may reap about 40–50 per cent of the total *dura* and millet harvests in good years, the turnover on crop markets is dominated by the other two subsectors, in particular by the mechanized rain-fed production.

Khartoum, the Northern and Nile Provinces, Port Sudan and the coastal areas get their supplies of grain from Kassala Province, the most important region of surplus, with Gedaref the centre of mechanized farming. The Blue Nile Province, another area with surplus, supplies Khartoum, the Gezira, the White Nile and parts of the southern provinces (mainly Bahr El Ghazal). The latter also get supplies from Upper Nile Province (especially the Renk area) which supplies grain to the western provinces. The years 1983–5 were marked by a deficit in grain production and supply resulting from the drought period during the first half of the 1980s and the immigration of refugees from neighbouring countries. According to FAO estimates, the grain deficit in Kordofan amounted to about 80 per cent in 1984; in 1986/7 the supply deficit had been diminished only to about half of the 1984 amount (El Sammani 1986).

Marketing and market policy Except for wheat cultivated in irrigated schemes, grain is not turned over on official markets. The marketing channels and institutions for traditional rain-fed cultivation are somewhat different from those of the mechanized and irrigated schemes. Smallholders are governed by the same conditions which are characteristic of the marketing of other cash crops (e.g. *sheil*, limited mobility, and shortage of liquidity). Most of the harvest from mechanized schemes is marketed by the farmers themselves who tend to be traders and transport entrepreneurs at the same time. Compared to the smallholders, they are in a position to make independent arrangements concerning place and time of sales. Mechanized rain-fed farming is favoured not only by the low tax burden and high subsidies (Oesterdiekhoff 1980, pp. 336–9; O'Brien 1985, p. 25); the ABS grants credits on favourable terms to cover most of the total production costs, but also provides support (especially storage and finance) to mechanized farmers for the marketing of grain.

Besides the traders who operate only on a local level, there are wholesale merchants who frequent the urban secondary markets and hold sway over the marketing system of a whole region. Their activities are not restricted by any official control and they are not obliged to keep accounts;

tax assessments are based on their own assessments of turnover and margins. A trader's chance to make high profits depends on whether he has sufficient capital and storage facilities to buy the crop immediately after harvest and store it for a period of up to six months. There are only small margins on day-to-day buying and selling prices, but the opportunity for larger profits results from seasonal price fluctuations. The price spread is not only due to actual shortages for remainders of last year's crop are always offered and the seasonal price escalation occurs independent of the respective harvest volume of the current year. It is, therefore, almost certain that manipulations by traders have contributed significantly to seasonal price increases (Louis Berger International 1983, p. 58).

Fluctuations in seasonal grain prices caused by seasonal peaks in supply and speculative practices, can gain ground and extend widely for they are fostered by short-term inelasticities of supply and demand to price movements, i.e. the fact that both supply and demand usually respond slowly and only weakly to price alterations. In the short-term, supply is the independent factor, changes in which—a shift to the right or left of the supply curve—cause considerable price responses. Even a small decrease in supply, produced deliberately, may result in comparatively great price jumps. This is an effective device for speculators and traders dominating the markets to appropriate monopoly rents and it is obviously used quite often: 'The element of speculation plays an important role in shaping stock policies and hence in determining prices' (Mohamed Farah Gibriel 1979, p. 20).

The Government is making efforts to gain more control on the grain markets and plans a substantial extension of public grain stores. The ABS has been ordered to store about 25 per cent of the total crop, i.e. about 1 million tonnes in 1986/7. For the project to be successful, the available storage facilities of 350 000 t must be supplemented by leased storage; this will prove to be a considerable obstacle in view of the notorious lack of suitable large stores. In all regions of the Sudan, smallholdings have storerooms which are well-suited to conserve the harvested crop over a long period with only minimum losses (Ibrahim 1987, pp. 31–2). By contrast, the storage of grain by traders and the Government is characterized by high losses after only a few months (20 and 50 per cent loss after six and twelve months, respectively). In addition to a moderate extension of the public storage capacities for a national security reserve, credit schemes to enable smallholders to increase their capacities for storage would be advisable. The deficient transport and communications system which complicates centrally-directed action is another reason for promoting local storage and thus avoiding local gaps in supply.

Price movements and price policy Grain prices are characterized by massive short- and medium-term fluctuations. The variations, often occurring suddenly, are recorded in the price statistics as monthly price jumps and annual average price movements. Since the beginning of the 1960s, annual prices show fluctuations amounting to 80 per cent. There was a moderate increase at the end of the 1970s, followed by rapid increases, exceeding overall inflationary tendencies, at the beginning of the 1980s.

Seasonal fluctuations follow a predictable pattern. The harvest season is defined as the period immediately after harvest (November to April), and at this time supplies are plentiful and prices low. Almost half of the turnover on secondary markets is concentrated during this period for many smallholders and farmers are interested in quick sales in order to discharge their credit liabilities to local money-lenders or banks (Antoun 1986, pp. 10–11). During the off-season (May to October) supplies are more limited and high prices prevail.

Seasonal price spreads are particularly pronounced on grain markets. In El Obeid price increases of up to 100 per cent and on the markets of Khartoum and Gedaref increases of 72–160 and 100–250 per cent, respectively, within a year have been recorded (Mohamed Farah Gibriel 1979, pp. 6 and 27–8). As mentioned above, it is highly probable that the traders whose profit margins are determined primarily by seasonal price spreads, manipulate markets and prices. Because considerable price jumps from one month to the next are a particular feature of grain markets, speculators can count on high profits within a very short time. It is a striking phenomenon that these fluctuations are larger in deficit areas (El Obeid, Shendi) than in those producing a surplus (Gedaref, Er Renk).

As the short-term fluctuations have no economic function and have a negative effect on producers and consumers, it is advisable to implement measures of market stabilization in order to level the price fluctuations and confine speculation. In this way seasonal price spreads will be reduced to the costs of storage. Medium-term price movements must, however, take effect because the supply is influenced by the previous year's prices (Mohamed Farah Gibriel 1979, pp. 17–19). Such a differentiated price policy requires a considerably improved marketing infrastructure, in particular, a net of adequate storage facilities at district, village, and household level, an efficient survey and communication system, and improved means of transport. In addition, it is necessary to find a price formula which takes into account the needs of consumers as well as the prerequisites of grain cultivation.

The special handling of the wheat market demonstrates that the Sudanese Government is prepared and in the position to intervene in grain markets. Subsidized wheat consumption is concentrated in urban areas and depends for the most part on imported supplies. It has increased four-fold since the beginning of the 1960s and has amounted to more than 600 000 t annually since the beginning of the 1980s. As the domestic production fell far short of this expansion of demand, the supply deficit was covered by rapidly increasing imports which have amounted to more than 400 000 t per annum since the beginning of the 1980s. For a short period in 1983–5, this import dependency intensified sharply due to supplies from food aid. In 1983/4, this amounted to 444 000 t of wheat (MANR 1986*b*, p. 76). However, the production gap is to be considered a long-term structural outcome of market and price policies.

As a result of the 'cheap bread policy', the real price of wheat bread decreased by an average annual rate of 2.5 per cent between 1970 and 1980. Even the administered price rise of 60 per cent in March 1982 merely re-established the level prevailing at the end of the 1970s (Franklin 1985, pp. 51 and 53). Consequently, between 1970 and 1982, the price ratio of sorghum to wheat bread increased annually by an average of 3.4 per cent. Thus the urban-elitist preferences for wheat bread were subsidized by the Government's price interventions. This is reflected by the gradual substitution of the traditional *kissra* bread made from sorghum, by wheat bread, and by the low demand for bread made from a mixture of traditional grain and wheat. Thus, the wheat price policy is contributing to the creation of 'a wheat import trap' which will grow as wheat becomes an indispensable ingredient of consumption.

Conclusions and recommendations

The characteristics of the current marketing system can be depicted as follows:

1. The marketing infrastructure is grossly deficient with regard to transport, storage, finance, and communication and market services are concentrated on a relatively small group of wholesalers and two parastatal companies, the Sudan Oilseeds Company and the Gum Arabic Company.
2. This situation and other difficulties which peasants encounter when taking up cash-cropping besides subsistence production, are responsible for their exclusion from official markets. Smallholders are confined to the realm of informal marketing involving low producer prices, scarcely influenced by price policies.
3. Price and market policies ignored the fact that the majority of producers are beyond official reach. They also failed to take account of parallel markets, which are far more important than formal markets in terms of traded volumes. With the gradual shift of market outlets from export to domestic demand, the scope for informal and illegal marketing and for shirking taxes and other market duties has increased considerably.
4. Viewed in a macroeconomic context, the marketing system is biased in favour of urban final demand and the export trade. In this respect,

> current marketing is no different from the colonial market system, both of which aim to syphon off agricultural surpluses to other (metropolitan urban, industrial, and administrative) sectors.

Market policies failed to adopt a different approach which would not only have met with strong resistance from merchants with significant political influence, but would also have required fundamental changes in overall agricultural policies. After more than a decade of stagnation and the severe set-back during the drought, Government has reached a point at which it must rethink conventional post-colonial attitudes towards agriculture. It has been reminded of the overwhelming economic importance of this sector. Although the Sudanese Government has started to restrict traders' market power at meat, grain, and oilseeds markets, as seen above, and to increase some floor prices, particularly for gum arabic, it remains to be seen whether these measures are the first steps of a comprehensive programme or merely a rough and ready repair.

The main elements which should be included in a new market policy approach are as follows. An extension of transport capacities is a prerequisite for any increase in marketed surpluses; in 1982 a bumper crop of groundnuts in the western provinces could not be conveyed for lack of transport (World Bank 1983, p. 53). Transport by truck is expensive but although transport by rail is cheap, it is unreliable and involves long delays. Transport deficiencies also contribute to high marketing margins.

There is common agreement that the problem of inadequate public storage capacities should be addressed by a regionally balanced extension programme (FAO 1986, p. 54; Louis Berger International 1983, pp. 134–6; Sudan Government 1986, pp. 37–8). Provision of storage facilities can be an important element of market power; combined with limited and monopolized transport capacities, it is responsible for excessive price fluctuations. A multi-system approach involving Government-controlled storage at district level to smooth out short-term price fluctuations and maintain supplies to domestic and export markets, public-controlled village stores, and traditional stores at household level could provide a maximum of benefits in terms of food security and price stabilization (Ibrahim 1987, pp. 32–4).

In view of the low income level of the majority of smallholders and their dependence on traders and moneylenders, the development of storage facilities at village and household level requires financial and political support. In some areas the ABS has launched credit schemes for smallholders who are members of co-operatives. The loans cover production costs, inventory, and marketing expenses. The crops are stored in warehouses provided by ABS and are sold late in the season when prices have risen. Currently, only groundnuts and sesame are included in this programme, but it can easily be extended to include more crops and regions (Louis Berger International 1983, p. 129). Marketing co-operatives should be introduced and promoted in collaboration with the ABS or a regional development institution so that financial and marketing support can be provided.

The subject of price policies is particularly ticklish for it touches on vested economic interests (Bates 1981, pp. 30–44). The price mechanism has been used by the Government as a means of resource transfer from the agricultural sector, as evidenced particularly by gum marketing. It works mainly to the benefit of the central budget and makes allowances for traders' interests. The procedure does not take account of important implications of agricultural pricing which have a bearing on smallholders' cash cropping. The 'top to bottom' price fixing should be replaced by a bottom up approach starting from the needs of smallholders.

The levels of nominal and real producer prices appear to be a random result of official price determination. Higher prices, which should be established gradually in order to avoid market disturbances, would be a prerequisite of any agricultural recovery programme. Higher prices are needed for gum arabic, in particular, to prevent the ecologically-valuable stands of *Acacia* spp. being cut down. Short-term fluctuations which are a burden to consumers and producers but a benefit to traders should be smoothed out. A stable and transparent price development might be more of a price incentive than a mere price increase, for it reduces the risks of cash-cropping for smallholders operating under difficult ecological conditions.

Agricultural recovery cannot be brought about merely by changes in marketing and market policies. Cultivation and animal husbandry are beset with numerous difficulties which require a different kind of remedy. Nevertheless, marketing will inevitably be an important aspect of policy considerations for new economic strategies. From a marketing point of view, it is advisable to pursue a higher degree of market integration within the scope of a regionally inward-looking development. Economic linkages within rural areas should be enhanced on a regional level without denying chances of interregional and also international specialization. A higher degree of processing of agricultural raw materials in the countryside must be achieved so as to generate more employment, additional income, and a better provision of manufactured produce which acts as producer incentive for cash-cropping smallholders. In this context, small-scale manufacturing should be promoted for it can intensify forward and backward linkages to the agricultural sector.

In order to implement rehabilitation and recovery programmes and to provide adequate producer incentives, agriculture needs a larger share of its surplus for intra-sectoral consumption and investment. Market channels should be redirected on a more horizontal course; their current flow is mainly vertical, i.e. leading from rural to urban markets. At this point, the matter becomes eminently political, as it touches on vested economic and political interests of the state, mercantile classes, and urban consumers.

References

AAAID (Arab Authority for Agricultural Investment and Development) (1982). *Feasibility study for a feedlot complex project in western Omdurman-Sudan*. AAAID, Khartoum.

Abbott, J. C. (1967). The development of marketing institutions. In *Agricultural development and economic growth* (ed. H. M. Southworth and B. F. Johnston), pp. 364–98. Cornell University Press/Ithaca Press, London.

Abbott, J. C. (1968). Marketing issues in agricultural development planning. In *Markets and marketing in developing economies* (ed. R. Moyer and S. C. Hollander), pp. 87–116. Richard D. Irwin Inc., Homewood, Illinois.

Ahmed Hamza Khalifa (1976). Marketing in the Sudan. In *An introduction to the Sudan economy* (ed. Ali Mohamed El Hassan), pp. 201–15. Khartoum University Press.

Ahmed Humeida Ahmed Ali (1984). *Credit institutions for smallholders in the traditional agricultural sector in west Sudan*. Faculty of Agriculture, University of Khartoum, Khartoum.

Ahmed Humeida Ahmed Ali and Zohair Mubarak Abdalla (1976). *Rural credit in the Sudan Gezira Scheme*, Department of Rural Economy, University of Khartoum, Research Bulletin No. 24. Khartoum.

Antoun, V. (1986). *Production of sorghum in Sudan*. USAID, Khartoum.

Bates, R. (1981). *Markets and states in tropical Africa. The political basis of agricultural policies*. University of California Press, Berkeley.

Beshai, A. A. (1976). *Export performance and economic development in Sudan 1900–1967*. Oxford University Press, London.

Blake, H. (1983). *Policy considerations in the production and marketing of gum arabic in Sudan*. USAID, Khartoum.

El Fatih Shaaeldin (1984). *The evolution and transformation of the Sudanese economy up to 1950*, Development Studies and Research Centre Monograph, No. 20. University of Khartoum.

El Fatih Shaaeldin and Siddig Umbada (1984). *Foreign exchange leakages in Sudan*, Economic and Social Research Council Bulletin, No. 100. Economic and Social Research Council, Khartoum.

El Haj Bilal Omer (1985). *The Danagla traders of northern Sudan. Rural capitalism and agricultural development*. Ithaca Press, London.

El Sammani, M. O. (ed.) (1986). *Kordofan rehabilitation and development strategy*, 3 Vols. Sudan Government/UNDP, Khartoum.

FAO (1982). *A policy and action plan for strengthening national food security in the Sudan*. FAO, Rome.

FAO (1986). *Technical co-operation programme. Agricultural policy development—Sudan. Development in the Northern Region of the Sudan*. Technical Report, FAO, Rome.

Farah Hassan Adam and Apaya, William Andrea (1973). Agricultural credit in the Gezira. *Sudan Notes and Records*, **54**. Khartoum.

Franklin, D. L. (1985). Political economy of agricultural pricing: The case of Sudan, unpublished paper.

GITEC (1980). Gum arabic development. Unpublished Final Report for EEC.

Ibrahim, Fuad (1987). Combating famine by grain storage in western Sudan. *Geo Journal*, **14**, 29–35.

Jones, W. O. (1972). *Marketing staple foods in tropical Africa*. Cornell University Press/Ithaca Press, London.

Keddeman, W. (1978). A survey of the possibilities for alleviating rural poverty. In *Employment, productivity and incomes in rural Sudan* (ed. Ali Abdel Gadir Ali and W. Kedderman), pp. 71–100. ILO, Geneva.

Louis Berger International, Inc. (1983). *Kordofan region agricultural marketing and transport study*. USAID, Khartoum.

Mahmoud, Fatima Babiker (1984). *The Sudanese bourgeoisie: vanguard of development?* Zed Books Ltd and Khartoum University Press, London/Khartoum.

MANR, Department of Agricultural Economics (1985). *Agricultural prices in Sudan: an historical review and analysis, 1970–1984*. Marketing Section, Department of Agricultural Economics, MANR, Khartoum.

MANR, Department of Agricultural Economics (1986*a*). *Agricultural commodity prices—1985 summary*. Marketing Section, Department of Agricultural Economics, MANR, Khartoum.

MANR, Department of Agricultural Economics (1986*b*). *Agricultural situation and outlook, annual report 1985/86*. Department of Agricultural Economics, MANR, Khartoum.

MNP (Ministry of National Planning) (1980/1). *Internal trade statistics and price indices, 1980/1*, Department of Statistics, MNP, Khartoum.

Mohamed Farah Gibriel (1979). *Factors contributing to dura price movement in the Sudan*, Economic and Social Research Council Bulletin, No. 78. ESRC, Khartoum.

Mohamed Farah Gibriel (1984). *Groundnuts production and marketing in Sudan*, Economic and Social Research Council Bulletin, No. 119. ESRC, Khartoum.

Mohamed Hashem Awad (1967). The economics of internal marketing in the Sudan. Unpublished Ph.D. thesis, University of London.

O'Brien, J. (1985). Sowing the seeds of famine: the political economy of food deficits in Sudan, *Review of African Political Economy*, **33**, 23–32.

Oesterdiekhoff, P. (1979). *Interne Vermarktungsbedingungen und Verteilung der Aussenhandelseinkommen*. Research Report, University of Bremen, Bremen.

Oesterdiekhoff, P. (1980). Der Agrarsektor des Sudan. In *Der Sudan. Probleme und Perspektiven der Entwicklung* (ed. R. Tetzlaff and K. Wohlmuth), pp. 257–384. Alfred Metzner Verlag, Frankfurt.

Oesterdiekhoff, P. (1983). New international economic order and small peasants' export production in least developed countries. Case study Sudan. In *The development perspectives of the Democratic Republic of Sudan. The limits of the breadbasket strategy*, Afrika-Studien, No. 109 (ed. P. Oesterdiekhoff and K. Wohlmuth), pp. 137–61. Weltforum Verlag, Munich.

Oesterdiekhoff, P. (1988). *Landwirtschaftliche Vermarktung und Marktpolitik. Strukturen, Defizite und politische Optionen im agrarischen Marketing von Regenfeldbau und Viehwirtschaft*, Sudan Economy Research Group Discussion Paper, No. 12. University of Bremen.

Reeves, E. B. and Frankenberger, T. (1981). *Socio-economic constraints to the production, distribution and consumption of millet, sorghum, and cash crops in North Kordofan, Sudan*, Department of Sociology, Report No. 1. University of Kentucky, Lexington.

Shaw, D. J. (1966). The effects of moneylending (*sheil*) on agricultural development in the Sudan. In *Agricultural development in the Sudan*, Papers for the 13th Annual Conference of the Philosophical Society of the Sudan, Vol. 2 (ed. D. J. Shaw), Appendix pp. D56–9. Philosophical Society of the Sudan, Khartoum.

Sidgi Awad Kaballo (1984). *The supply response of traditional oilseeds producers in Kordofan*, Economic and Social Research Council Bulletin, No. 117. ESRC, Khartoum.

Sudan Government (1986). *Strategy for development of rain-fed agriculture. Marketing, pricing and incentives*. Sudan Government, Khartoum.

Sudanow (1986). Ministry of Culture and Information, Khartoum.

Wilmington, M. W. (1955). Aspects of moneylending in northern Sudan. *Middle East Journal*, **9**, 139–46.

World Bank (1981). *Accelerated development in sub-Saharan Africa: an agenda for action*. World Bank, Washington D.C.

World Bank (1983). *Sudan—pricing policies and structural balances*, 3 vols. Country Programs Department, World Bank, Washington D.C.

World Bank (1987). *World development report*. World Bank, Washington D.C.

Zohair Mubarak Abdalla (1974). *Marketing of livestock in the Sudan*. Department of Rural Economy Research Bulletin, University of Khartoum, Khartoum.

21

Agricultural education, research, and extension

W. I. ROBINSON

'Not at the mouth of his clean lipped guns shall ye learn his name again
But letter by letter, from Kaf to Kaf at the mouth of his chosen men
He has gone back to his own city, not seeking presents or bribes
But openly asking the English for money to buy you Hakims and Scribes'

Kipling (1898)

Thus spoke a 'Mohamedan schoolmaster' in the Bengal Infantry stationed in Suakin upon hearing that Kitchener was collecting money in England to build a College in Khartoum. The schoolmaster may have been a figment of Kipling's imagination but the occasion certainly was not, for Kitchener's initiative led to the opening of Gordon College in Khartoum in 1902. Gordon College, later to become Khartoum University, is really the starting point for both the themes of this chapter. Firstly, its foundation heralded what was to become known as the 'educational experiment in the Anglo-Egyptian Sudan' (Currie 1934) an experiment which was to shape the current education system of Sudan and have a fundamental effect on education development throughout Africa.

Secondly, the individual activities of the first Director of Education brought about the beginnings of agricultural research in Sudan which resulted in the creation of a centre which, at its zenith, was the foremost agricultural research centre in Africa.

The objectives of the educational experiment were to provide the right amount of education through: 'The diffusion among the masses of the people of education sufficient to enable them to understand the machinery of Government, particularly with reference to the equitable and impartial administration of justice' and . . . 'The creation of a small administrative class capable of filling many Government posts, some of an administrative, others of a technical nature'.

Laudable as they appeared in 1900, these objectives contained no plans to develop that nebulous commodity that forms the ligaments and tendons of any corporate state, namely middle level management or the educated artisan. In other words there were no plans to develop a group with the administrative ability to organize its own enterprises outside of the government structure. The need for competent artisans was, however, identified as a third objective and two instructional workshops of a 'severely practical' nature were established in Khartoum and Omdurman to provide artisans for government service. The strategy was so successful that in 1942 the educationalist, Humphrey Bowman (Bowman 1942) wrote that: 'as a result of the policy followed over a number of years the Sudan has been able to furnish itself in all its lower and many of its intermediate grades, with a purely indigenous Civil Service' and to continue . . . 'the half-educated, unemployed class, so prevalent in Egypt and India may be said to be almost non-existant'.

This ill-judged statement says a great deal about the attitudes towards education and development at that time, attitudes which do not seem to have progressed since Kitchener's day. The statement also contains an explanation for some of the sadder aspects of Britain's educational legacy to Sudan. Under this narrow system, secondary education was reserved for a selected few who were destined for government service. Secondary education, and later tertiary education, provided the only routes into government service and government service became the only option after secondary and tertiary education. Once these interdependencies became established, they were equally effective in all disciplines. Even vocationally-orientated institutes like agricultural schools followed the lead established by the instructional workshops and produced workers for the Civil Service. As a result of the policy, the agricultural industry in Sudan has evolved dichotomously, an essentially illiterate agricultural workforce farming its own holdings or labouring for the *jallaba* (merchant barons), while a cadre of government employees who are essentially out of touch with the farming community, staff the ministry offices, colleges, and research institutes.

The agricultural ligaments and tendons that could link the farming muscle to the administrative skeleton are missing in the Sudan. Meanwhile in the rural areas of India 'the half-educated, unemployed class' so despised by Bowman, has provided the missing link which has led to economic recovery in India's agricultural industry. Had such a group, that is, an educated peasantry, been present in rural Sudan, the polarization of action and thinking may not now be so obvious.

Once the absence of a technician class was identified as a hindrance to development, intermediate colleges, designed to provide the training for middle level managers and technicians, were created. True to the established tradition, these were geared to providing assistants for government service. The courses and curricula, modelled on western concepts, are so far removed from the reality of Sudanese agriculture that the technicians trained in this way consider job opportunities in agricultural areas away from the Wad Medani–Khartoum axis to be 'hardship' postings. In the late

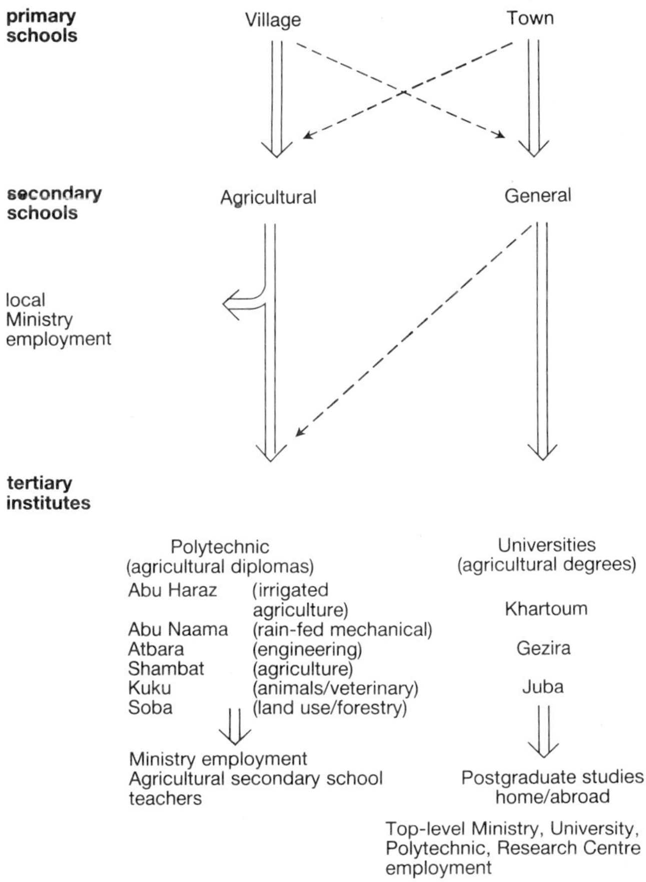

FIG. 21.1. Agricultural education: main channels of student development.

1980s, Sudan's agricultural education system produces 700 graduates and diplomats annually through the institutions shown in Fig. 21.1. Very few of these people find their way into practical agriculture. The traditional sector of agriculture is not sufficiently attractive for students who have spent nine years or more undergoing formal education and the mechanized private sector is dominated by merchant barons who recruit their technicians at a young age, direct from the villages. The only opportunities open to the educated agriculturalist appear to be in government organizations which are ill able to afford a steady recruitment of 700 per year. Projects for improving the quality of training for graduates by improving teaching facilities, equipment, and programmes in the agricultural institutions have been initiated, but the question of how to utilize graduates most effectively remains unanswered. The greater proportion of students receiving agricultural education are men, many of whom began their education in the village primary schools and so understand village dynamics. Unfortunately, the women who receive agricultural education usually come from the progressive households of the urban elite and are likely to be the least suitable candidates for roles in participatory development schemes in remote areas. Yet it would appear from field observations that the massive harvests of cotton, sorghum, and sesame rely very heavily on women labourers. Peasant smallholdings run on a family basis, always depend on female labour at crucial times of the year, but are now depending increasingly upon female management, as the rural exodus of able-bodied males continues to grow. Women would benefit from more specific agricultural education or training, rather than the odd course in home economics organized by urban-orientated supervisors. Their male counterparts would also benefit from agricultural education courses which reflected the characteristics of villages and their agricultural production systems as well as the major production systems.

The most immediate way to fill the middle level management gap has been identified as the provision of training to the existing workforce. Various training projects sponsored by bilateral and international agencies have been implemented to meet specific demands in a well-planned fashion. However, there is no indication of the continuity of such training schemes once the international support ends. In southern Sudan, the Amadi training centre has embarked on a desperate search for funds in order to sustain its training programme for workers in rural development. Without such support not only will the output of its trainees cease but the infrastructure established will follow the fate of so many other project buildings in Sudan, collapsing through lack of maintenance and appropriate use. The project approach to training leaves much to be desired regarding continuity and the evolution of an appropriate system. Furthermore, externally sponsored *ad hoc* courses are hard to build into any effective training programme. What is needed is a clear interdependent policy based on various trades and skills required in the urban and rural areas (El-Jack and Taha 1977). A centrally-controlled national programme is unlikely to be the answer; co-ordinated programmes at the regional and district level with enough flexibility to allow local variations to be accounted for would be more likely to succeed.

The separate development of agricultural extension programmes has, by and large, been reserved for the main schemes and projects. Until recently, the inspectors of the schemes were also thought of as extension officers. However, there is a world of difference between policing strict land use policies and teaching improved practices to farmers who have freedom of choice.

Within agricultural development projects more emphasis is placed on the development of the agricultural extension worker's communication skills through investment in sophisticated equipment and training packages, rather than through the development of the agent's practical agricultural skills. Undoubtedly the 'medium' is important, probably as important as the 'message', but in the case of agricultural extension, successful communication of the message has more to do with the agricultural ability (skills and understanding) of the agent than with media packaging. The content of the message has to be understandable and applicable, and it is up to the research divisions to see that this is the case.

Weak structures and poor communication between education, training, research, and extension have been identified as factors contributing to the noted lack of success of agricultural projects (Doran 1980). However, the difference in objectives between the farmers and the planners is much more likely to be the main cause. Perhaps the key to a new beginning has already been found. Investigative research into environmental and social issues related to agricultural production is an integral part of the research approach adopted by the Western Sudan Agricultural Research Project (WSARP). This localized research approach is not without its critics who consider it to be less cost-effective than investigation into more universal issues. Yet the need to look at local issues was the spark that created Sudan's first real agricultural research experience.

Whilst it is fair to say that papers of scientific interest were produced by individual investigators on subjects as diverse and as currently pertinent as the *Acacia* species of the Nile Region (Schweinfurth 1867), water resources of the Nubian Desert (Aston 1897), and the acclimatization of domestic animals (Schweinfurth 1882), such activities remained unco-ordinated until the establishment, in swift succession, of the Wellcome Research Laboratory in Gordon College and the Shambat Research Farm, where botanic and agronomic experiments were to be conducted. The Department of Education established the Shambat Research Farm in 1904, a mere fifteen years after Sir John Bennet Lawes placed his laboratory and experimental fields at Rothamsted, England, into the hands of trustees who were charged with the responsibility of continuing the tradition of experimentation on subjects designed to bring about the financial improvement of the English agricultural industry. The Shambat Farm also seems to have been established due to the foresight of one man, the Director of Education, James Currie, who is said to have been dissatisfied with progress made in the Department of Agriculture at the time, and so determined to undertake more applied agricultural research within his department as well as the studies in the Wellcome Laboratories (Bowman 1942).

During the first seven years at Shambat, cotton growing was shown to be a financially viable proposition. Following the commercial success of the first field trials based on Shambat's findings and conducted on 60 ha of irrigated land at Tayiba in 1910/11, the Gezira Scheme which was ultimately to become what has been called 'the largest single farm in the world', was started.

The creation of Sudan's major agricultural research station in Wad Medani at the heart of the Gezira Scheme followed soon after the first cotton field trials and this has remained the headquarters of agricultural research operations ever since 1918, the year of its official opening. In keeping with its integral role in the agricultural development conducted between the two World Wars which was concentrated almost exclusively on Gezira, the installations in Wad Medani became among the finest in Africa. The need to investigate other regions resulted in the establishment of two other centres at Kadugli in 1935 and at Yambio in 1948. For what would appear to be logistical and political reasons, neither station had as great an effect on agricultural events as the original Gezira/Shambat investigations.

Just prior to Independence the need to diversify and to investigate alternative systems outside the Gezira Scheme, but within range of the existing infrastructure, prompted the establishment in 1952 of the Kenana Research Station at Tozi, with a mandate to investigate rain-fed agricultural systems. By the beginning of the 1960s, as a result of the post-Independence developmental emphasis on irrigated agriculture and in a new wave of activity, research stations and sub-stations were created at Hudeiba (1962), Khashm el Girba (1960–3), Guneid (1963), Sennar (1963), and Maatug (1963); also in 1963, the Kenana Station was moved to Abu Naama, where water for irrigation was more readily available (Salih, 1990). With the temporary closure of Yambio in 1964, the then Agricultural Research Division of the Ministry of Agriculture, Food and Natural Resources (which became the Agricultural Research Corporation in 1967) concentrated almost all of its activities in and around the Gezira Scheme or at Hudeiba, a similar site to Gezira at the confluence of the Nile and Atbara Rivers.

In 1975, as part of a rationalization process reported by Idris (1975), the semi-autonomous Agricultural Research Corporation (ARC) assumed responsibility for: the re-opened Yambio Research Station; the Forestry Research Institute at Soba; the Food Research Centre, Khartoum North; and the Wildlife Research Centre, Khartoum.

Research activities in range and pasture research at Ghazala Ghawazat, fisheries and marine biology at Port Sudan, and gum arabic at El Obeid were also the responsibility of the ARC. This left the remaining ministerial research bodies, the Veterinary Research Institute and the Animal Production Research Centre, under the supervision of the Animal Resources Division. The location of existing research centres is shown in Fig. 21.2.

Veterinary research has a long and prestigious history in Sudan. It began in 1913 as part of the veterinary services established initially for the welfare of introduced and indigenous, domestic and military animals. Following sixty years of diverse veterinary services conducted by the Veterinary Department, the existing Institute in Soba was established to conduct research, identify specimens, and produce vaccines. Five sub-centres, located at Malakal (Southern Region), Nyala and El Obeid (Western Region), Sennar (Central Region), and Kassala (Eastern Region), each with field laboratory facilities were charged with the responsibility of directing regional veterinary operations, particularly conducting disease surveys. In 1986 veterinary services became the responsibility of the newly-formed Ministry of Animal Resources, thus reintroducing the administrative structure which operated from 1950 to the early 1970s.

The line between veterinary and animal production research responsibilities in Sudan seems always to have been blurred. Initially the Veterinary Research Division conducted all animal-oriented research within the Ministry of Animal Resources. Then, research on pasture and irrigated forage for dairy and beef production was split off to the Animal Production Research Administration. Nutrition and metabolic analyses seem to have been retained by the veterinary group. The rangeland and pasture research, taken over by the ARC in 1975, was historically part of the responsibility of the old Ministry of Animal Resources, through the activities of the Animal Production Research Administration, Khartoum. The composite research responsibilities of the Animal Production Research Administration included: Ghazala Ghawazat Research Station, Darfur; Um Benein Research Station, Blue Nile Province; National Dairy Research Centre, Gezira; Atbara Research Station, Nile Province; El Huda, Gezira; and Kuku Central Animal Production Research Station, Khartoum.

Of the above centres, only Ghazala Ghawazat, was established to undertake research on traditional rangeland methods of production; the others were created to carry out research on problems associated with the modern livestock sector. This incorporated the use of irrigated forage and agronomic by-products by imported exotic or improved Kenana or Butana cattle, and intensively-fattened sheep.

The latest developments in research organization tend to obscure the picture even further. Following Ministerial recognition that the farmers and pastoralists of western Sudan had not benefited from previous research activities, a reorganization of administrative responsibilities was carried out in 1979 resulting in the establishment of the Western Sudan Agricultural Research Project (WSARP). As a result of the reorganization, the Kadugli, El Obeid, and Ghazala Ghawazat Research Stations were grouped with a new station in El Fasher and all underwent extensive rehabilitation and expansion. With the help of specialists, funded by USAID, a fully integrated research programme including rangeland management, agricultural engineering, agronomy, horticulture, pest management, agricultural economics, social science, animal production, and health, was prepared and adopted. Although under the ARC umbrella, WSARP was essentially autonomous, with its own budget, administration, and publications department.

In Sudan, agricultural research is also undertaken by a variety of non-ARC or independent bodies. These include the Universities of Khartoum,

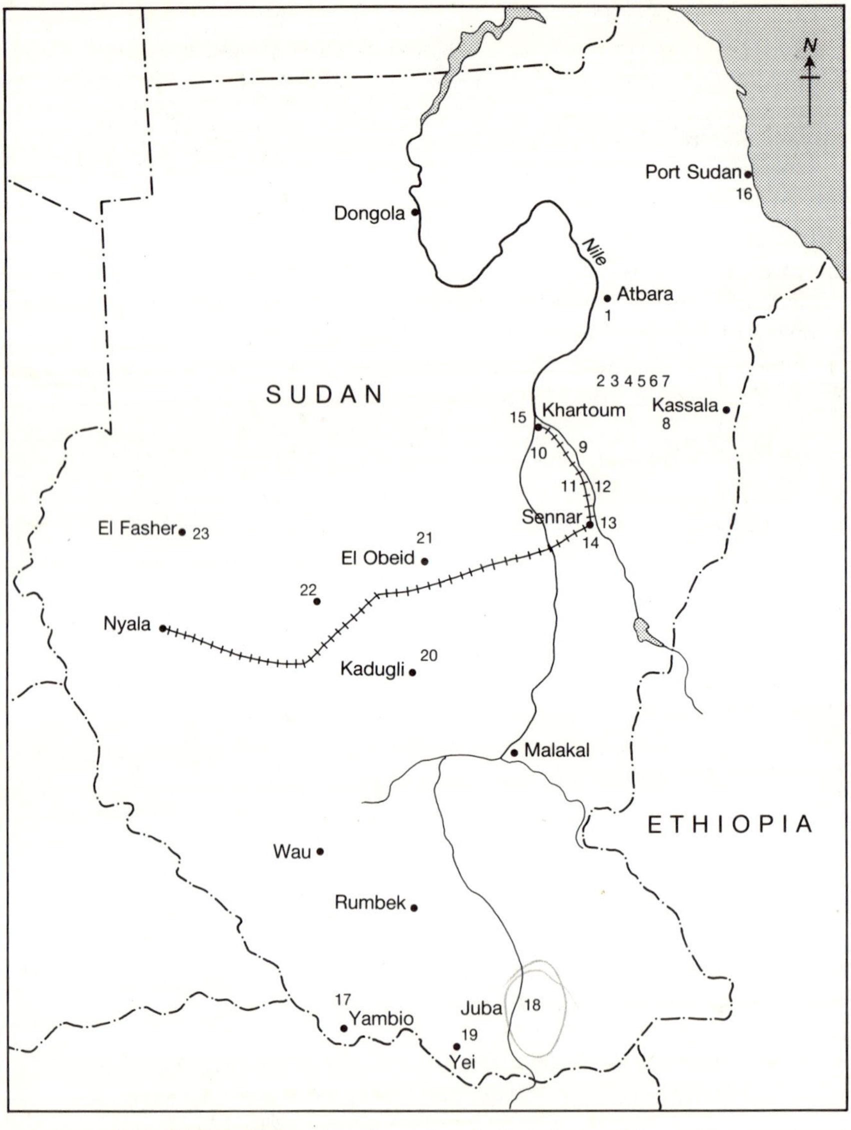

FIG. 21.2. Sudan: location of agricultural research centres.

Gezira, and Juba, and research units in various development projects. Of the former group, the University of Khartoum has by far the longest history. The School of Agriculture was established in 1938 and became a Faculty in 1956. About this time, academic staff began contributing to agricultural research. The Veterinary School followed an identical pattern of development. Since the 1950s, research contributions from the two faculties have covered a very broad spectrum largely as a result of the

Legend for map Figure 21.2

Name	Subjects
1. Hudeiba Research Centre	General agriculture
2. Shambat Research Centre	General; cotton breeding
3. University of Khartoum	General agriculture
4. Soba	Veterinary
5. Soba	Forestry
6. Wild Life Research Centre, Khartoum	Ecology
7. Food Processing Research Centre, Khartoum	Nutrition and food processing
8. Khashm el Girba	General agriculture
9. Guneid	Sugar cane
10. Maatug	General agriculture
11. Gezira	General agriculture
12. University of the Gezira	General agriculture
13. Sennar	General agriculture
14. Um Benein	Animal production
15. Animal Research Centre, Khartoum	Animal production
16. Fisheries Research Centre, Khartoum	Fisheries and marine biology
17. Yambio	Rain-fed agriculture (south)
18. University of Juba	General agriculture (south)
19. Yei	Rain-fed agriculture (south)
20. Kadugli	Rain-fed agriculture (west)
21. El Obeid	Rain-fed agriculture (west) and gum arabic
22. Ghazala Ghawazat	Animal production
23. El Fasher	General agriculture

Source: Robinson (1987).

members of staff following their own particular interests. The Department of Botany and the Institute of Environment Studies have undertaken research projects and resource inventory preparations related to crop production, plant pathology, ecology, and desertification. The Hydrobiological Research Unit and the Department of Social Anthropology and Sociology have also contributed to what could be classified as agricultural research, through the studies of individual members of the two departments.

The Universities of Gezira and Juba were established in 1976 and 1977, respectively, and although, in theory, the agricultural problems tackled should be those facing the communities served by the Universities, the same approach of individual workers following particular interests pertains.

From Currie's first experiments to the present day, agricultural research in the Sudan has had a complex evolution. The Ministry recognized that some form of co-ordination was required and this prompted the establishment of the National Council for Research (NCR) in 1970. The Council is an autonomous body consisting of four specialized councils supervising agricultural, medical, scientific and technological, and social and economic research. The role of the council supervising agricultural research, the Agricultural Research Council (ARC), has been to influence research through the allocation of Sudanese research funds to individuals or institutions. The ARC has seven committees whose members are drawn from various constituencies including the private sector, scientific associations and departments. The seven committees are: Planning and Co-ordination; Natural Resources Research; Plant Production Research; Animal Production Research; Food Processing and Technology Research; Zoonoses Research (with Medical Council); Agricultural Socio-economic Research (with Social and Economic Research Council).

This ARC body should not be confused with the Board of Management of the Agricultural Research Corporation, which is also known as the Agricultural Research Council. This Board of Management is made up of the managing directors of all the irrigation schemes, representatives of the Universities and the other Agricultural Corporations, the Ministries of Finance and Industry, and the ABS. The Board meets at least twice a year and plans the agricultural research policy for Sudan, approves the ARC budget and appointments, and reviews the work of the Technical and Administrative and Finance Committees of the Agricultural Research Corporation.

Apart from internally funded and organized research, a further phalanx of research workers has been provided by International organizations and overseas groups working in Sudan. In some instances these external groups work in conjunction with Sudanese Institutions but often, either as individuals or groups, they pursue their own academic interests. The research activities of the Food and Agriculture Organisation (FAO) (pest management, agricultural planning and statistics, and fertilizers), the International Centre for Agricultural Research in Dry Areas (ICARDA) (faba beans), the International Crops Research Institute for the Semi-Arid Tropics (ICRISAT) (sorghum and millet), the Overseas Development Administration (ODA) (forestry and animal production), and the Arab Centre for the Studies of Arid Zones and Dry Lands (ACSAD) (animal production and economic development) are fully integrated with the Agricultural Research Corporation's activities, but many externally-funded groups, including chemical companies and project groups, seem to work independently on their own programmes. Links are also maintained with the Ford Foundation, the International Atomic Energy Commission, the Overseas Development and Natural Resources Institute (ODNRI)), and a multitude of Universities (MAFNR 1977).

The need to improve communications between the diverse groups operating in the same field was recognized by the National Council for Research in 1982. During that year a register of research in progress 'in Sudan, about Sudan or by Sudanese outside the country' was drawn up with the assistance of UNESCO. The statistical analyses included in the National Register of Current Research are summarized in Table 21.1.

TABLE 21.1. Summary of current research in the Sudan, 1982/3

Field of science	% contribution	Total	Number of research project locations					
			Khartoum	Central	North	East	South	Foreign
Agricultural science	61	775	205	382	133	25	5	25
Social	12	155	95	4	—	—	3	53
Medical	9	114	92	12	—	—	5	5
Engineering	8	104	78	6	—	—	1	19
Maths and national science	6	71	44	1	—	2	—	24
Arts	4	53	42	—	—	—	—	11

Source: National Research Council (1984).

The concentration of current research in the agricultural sector is clear but there is a very uneven distribution of research across the country. The locations of existing research centres are given in Fig. 21.1, and the organigramme presented in Fig. 21.3 is an attempt to represent the current national research interests. It demonstrates clearly the sectional and fragmented nature of the research structure, but shows how the formation of WSARP reflects a will to integrate the efforts of bodies working in the same geographical area. The fact that the co-ordinating council (not shown in Fig. 21.3) is split by specialism into seven committees makes an integrated approach more difficult to achieve.

Studies in or about the Sudan undertaken since 1900 have been analysed by discipline and by area over consecutive three-year periods up to 1986 (Robinson 1987).

A summary graph presented in Fig. 21.4 shows the number of agriculturally related studies undertaken in each zone since the turn of the century.

For studies applicable to the Sudan as a whole, or unclassified, there has been a continuous low level of interest gradually mounting until the late 1950s and Independence, when an explosion of interest was detonated. The number of studies continued to expand until the early 1980s when, by all accounts, the resources all but ran out. This resulted in the curtailment of field trials, printing and publication ceased, research proposals were not implemented, and research facilities were no longer maintained. Low salary levels have also forced research workers to seek alternative employment elsewhere or to supplement their employment in non-study areas in order to maintain a reasonable standard of living.

For the north and east and the Gezira zones of the Sudan, the graphs exhibit very similar patterns as a result of similar circumstances. The north and east graph represents the research activities in the faculties of the University of Khartoum, as well as the other research stations in the area (Fig. 21.2). The Gezira graph includes the activities of the ARC HQ station at Wad Medani plus Khashm el Girba, Sennar, Um Benein, Tozi, Guneid, and the University of Gezira all situated on the central clay plain. The increased activity in the early 1960s highlights the establishment of the majority of stations at that time.

In the South, the area with the greatest natural agricultural potential, there was no such increase in research activity. The establishment of a research station at Yambio after the Second World War can be detected in an increase in studies undertaken in the South between its opening in 1948 and its temporary closure in 1964. An increase in reported studies from 1975 coincides with the reopening of Yambio, a spate of project activities and the start of the Jonglei Canal. These studies, tended to drop off in 1984, but their decline is more likely due to the new civil war in the area than to a lack of project funds.

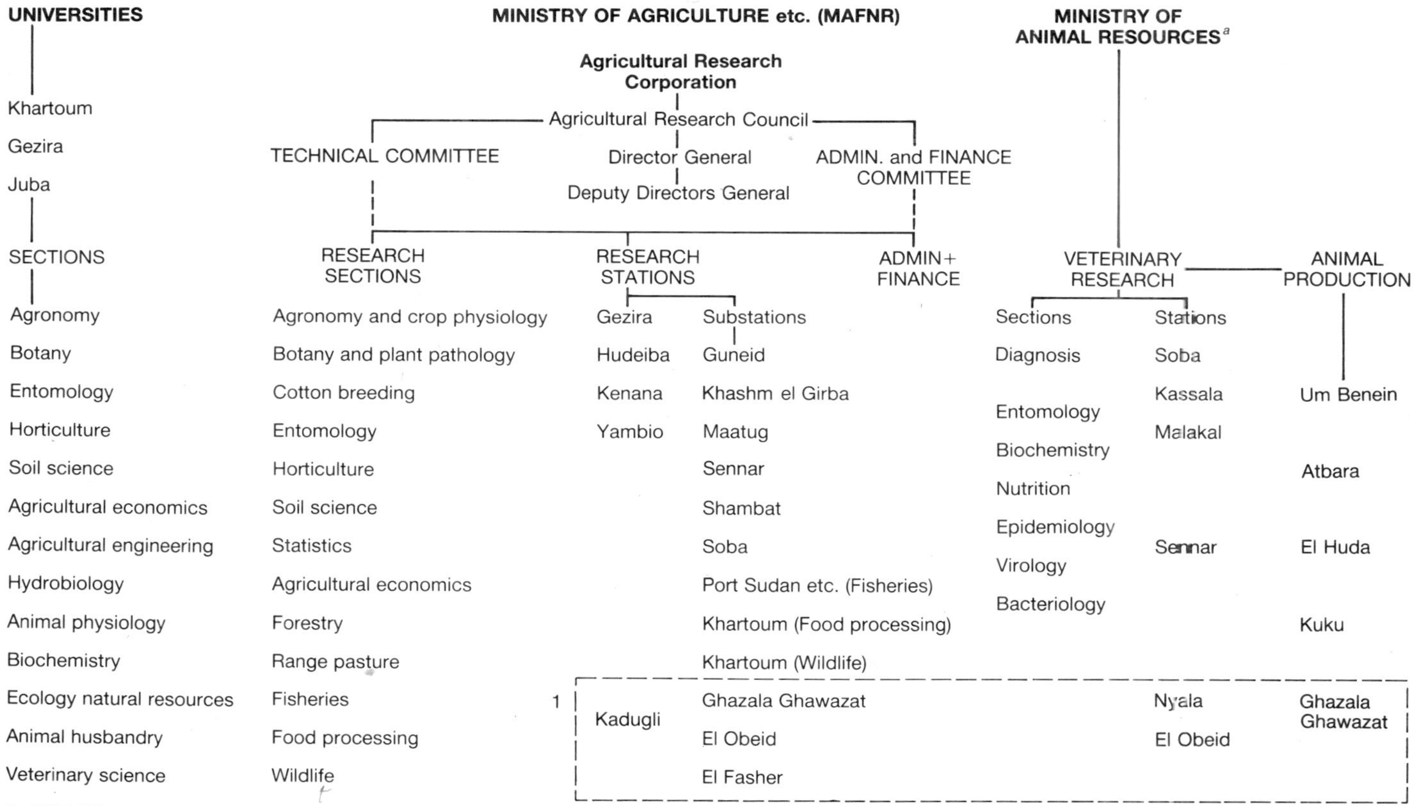

1=WSARP

[a] Regrouped with Ministry of Agriculture in 1990

FIG. 21.3. Sudan: agricultural research structures, 1986.

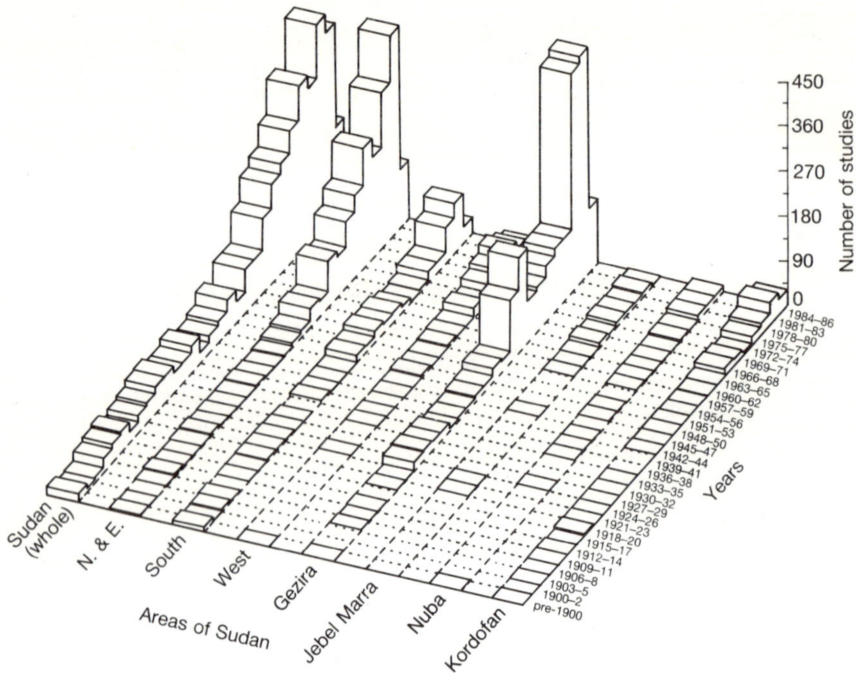

FIG. 21.4. Sudan: studies related to agriculture since 1900.

Areas of Kordofan close to the Gezira Scheme have attracted investigators since the turn of the century, but even including the work done in the Nuba Mountains area from 1935 onwards, (graphed separately) there has been only a sustained low level interest until the establishment of WSARP in 1979. As Kadugli was the first of the WSARP centres to become operational, most studies were based there and so relate either to Kordofan or the Nuba Mountains rather than to Darfur (west), although this will be redressed with the opening of El Fasher Station in 1989/90.

Exploratory visits into the west of Sudan are reflected by periodic recorded studies becoming more consolidated after the Second World War. Pre-project investigations form the basis of the reported work in the 1960s in the graphs for Jebel Marra and the west. Thereafter, project analysis and desertification studies have been conducted regularly.

Increasingly, research activities in Sudan have become more academic reflecting the disparity between the two sides of the agricultural industry mentioned earlier. It would seem that these divergent paths have to be reunited if the state-run services are to have any real impact on agricultural development.

The notable success of Sudanese agricultural research in the field of cotton occurred when there was a very close relationship between research and commodity development in commercial, academic, and administrative areas. Within the present climate in the Sudan, the challenge is to recreate such links for each commodity/system, including the rain-fed production of oilseeds and cereals and the traditional methods of livestock production.

The next few years will be crucial to the survival of the WSARP experiment. If it survives and research is brought more closely into line with farmers' practical needs, then perhaps there is a chance that a rurally-orientated agricultural education system will also emerge in due course.

References

Kipling, R. (1898). Kitchener's School. In *Rudyard Kipling's Verse*, 1919 Vol. I, pp. 267–9. Hodder & Stoughton, London.

Aston, E. (1897). Water analysis of the Nubian Desert. *Journal of the Geological Society*, **4**, 19.

Bowman, H. (ed.) (1942). *Middle East window*. Longmans, Green & Co. Ltd., London.

Bowman, H. (1942). The Wellcome Laboratories. In *Middle East window* (ed. H. Bowman), pp. 96–101. Longmans, Green & Co. Ltd., London.

Currie, J. (1934). The Educational Experiment in the Anglo-Egyptian Sudan, 1900–1933. *Journal of the African Society,* **33**, 361–73.

Doran, A. (1980). Agricultural extension and development: the Sudanese experience. *British Society for Middle Eastern Studies*, **7**, 39–48.

El-Jack, A. H. and Taha, A. R. E. A. (1977). *Vocational training and economic development in the Sudan*, Monograph Series No. 2. Development Studies & Research Centre, University of Khartoum.

Idris, H. (1975). *Agricultural research capability and scope in Sudan*. National Council for Research, Khartoum.

MAFNR (Ministry of Agriculture, Food, and Natural Resources (1977). *The Joint Team Report*. Ministry of Agriculture, Food and Natural Resources, Khartoum.

Natural Resources Council (1984). *National register of current research*. NRC, Khartoum.

Robinson, W. I. (1987). *A review of agricultural studies of Sudan*. Centre for Arid Zone Studies, University College of North Wales, Bangor, UK.

Salih, M. O. M. (1990). *Long term experiments at Tozi*, ARC Animal Report Gezira 1961/62.

Schweinfurth, G. (1867). Enumeration and description of the *Acacia* species of the Nile Region. *Linnaea*, **35**, 7–9.

Schweinfurth, G. (1882). Concerning acclimatization of various domestic animals in the Equatorial Nile Region. *Ausland*, **55**, 13–14.

22

Agricultural machinery and equipment supply

A. H. ABDOUN

The early mechanization of agriculture

Mechanical power was first introduced in the agriculture of the Sudan during the early 1920s at the time of the establishment of the Sudan Plantation Syndicate. This syndicate was pioneering the use of a plough tackle unit consisting of two 60 hp engines pulling a combined cultivator or multiple ridger.

Later in the mid-1940s the increased use of machinery was associated with the need to produce more food during the Second World War and the need to expand agriculture into vast areas with the sole purpose of satisfying the increasing demand for food. At this time a 50 hp tractor and a wide-level disc harrow with seederbox was introduced from Canada. By the 1950s, the private sector had started to invest in machinery; this marked the real start of crop production using machinery in the central rainlands.

The early introduction of large-scale machinery in the central rainlands caused no significant disturbance for traditional farmers, but led to a horizontal expansion policy and the opening up of agricultural land in conjunction with greater use of machinery. Initially, land was distributed to private sector operators with a subsidy to procure medium-range tractors of 50–65 hp. Later on an increase in power utilization was considered feasible and, more recently, there is the trend towards much higher horse-powered tractors ranging to 200 hp or more in certain large-scale investment enterprises.

The traditional sector and intermediate technology

Increasingly, the role of the traditional farmer in the national economy has been realized and fully appreciated. The traditional sector provides a means of subsistence living for a large proportion of the population, but also produces a large share of the production of cash crops. The percentages of different crops grown in the traditional, mechanized rain-fed, and irrigated sectors is shown in Table 22.1. In the last decade, the traditional sector has suffered very badly from the effects of drought, which have affected all African countries to a varying degree.

TABLE 22.1. Percentage of crop production in the traditional rain-fed sector

Crop (%)	1980/1–84/5			1985/6			1986/7		
	I	M	T	I	M	T	I	M	T
Sorghum	4	28	18	5	39	15	4	40	14
Millet	0	0	18	0	1	18	0	0	18
Sesame	0	3	9	0	6	6	0	6	7
Wheat	2	0	0	2	0	0	1	0	0
Groundnuts	2	0	10	1	0	4	1	0	5
Cotton	5	0	1	3	0	0	4	0	0
Total	13	31	56	11	46	43	10	46	44
Country arable land ('000 ac)		16310			21645			20370	

I: irrigated; M: mechanized rain-fed; T: traditional rain-fed.
Source: derived from Statistics Section, Department of Agricultural Economics Ministry of Agriculture and Natural Resources (1987).

Although some experts have considered rapid changes from primitive cultivation to more advanced agriculture to be appropriate, the concept of intermediate technology, which has not attracted much interest in the Sudan, has significant advantages for the traditional farmer.

Some emphasis has been placed on developing practical and simple machines and on developing programmes which would bring about gradual mechanization of agriculture without causing drastic changes to the existing practices. With this aim, the use of 'power' might be categorized as follows:

1. Locally made hand tools, whose promotion is regarded as fundamental for the so-called subsistence agriculture, which is geared to the production of staple grain crops.
2. The use of simple animal-drawn implements which were originally used primarily in the northern part of the country.
3. The use of high-powered machinery over large areas with the view to modernizing agriculture.

Hand tools

In the traditional sector agricultural operations are carried out with simple hand tools of a type that have been used for centuries without being improved significantly. These tools are simple and practical for the farmer to use but their use is affected by the scarcity of labour (Table 22.2). Using existing hand tools, a man might require about 30 hours to complete a task which could be undertaken by one hp hour of machine power. There is a great need for hand tools to be technically efficient, robust, and instantaneously available to farmers at comparatively cheap prices.

TABLE 22.2. Labour requirement for traditional hand cultivation

Operation	Time	% Available labour	
		Light soil	Heavy soil
Clearance and land preparation	May–June	28	95
Sowing and land preparation	June–July	88	100
Sowing and weeding	July	100	100
1st weeding	July–August	100	100
2nd weeding	August	100	100
3rd weeding	September	100	100
Harvesting	October	71	38

Source: Silsoe College (1987).

In order to improve the local manufacture of hand tools, certain international organizations have proposed the establishment of workshops in different parts of western Sudan. It has been found that certain foreign imports are too expensive for traditional farmers to purchase.

Most of the hand tools are used for soil loosening and partial inversion of soil, or for ridging and seed-bed preparation. They are well adapted to this sort of work in most of the regions but are generally made of relatively short-life materials.

Tools for hoeing (Fig. 22.1) are common in most of the regions depending on the process and mode of cultivation. Each tool has the appropriate handle for a stooping or standing position. A sowing stick (Fig. 22.2) is used by one person but has been abolished gradually in many areas due to the quick progress of mechanization. Tools for levelling (Fig. 22.3) can be pulled by one or two persons. The function of such tools is either to level small plots or to raise banks and borders around small areas. Tools for harvesting (Fig. 22.4) resemble curved knives but are serrated on the inner side. These are used for cutting off heads of sorghums ready for threshing and for cutting alfalfa. Local tools for cultivation (Fig. 22.5), are mainly for

'Long-handled type'

'Short-handled type'

FIG. 22.1. Local hoeing tools.

'Sowing stick'

'Sowing rod'

FIG. 22.2 Local sowing tools.

FIG. 22.3. Local levelling tool.

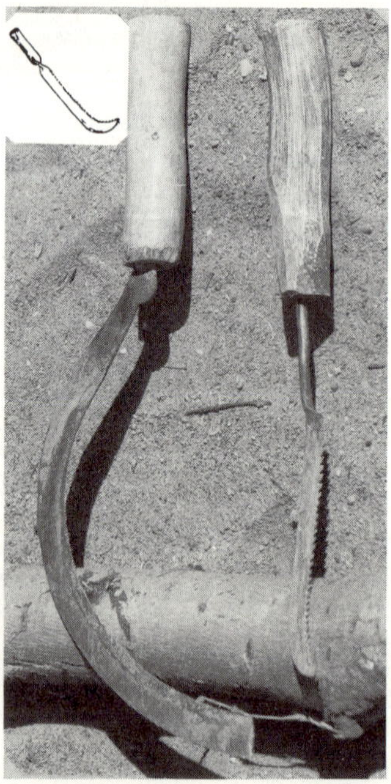

FIG. 22.4. Simple local harvesting tools.

'Long-handled tool' 'Short-handled tool'

FIG. 22.5. Local cultivation tools.

weeding within the rows. The handles of these tools are made from branches of trees or shrubs.

Animal-drawn implements

As the reduction of the fallow interval in shifting cultivation can cause a significant loss of soil fertility, can increase weed growth, and can lead to soil erosion, some traditional farmers have tended to use animal power in the fields. A family working with hand tools can prepare and cultivate only two hectares of land, whereas one using animal-draught power can work more than four hectares. However, the need for weed control, whether mechanical or chemical, is a constraint where a larger area is cultivated.

The use of animal-drawn implements was first common in the northern regions of the Sudan (Fig. 22.6). In the western regions it became a new phenomenon, but there were no artisan blacksmiths to service the implements. Later, a project was initiated and some field implements for the traditional farmer were manufactured locally. The idea was pioneered in some neighbouring countries in Africa where oxen pulled ploughs, cultivators, and small simple seeders which were made in local workshops by exploiting any available local materials (Fig. 22.7).

FIG. 22.6. Traditional ploughing by oxen in northern Sudan.

The camel could play a major role in providing animal power in remote areas, but the most difficult practice is to train, maintain, and keep the animal fit for work. This could be a great burden for the farmer through the hard dry season. Hiring of camels might provide financial benefits for the owner and potential benefits to the farmer.

In a recent study, Silsoe College (1987) conducted a survey on an animal-drawn mouldboard plough in order to evaluate its place in the farming system. The original artisan plough had been improved so that it could be converted into a toolbar (Fig. 22.8). A further study of the Nuba

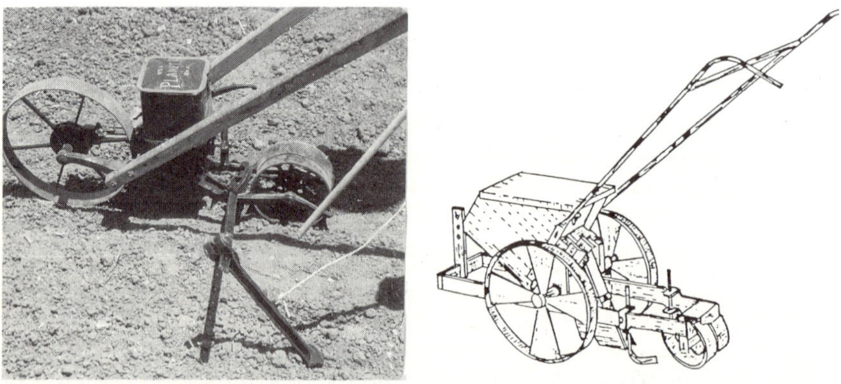

'Hand-driven planter' 'Seeder locally modified'

FIG. 22.7. Seeding equipment.

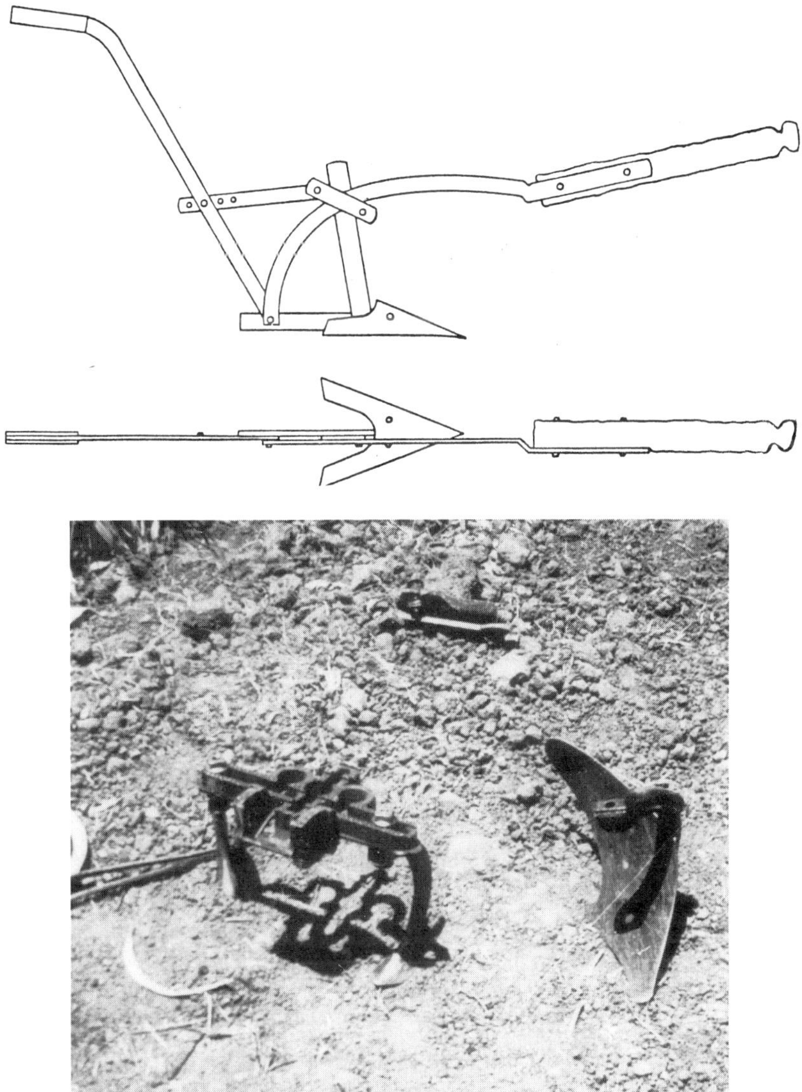

FIG. 22.8. Ploughing equipment.

Mountain hoe (Fig. 22.9) indicated that all parts with the exception of spring tines could be made locally.

A major breakthrough in rural productivity could be achieved by the promotion of animal-drawn implements for cultivation. However, in some remote areas many farmers were anxious to solve their problems on their own lest they become dependent on the results obtained with new implements. In the Silsoe College study 40–91 per cent of farmers felt that

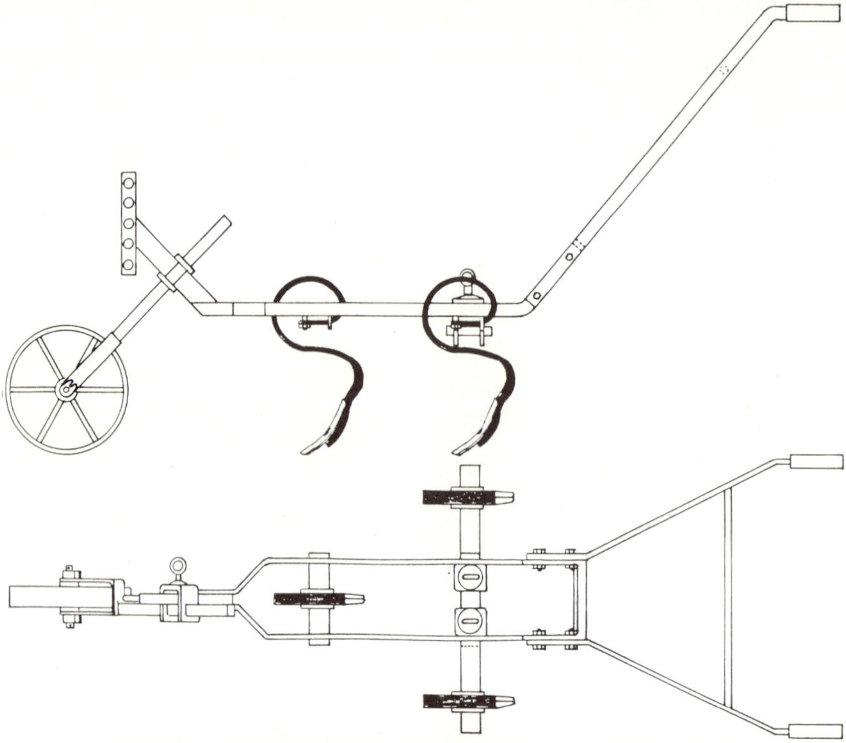

FIG. 22.9. Simple hoeing equipment.

ploughing improved moisture infiltration of the soil and 72 per cent agreed that ploughing made weed control easier.

Mechanical power technology

As in other developing countries, the mechanization of agriculture in the Sudan has been selective, only partial, and it has had international involvement. Moreover, the adoption of mechanized tillage operations has frequently involved the introduction of inappropriate techniques applicable in temperate countries.

The size of tractor and attachments should be related to the amount of work as dictated by the timeliness and cost of operations. Table 22.3 gives an indication of the difference in performance of traditional cultivation compared with mechanized operations. The total operating costs decrease with the number of hours worked because the fixed costs per unit remained relatively constant. Multi-purpose machinery would give maximum utilization of machinery.

TABLE 22.3. Comparison of costs of operations (£S)

Operation	Time to perform task ($h\ ac^{-1}$)		Remarks
	Mechanized	Traditional	
Ploughing and harrowing	1.10	25.00	
Levelling	0.40	8.10	
Sowing	0.30	8.00	
Weeding	0.30	10.00	6 persons for
	0.25[a]	1.30[b]	2 days in Traditional

[a] Field sprayer.
[b] Motor sprayer with long hoses.
Source: AOAD (1975).

According to Sheikh *et al.* (1979), the operating costs per acre with a 12 hp power tiller and an 18 hp 4-WD garden tractor were three- and two-fold greater, respectively, than with a 47 hp general purpose tractor. Although the largest general purpose tractors were found to be the most expensive to operate on a per hour basis, they were cheaper in terms of cost per hp. It would be feasible to use larger tractors whenever sufficient work was available (Moses and Frost 1965). The range of mechanical power utilization in the Sudan and the extent of experience are outlined below.

Small field tractors

The introduction of small field tractors and their attachments was a response to the requirement for low-powered mechanization for cultivation, but they brought little progress. The adoption of such tractors was hindered by farmers' reluctance to possess tractors of less than 35 hp for private use. Early in the 1970s, 30 hp Bulgarian track layers were imported and distributed to different parts, but after-sales services were poor. The spread of small farm tractors was also hampered by the negative effect of ploughing in the cement-like soils found in many areas.

Single axle tractors (Fig. 22.10) were also tried, but their general performance was not so good in the heavy clay soils as in the light silty soils of the irrigated areas. Wheel slippage was a common problem and this lowered the field efficiency.

General-purpose tractors and implements

The implicit assumption that farmers will adopt specific equipment if it is available in the Sudan, might indicate that either the farmer is not in a

FIG. 22.10. Single-axle walking tractor on trial in the field.

position to take an economic decision, or that he is convinced that the adoption of the machine will be profitable for him. The present situation in the Sudan is such that procurement of machinery by the individual farmer is debatable because of the burden to him of the outstanding prices of tractors and implements, together with their increasing running costs. In spite of the expensive inputs productivity is still low.

Machinery requirements in rain-fed areas

Since 1944, in rain-fed areas the traditional requirements had been confined to the possession of 2-WD tractors of 50–75 hp, a wide-level disc harrow with seederbox, and a trailer. This machinery is suitable for permanent farming in the mechanized crop production areas, which were divided into holdings of about 1500 acres each.

The total rain-fed area is estimated to be approximately 18.24 million acres of arable land, which includes both traditional and mechanized farming systems (Table 22.4).

A rough estimate of the total mechanical power requirement for the rain-fed land can be worked out by taking into consideration the rate of 0.6 hp per ac as an appropriate standard for tractors. The need to cover an area of about 8.22 million ac in the mechanized sector would be within the range of 21 800 tractor units, where the engine power was 90–100 hp. Based on the same rate, the calculated requirement would be about 23 870 tractor units to modernize traditional agriculture.

However, the present level of machinery is not commensurate with the

TABLE 22.4. Areas of crops (1980–7) ('000 acres)

	5-year average (80/1–84/5)			1985/6			1986/7		
		rain-fed			rain-fed			rain-fed	
Crop	I	M	T	I	M	T	I	M	T
Sorghum	628	4639	2869	1123	8537	3215	827	8190	2793
Millet	16	6	2867	22	126	3980	15	71	3590
Sesame	—	588	1424	—	1178	1340	—	1296	1316
Wheat	293	—	—	360	—	—	268	—	—
Groundnuts	282	—	1699	146	—	805	232	—	949
Cotton	860	36	103	753	2	58	782	2	37
Total	2079	5269	8962	2404	9843	9398	2124	9559	8685

Average of crops for the 7 years (1980–7) ('000 ac): I 2200; M 8220; T 9000.
Source: Department of Agricultural Economics (MANR) (1987).

TABLE 22.5. Imports of tractors and implements (1976–86)

Year	Tractors *a*	Tractors *b*	Disc harrow	Disc plough	Combine	Planter	Wide level disc harrow	Ridger
1976	1084	1722	132	292	182	12	97	500
1977	290	792	56	—	106	—	75	90
1978	1033	852	NA	416	30	13	132	100
1979	533	950	NA	—	—	—	—	—
1980	833	1500	99	654	304[a]	27[a]	3	163
1981	2183	1900	NA	NA	NA	NA	NA	NA
1982	2348	900	—	—	150[a]	NA	NA	NA
1983	329	950	—	—	199[a]	28[a]	NA	NA
1984	471	1387	—	—	239[a]	—	NA	NA
1985	88	600	—	—	1[a]	1[a]	NA	NA
1986	587	—	—	—	6[a]	2[a]	NA	NA

Source: [a] Department of Agricultural Economics (MANR) (1987).
[b] ABS (1987).
NA = details not available.

huge land endowment (Table 22.4), particularly the commanding sub-sector of mechanized farming which constitutes about 43 per cent of the cropped area. This absorbs more than 60 per cent of the tractors and 80 per cent of the wide-level disc harrows (Fig. 22.11).

In a recent survey, conducted by the Development Studies and Research Centre of the University of Khartoum (1985), it was established that 34 per cent of the farmers in the Habila area, 96 per cent in the Damazin area, and 65 per cent in the Gedaref area utilized their tractors as a means of transporting labour, water, and fuel to the production camps. This represented a massive waste of power during the season.

Machinery requirements in irrigated areas

In irrigated areas, the cropping system usually depends on the variety and intensity of the crops grown. The average amount of cultivable land is estimated to be equivalent to about 2.2 million acres and this covers a number of food and cash-crops (Table 22.5). Assuming a rate of 1.75 hp per ac as an appropriate standard for mechanizing the irrigated areas which experience diversified and intensive cultivation, an estimated requirement of about 21 000 tractor units, each of 75 hp would be appropriate.

Comparing the number of tractors and selected implements in the big

(a)

(b)

(c)

FIG. 22.11. Machinery used in rain-fed areas, (a) land preparation, (b) sesame harvesting, (c) sorghum harvesting.

TABLE 22.6. Total number of tractors and selected implements in Gezira (1975)

Item	Gezira and Managil	Mechanization Centre
Agricultural tractor	1687	2007
Crawler	66	78
Rotary cutting blade	58	70
Ridger	2621	1502
Offset disc harrow	22	725
Wide level disc harrow	67	363
Fertilizer distributor	70	—
Planter	102	894
Inter-row cultivator	74	744
Rotavator	30	79
Digger shaker	56	761
Groundnut combine harvester	120	371
Wheat combine harvester	504	395

Source: Affan Agricultural Company (1976).

agricultural corporations such as Gezira and Rahad (Tables 22.6 and 22.7), it is evident that the acquisition of appropriate tractors and implements in the irrigated areas has been unbalanced and often not relevant to requirements. In the Rahad Agricultural Corporation many tractors and imple-

TABLE 22.7. Tractors and combine harvesters by make reaching local markets (1956–75)

Tractors			Combine harvester	
Make	Quantity		Make	Quantity
	Gezira	Managil		
Nuffield	56	221	Massey Ferguson	11
Massey Harris	7	4	John Deere	27
Massey Ferguson	567	245	Claas	429
Ford	47	28	Iskafour	37
Fordson	49	—		
Rackofica	8	4		
Zadgar	31	4		
Belarus	12	—		
John Deere	8	—		
Zetor	7	6		
Farmall IH	20	3		
Others	43	130		
Total	855	645		504

Source: Affan Agricultural Company (1976).

TABLE 22.8. Farm machinery in the Rahad Agricultural Corporation

Tractors	Total	Working	Not working	Written-off
Massey Ferguson				
2720	32	32	—	—
290	64	64	—	—
185	374	258	6	110
Ford				
9700	8	3	4	1
6600	10	5	5	—
5000	30	4	—	26
International Harvester				
1086	28	16	10	2
Total	546	382	25	139
Implements				
Heavy disc harrow	44	42	2	—
Light disc harrow	200	115	42	33
Land leveller	60	—	41	19
Ridger	297	190	107	—
Planter	200	—	50	150
Fertilizer attachment	168	—	168	—
Inter-row cultivator	50	50	—	—
Sprayer	20	—	10	10
High-clearance spray	10	5	5	—
Fertilizer distributor	100	100	—	—
Harvesting Machines				
Groundnut digger	246	106	64	76
Groundnut combine harvester	182	100	25	57
Cotton picker	50	50	—	—
Pick-up baler	50	50	—	—

Source: Rahad Agricultural Corporation (1985).

ments were either not working or written off (Table 22.8). Examples of the mechanization of some crops in irrigated areas are shown in Fig. 22.12.

Pattern of supply and marketing of agricultural implements

Importation of tractors was first known as early as 1925 and the quantity of imports rose with the increase in agricultural activities in the Sudan. The numbers of tractors imported over the last fifteen years (Table 22.9) show a

(a)

(b)

FIG. 22.12. Mechanization of some crops in the irrigated areas, (a) groundnut harvesting, (b) kenaf cutting.

TABLE 22.9. Tractors imported into the Sudan (1970–85)

Year	Tractors (selected known makes)										
	Massey Ferguson	Ford	International Harvester	Leyland	Belarus	John Deere	Steyer	Zetor	Ebro	Others	Total
1970	253	80	115	198	175	—	—	450	—	—	1271
1971	—	—	75	109	571	32	—	—	13	—	800
1972	—	5	50	—	75	—	—	—	—	—	130
1973	1105	—	100	185	635	—	15	—	—	—	2040
1974	348	400	50	532	—	587	130	—	—	—	2047
1975	21	—	100	250	52	110	130	—	16	—	679
1976	1522	—	50	50	—	60	40	—	—	—	1722
1977	312	—	35	245	—	—	200	—	—	—	792
1978	—	141	—	190	—	500	4	—	17	—	852
1979	250	200	—	—	—	—	—	—	—	500	950
1980	300	200	500	—	—	—	—	—	—	500	1500
1981	1300	300	—	—	—	—	—	—	—	—	1600
1982	500	400	—	—	—	—	—	—	—	—	900
1983	500	250	200	—	—	—	—	—	—	—	950
1984	600	300	300	—	—	187	—	—	—	—	1387
1985	500	50	200	—	—	—	—	—	—	50	800
Total	7511	2326	1775	1759	1508	1476	519	450	46	1050	18420

Source: Agricultural Bank of Sudan (1986).

TABLE 22.10. Combine harvesters imported into the Sudan (1970–85)

Year	Claas	Massey Ferguson	CKM4	John Deere	Fahr	Others
1970	35	18	50	10	—	—
1971	45	30	60	—	—	—
1972	50	—	—	50	—	—
1973	50	—	—	—	—	—
1974	80	—	—	—	—	100 JF
1975	120	—	—	50	—	—
1976	170	—	—	—	20	—
1977	100	—	—	—	—	20 NH
1978–85	630	—	—	—	—	—
Total	1280	48	110	110	20	120

Source: Agricultural Bank of Sudan (1986).

significant, but not that regular annual flow. Imports of combine harvesters favoured the make 'Claas' (Table 22.10).

In recent years consumer satisfaction has been low due to many factors, but mainly the scarcity of machines in the free market. In earlier years, credit was available from suppliers at an interest rate of 5 per cent on signing the contract, 15 per cent as COD, and 85 per cent paid over five years. Such easy terms of payment no longer exist. Manufacturers refrained from financing because of the difficult financial situation in the country and the local and foreign suppliers started to rely on foreign aid and grants.

Tractors and implements

The bulk of demand for tractors is from individual farmers. The remaining demand originates from public agricultural institutions. In former years the farmer had every opportunity to choose the makes and models of the machinery he required. Nowadays a farmer's choice is severely restricted due to successive devaluations, economic inflation, and certain foreign exchange barriers.

In the recent survey by the Development Studies and Research Centre of the University of Khartoum (1985), it was found that 52 per cent of farmers in the Habila area claimed that they were unable to possess the tractors of their choice. This was due to the difficulty of attaining credit facilities from any source other than the Agricultural Bank of Sudan (ABS) which confines farmers to specific types of tractors. In the Damazin area of mechanized farming, 65 per cent of farmers were happy with their choice

TABLE 22.11. Ownership of machinery in rain-fed mechanized agriculture

Type	% Machinery used			
	Owned (%)		Hired (%)	
	Gedaref	Others	Gedaref	Others
Tractors	95	85	5	15
Wide level disc with seederbox	92	82	8	18
Combine harvester	40	16	60	84
Trailers	100	91	0	9

Source: Abdel Nour (1985).

of machinery. In Gedaref and other areas of mechanized rain-fed agriculture, most of the tractors used were owned by individual farmers; the remainder were hired. Combine harvesters were usually hired (Table 22.11).

Between 1970 and 1985, tractor prices doubled almost every five years (Table 22.12). Farmers were no longer able to afford to purchase with cash from the dealer, but had to approach the ABS, which would not provide credit facilities for all.

TABLE 22.12. Relationship between tractor prices and sorghum prices

Year	Tractor (£S)	1 bag sorghum (£S)
1970	1800	1.80
1975	3012	3.00
1980	17000	20.00
1985	32000	100.00

Source: Abdel Nour (1985).

Discarding the inflation since 1975, the price of a tractor remained equal to the price of about 1000 bags of sorghum until 1980. In 1985, the tractor price took a sharp decline relative to sorghum.

Animal-drawn implements

The promotion of animal-drawn implements provides an opportunity for increasing the production of crops in rural areas. The increasing demand for such implements has been confirmed by the results of a study in the Nuba Mountains area (Silsoe College 1987) to assess the feasibility of local manufacture of agricultural implements (Table 22.13).

TABLE 22.13. Implements sold under credit in the Nuba Mountains Development project

Item	Quantity				Total
	1982/3	83/4	84/5	85/6	
Ploughs	62	390	462	800	1714
Hoes	241	229	442	367	1279
Seeders	168	97	114	57	436
Lifters	86	39	4	31	160
Ridgers	58	44	28	52	182
Carts	204	16	—	138	358

Source: Silsoe College (1987).

Distribution and sales of machinery

Before 1969, farm machinery was imported by different private dealers, who represented sales agents of the manufacturers. Due to the policy of nationalization in the early 1970s a number of companies stagnated, but Massey Ferguson products and Claas combine harvesters were not drastically affected because they were well established and had a sound reputation among the farmers. Moreover, the continuous flow of machinery and spare parts into the country suffered a serious setback.

Distribution

Although the distribution of machinery has been affected by the rigid foreign exchange restrictions, some companies have been quite influential in terms of market supply and distribution. More than fifteen makes of tractor and seven makes of combine harvester were imported and reached the local markets between 1970 and 1985 (Tables 22.9 and 22.10). The most common and popular were Massey Ferguson and Nuffield, followed by John Deere, Ford, and International Harvester. These types accounted for about 80 per cent of total tractor imports. The Claas combine harvester dominated the combine harvester market, accounting for 76 per cent of imports (96 per cent in 1976). Companies able to provide spare parts had a significant market advantage because they were able to provide reasonable after-sales services.

The ABS has a significant effect on the distribution of machinery because of its easy access to foreign exchange links at the official level. The ABS can procure machinery on a wholesale basis and sell it on credit terms to the farmers. This is not done by other importing agents.

Imports and sales

Through its privileges and sales facilities, the ABS can command about 40–47 per cent of the total tractor imports (Table 22.14) and about 44 per cent of the wide-level disc imports.

Imports vary from year to year, particularly for the ABS, but the remaining imports were shared by the various dealers, who try to effect sales to farmers by offering low prices. The problem for farmers is the poor availability of after-sales services.

In recent years individual salesmen have appeared in the tractor market. These people, often working abroad, endeavour to profit by investing money in tractor transactions. They rarely provide spare parts and they do not provide any credit facilities. Such deals are, therefore, very detrimental for farmers.

TABLE 22.14. Tractor imports (1974–8)

Year	Importers		Total
	ABS	Dealers	
1974	1387	687	2074
1975	—	652	652
1976	1000	722	1722
1977	500	771	1271
1978	200	652	852
Total	3087	3484	6571
Average	617	697	1314

Source: ABS (1984).

Credit facilities

The actions of some dealers have contributed to price escalations which, in certain cases, exceed the black market levels, particularly for spare parts. To some extent the excessively high prices led to a fall in sales. Dealers could not offer credit to farmers who were unlikely to be able to make regular repayments and the unpredictable crop yields further diminished the chances of repayment. However, dealers who were unable to sell their commodities to farmers did not give up their trade activities because in times of shortage the ABS tended to absorb their stock. This created more problems with after-sales services, especially in remote production areas. The terms of credit offered by the ABS include a down-payment of 25 per cent before the credit is granted, followed by equal annual instalments paid over three years for the remainder. These terms are effected provided that the ABS is given some sort of security such as a post-dated money order, personal guarantee, or mortgage.

Industrial policy and strategy for machinery supply

Agricultural products make a significant contribution to the national economy (Table 22.15). However, industrial production was regarded as a means of attaining high incomes and hence raising living standards.

Local manufacturing of agricultural equipment, particularly concentrated in the capital, Khartoum, has been the subject of some speculation and a policy of producing simple implements by artisans was adopted due to the absence of local components for more sophisticated machinery manufacture. No work took place on the development of small machinery manufacture except for some private attempts which failed to satisfy consumption, and did not serve industrial needs for many reasons including the lack of qualified personnel.

In the western regions of Sudan, where there are no artisan blacksmiths, great potential existed to develop rural industries and there have been some developments in the design and manufacture of local implements. However, farmers need to be trained to use and maintain such implements.

The policy to create facilities to manufacture field equipment, which was initiated during the 1977–83 Development Plan, became more problematic with Sudan's increasing debts crisis. The aim had been to boost local

TABLE 22.15. Distribution of GDP by selected sector (1971–81)

Year	% at Price 1975					
	Agriculture	Mining	Manufacture	Utilities	Construction	Services
1971	33.8	0.2	12.0	3.5	4.0	46.0
1972	34.3	0.3	8.8	2.6	3.2	50.8
1973	33.7	0.5	15.3	1.8	4.5	44.2
1974	37.4	0.3	9.5	1.7	4.2	46.8
1975	36.1	0.3	8.9	1.6	5.1	47.8
1976	29.2	0.3	7.9	2.2	5.0	54.9
1977	31.2	0.3	7.3	3.1	4.7	54.3
1978	33.1	0.3	7.0	2.2	4.1	53.3
1979	28.5	0.3	7.0	2.3	4.9	56.6
1980	27.5	0.3	7.7	2.3	4.8	57.2
1981	27.4	0.3	7.8	2.4	4.8	57.2

Source: Statistics and Survey Unit, UNDP (1982).

participation in the national industrialization programme and to cater for the expansion of the Khartoum foundry to meet market requirements.

Farm machinery—policy and strategy

To meet farmers' requirements agriculture needs to form an integral part of the country's overall development plan and there should be comprehensive plans directed towards an organized nationwide strategy. Mechanization in the Sudan had been on a project-by-project basis which has proved unsatisfactory. Unless more efforts to support overall nationwide programmes are exerted, the unco-ordinated approach will continue to hinder realistic solutions to outstanding problems.

There is a need to manufacture large numbers of tractors and implements in the Sudan, for the potential requirements are far higher than the existing annual imports. Therefore, strict regulations are required to solve the problem of currency resources. With the exception of a few formalities, there are no legal restrictions on the importation of machinery.

It is necessary to concentrate on the proper training of operators and mechanics so as to reduce damage and misuse of machinery, thus increasing its life time. In the case of the Sudan, it is essential to reduce current expenditure in foreign currency on procuring more and more new machinery before the life span of the existing machines has run out.

The need for reliable after-sales services is recognized by farmers. The junk piles of machinery in many corporations are due primarily to the lack of properly scheduled preventive maintenance and satisfactory repair. As existing workshops are inadequately equipped, the meagre facilities, scarcity and poor quality of lubricants have contributed to the failure of agricultural progress. With the scarcity or unavailability of spare parts, and the high costs, it has proved impossible to establish better after-sales services.

Seldom is the most suitable machine available to the ordinary farmer. He is forced to purchase machinery that is excessively expensive, or is incapable of performing its function to his satisfaction under the local conditions. To eliminate this discrepancy, a well-equipped centre, the Tambul Training and Testing Centre, sponsored by the French Government, was established to carry out proper testing and evaluation of imported machinery. This centre makes a significant contribution to the monitoring of the various types of machinery permitted into the country. As a sound strategy to revive the real demand for machines and to sustain its continuity, testing safeguards the flow of suitable and technically sound types of machinery which comply to the specifications designed to suit the local human and environmental conditions of the country.

References

Abdel Nour, B. O. (1985). *Marketing and distribution of farm equipment in the Sudan*. Unpublished paper presented at the Farm Machinery and Equipment Workshop. Joint Workshop between ILO and the Ministry of Agriculture and Natural Resources, November 1985.

ABS (Agricultural Bank of Sudan) (1984–7). *Annual reports, 1983–6*. ABS, Khartoum.

Affan Agricultural Company (1976). *Farm mechanization in the Gezira*. Feasibility study by Affan Agricultural Company, Consulting Engineers, Khartoum.

AOAD (Arab Organization for Agricultural Development) (1975). *Complete survey for mechanizing agriculture in Arab countries*. AOAD, Khartoum.

Department of Agricultural Economics, MANR (1987). *Agricultural situation and outlook. January monthly report*. Department of Agricultural Economics, MANR, Khartoum.

Development Studies and Research Centre (1985). *Survey report*. DSRC, University of Khartoum, Khartoum.

Moses, B. D. and Frost, K. R. (1965). *Tractor selection and management in farm power*. John Wiley & Sons Inc., New York.

Rahad Agricultural Corporation (1985). *Annual report: agricultural practices*. Rahad Agricultural Corporation, Sudan.

Silsoe College (1987). *A study to assess feasibility of local manufacture in Sudan*. Final Report, Animal-drawn implements. Unpublished report for EEC Delegation, printed in Khartoum.

United Nations Development Programme (1982). *Statistics and Survey Unit Report*.

Further reading

Abdel Rahman, H. A. and Harbi, E. I. (1984). *Agricultural mechanization in the Sudan*. Unpublished paper presented at the 2nd International Conference on Agricultural Mechanization, Amsterdam.

Arab Fund for Economic and Social Development and Industrial Development Centre for the Arab States (1978). *Preliminary study on the agricultural machinery and equipment in the Arab region*. Arab Fund for Economic and Social Development, Khartoum.

AOAD (1979). *Future of food economics in Arab countries (1975–2000), Part I. Food production*. AOAD, Khartoum.

Bansal, R. K. and Srivastava, K. L. (1981). Improved animal drawn implements for farming in semi-arid Tropics. *Agricultural Mechanization in Asia, Africa and Latin America*, **12** (2), 33–8.

Bansal, R. K. and Thierstein, G. E. (1982). Animal drawn multi-purpose tool-carrier. *Agricultural Mechanization in Asia, Africa and Latin America*, **13** (4), 27–36.

Bell, R. D. (1983). *Optimum tillage*. Report of the 5th Session, FAO Panel of Experts on Agricultural Mechanization. FAO, Rome.

Bukhari, S. B. (1982). Cost of owning and operating tractors. *Agricultural Mechanization in Asia, Africa and Latin America*, **13** (4), 61–7.
Department of Agricultural Economics, MANR (1984). *Yearbook of agricultural statistics*. Department of Agricultural Economics, MANR, Khartoum.
Department of Agricultural Economics, MANR (1986). *Agricultural situation and outlook. Annual report 1985/1986*. Department of Agricultural Economics, MANR, Khartoum.
Mahmoud, E. S. (1982). Mechanization for modern and traditional agriculture in Sudan. *Agricultural Mechanization in Asia, Africa and Latin America*, **13** (1), 37–40.
Sheikh, G. S., Sial, J., and Afzal, M. (1979). Comparative performance of two wheel and four wheel tractors. *Agricultural Mechanization in Asia, Africa and Latin America*, **10** (2), 55–8.
Tothill, J. D. (1954). *Agriculture in the Sudan*. (2nd edn). Oxford University Press, London.

23

National policies for agriculture

K. WOHLMUTH

Introduction

The economic crisis which has been affecting the Sudan since 1977/8 is having a particular effect on the agricultural subsectors (irrigated agriculture, mechanized rain-fed farming, traditional rain-fed farming, livestock-raising, and forestry) and also on the industrial sector of the country. These subsectors are suffering from the repercussions of the most severe economic crisis since Independence (1956), a crisis which is the result of unfavourable world market trends, over-ambitious development programmes in the 1970s, and a failure of policy reform at the macro- and sectoral-levels in the 1980s. The worsening balance of payments situation, the increasing indebtedness, the failure to come to terms with the International Monetary Fund (IMF), the banks and other donors, the decrease of official transfers from Sudanese workers abroad, the decreasing level of foreign aid, the increasing pressure on the Sudanese pound (£S), and the capital flight go parallel with unfavourable cotton marketing conditions and unstable internal political developments (Sudanow 1987*a*, *b*; Ottaway 1987).

The IMF and the World Bank are exerting strong pressure for policy reform on the Sudanese Government, but up to now the results of policy reform measures for the productive sectors and at the macro-level are not at all impressive. In 1980, the Sudanese Government presented a recovery programme to the donors of aid in order to regain confidence in international finance circles. The programme aimed at adjustments of the price structure, demand control, institutional reforms, especially at the level of government administration, restoration of the economic functions of the market (and privatization), and rehabilitation of agriculture and industry. Some progress in rehabilitation and policy reform was made, especially in 1982/3, when cotton exports were restored to the 1975/6 level and budget and balance of payments figures improved (World Bank 1985, p. IV). But the trend did not last long and the deadlock of negotiations with the IMF and the unsolved question of rescheduling the debt has had an increasing effect on the productive sectors via scarcity of hard currency, inflation, overvalued currency, price distortions, and capital flight. The major issues of economic management are now, in 1989, still unresolved.

Government action in agriculture has been limited mainly to the irrigated subsector, the main focus being the change of production relationships between the tenants and the Sudan Gezira Board management. The unfavourable world market situation for cotton, the continued overvaluation of the currency and the many other weaknesses of irrigated agriculture (such as ageing installations) have aroused the planners, for traditionally this sector was the main source of foreign exchange and government revenues. Important markets for industrial expansion are also located in the Gezira. Although, more recently it is acknowledged that the rain-fed sector 'shows social rates of return that are generally higher than for irrigation' (World Bank 1985, p. 14), this wisdom is not reflected in policy changes in the rain-fed farming and livestock sectors. The rain-fed sector has a higher share of the Gross Domestic Product (GDP) than the irrigated sector and generates higher net foreign exchange earnings. Therefore, a drastic policy change could contribute to the enhancement of growth and employment, basic needs provision and income redistribution, as well as the elimination of inflation and balance of payments imbalances.

In 1986, the Sudanese Government launched a major study programme entitled: *Strategy for Development of Rain-fed Agriculture* (Republic of the Sudan 1986), which emphasizes the potential of the rain-fed sector, describes the objectives of a strategy for its improvement and the basic conditions for success.

The strategy aims at food self-sufficiency and regional food security; production of surpluses for export; increased incomes for producers and sustained employment for the rural population; conservation, rehabilitation, and improvement of the natural resources; encouragement of private investments and initiatives for increased production by providing infrastructure, services, and policy support. It is a far-reaching programme and, unfortunately, up to now, no decisions have been taken as to how to implement it. This can be seen from the development budget for the fiscal year 1986–7 (MFEP 1986*c*).

It is very unfortunate that, at present, the dynamics of the Sudanese economy centre exclusively on services and the informal sector. The informal sector absorbs ten per cent of the labour force, the formal (mostly public and urban) sector, and the irrigated agricultural sector each absorb ten per cent of the labour force, whereas the rain-fed farming and livestock-raising sector with 70 per cent of the labour force is stagnating or in decline (ILO 1987, p. 41).

Evolution of national policies for agriculture in the Sudan

Four periods in the evolution of national agricultural policies may be distinguished for the Sudan (Oesterdiekhoff 1980; FAO 1986; Eltom 1986; Salam 1986): the colonial period; the post-independence period (1956–69); the 1970s including Numeiri's May Regime and the period of the 'bread basket' strategy; and the period of recovery and rehabilitation in the 1980s.

In the first period, agricultural policy was characterized by the promotion of cotton growing, concentrating on the Nile Basin while disregarding agricultural production in other areas of Sudan. In British colonial times the natural and human resources of other areas remained undeveloped, although the hinterland of Sudan was important as a source of cheap labour for the Gezira Scheme. Agricultural development took place only to a very limited extent in other areas but the so-called livelihood schemes in the Northern Region, the Nuba Mountains, the Gash and Tokar Deltas, and the Zande Scheme in the south should be mentioned. Some of these ventures were even cash- and export-oriented. During the Second World War mechanized rain-fed farming began in order to provide food for the army.

In the second period (1956–69) the colonial-type agricultural policy continued basically unchanged. The Gezira Scheme was nationalized and enlarged considerably with the opening of the Managil Extension and, for the first time, there was an attempt to change Sudan's production and trade structures with measures to intensify and diversify production towards wheat and groundnuts. Some railway development in this period allowed for the expansion of traditional cash-crop production, an increased monetization of the livestock economy, and the overall expansion of food production. Mechanized farming expanded considerably as did the agro-industries, particularly flour mills, textiles, sugar, and vegetable oils. The national policies on agriculture were implemented as the *Ten Year Plan of Economic and Social Development 1961/2–1970/1* (Ministry of Finance and Economics 1961) which emphasized only the modern irrigated and mechanized farming sectors. Although the plan was abandoned in 1964, the emphasis on large-scale irrigation and mechanized farming continued. Private investments into large-scale mechanized farming were encouraged and a Mechanised Farming Corporation (MFC) was set up in 1968. Already, unplanned and irrational expansion of mechanized farming into the central clay plains had caused serious deterioration of soils and pastures.

In the third period, starting with Numeiri's May Revolution in 1969, the colonial bias against traditional agriculture continued unchanged. The unplanned expansion of mechanized farming and the neglect of traditional agriculture led to productivity decreases in rain-fed farming and to enforced migration of labour to modern sector agriculture and to urban areas. The nationalization of the private cotton pump schemes brought a further decline in productivity. However, agro-industries were expanding and diversifying considerably and the Soviet-inspired *Five Year Plan 1970/1–1974/5* reinforced the bias towards modern agricultural subsectors and the promotion of agro-industries. The plan was extended essentially unchanged until 1977. Agricultural policies of the time were characterized by the focus on food import substitution via the horizontal agricultural expansion route; new areas of cultivable land were opened up instead of utilizing already used land more intensively.

However, it was during this period that the neglect and the degradation of the traditional sector became a policy issue. The *Six Year Plan of Economic and Social Development 1977/8–1982/3* (MNP 1977) was the first to recognize explicitly the importance of the traditional sector for Sudan's development, but the allocation of public funds did not reflect the new insight. A new strategy, the parallel development of modern and traditional subsectors was emerging. This should allow both subsectors to be sustained and should enable income redistribution and equity objectives to carry more weight. Influential in this policy change was the mission report of the International Labour Organization (ILO 1976), published at the request of the Sudanese Government. The report discussed several strategies to correct the dualism between the high-income irrigated and mechanized rain-fed agricultural sectors and the low-income, traditional agricultural and livestock-raising sectors. The first strategy was to bring people from the low-income to the high-income regions by allocating them tenancies in newly-irrigated schemes. Although the Rahad Scheme, a new irrigation project, was a step in this direction, the strategy was not a viable solution to Sudan's agricultural development problems because of its limited capacity to absorb even the expected population increase. The second strategy was to establish modern production units in low-income areas. The expansion of mechanized farming was operating in this direction and it was argued convincingly in the report that the system promoting large farming units of 1000 feddans (420 ha) or more may benefit only well-off investors. Such a system was unlikely to contribute towards reducing the income gap.

The third strategy was, at that time, recommended to the Sudanese Government. It was proposed to develop traditional agriculture and livestock-raising in the low-income regions, while continuing to promote the growth of modern agriculture in order to generate the savings necessary to develop the traditional sector. The Sudanese Government did not follow this recommendation and, in the following years the traditional sector even contributed to the development of modern sectors by providing cheap labour and by suffering heavy direct and indirect taxation. Cash-crop producers received prices far below the minimum of floor prices announced by the Sudanese Government. The result was a highly unbalanced development of agricultural subsectors, which led to further erosion of the production potential of traditional agriculture. The ILO's strategy for the transformation of the traditional agriculture was to be implemented during two distinct time periods: the 1975–85 period was for creating the preconditions for a take-off in the traditional sector, whereas the 1985–95 period was for massive investment in traditional farming comparable to the investment effort in modern farming. The strategy would have required a massive increase in the historically meagre allocation of public funds to traditional agriculture.

While rejecting the ILO strategy recommendations for traditional agriculture, the Sudanese Government actually increased its development expenditure enormously from 1973 in order to implement the 'bread basket' strategy, an ambitious programme for the modern agricultural subsectors and agro-industries (Awad 1983; Oesterdiekhoff and Wohlmuth 1983*a*, *b*). This programme aimed at the diversification of Sudan's production and trade structures to make the Sudan an important supplier of food to Arab countries. Huge production increases and large-scale processing of local raw materials were envisaged. Between 1972 and 1975, development expenditure increased from £S29.6m. to £S102.4m. (Wohlmuth and Hansohm 1984, p. 46). This policy was laid down in the *Six Year Plan*, and an annual planned growth rate of 6.5 per cent for agriculture and of 9.5 per cent for manufacturing and mining was envisaged. The Government assumed that vast, untapped natural resources were waiting to be exploited and that horizontal (land-consuming) agricultural policies were justified. The objectives of agro-industrial expansion were both import substitution (in sugar, textiles, and clothes) and export substitution (cotton by yarn and clothes, hides and skin by leather, and live animals by meat). From the outset, the strategy was limited in terms of creating domestic intersectoral and intra-sectoral linkages, for the intention was to replace raw material exports with exports of semi-finished products. In order to finance this leap-forward strategy the Government was relying on external resources for 52 per cent of the total (public and private) planned investments. About 88 per cent of the planned investment in agriculture was scheduled directly for modern sectors (irrigated projects; promotion of new export crops such as coffee, tea, and rice; and modern livestock ranging), and there were other expenditures for agricultural services again biased towards the modern sector. By contrast, projects aimed directly at traditional agriculture received only 3 per cent of the total investments scheduled for agriculture (Wohlmuth and Hansohm 1984, p. 30) and even this was used not for the improvement of existing farming systems, but for the introduction of modern mechanization programmes for smallholders into traditional rain-fed areas.

Huge increases in annual production were expected from this strategy, 12.0 per cent for sorghum, 12.6 per cent for wheat, 18.9 per cent for medium-staple cotton and 11 per cent for groundnuts. More important, production increases were expected to result primarily from area expansion rather than from higher yields (Wohlmuth and Hansohm 1984, p. 31). Horizontal expansion became the preferred strategy for rapid increases in agricultural production. Planned public investment remained concentrated on modern sector projects in eastern-central Sudan, whereas private investment was expected for development in new areas such as the western savannas and the southern Funj. The ecological impact of this strategy was not given any consideration at that time. A similar bias against the traditional sector and towards the horizontal expansion of mechanized farming was implicit in the Ministry for Agriculture's *Food Investment Strategy 1977–1985* (MAFNR 1977). This strategy aimed for food self-sufficiency not by developing peasant farming systems, but through large-scale infusions of private capital into mechanized farming.

Another important input to the horizontal agricultural development strategy was a document entitled: *Basic programme for agricultural development in the Democratic Republic of the Sudan, 1976–1985* (Arab Fund for Economic and Social Development 1976). This was the most comprehensive and ambitious formulation of the objectives and the programme of the 'bread basket' strategy. In September 1986 the Sudanese Government prepared documents for a meeting of the Arab Ministers of Agriculture in Jordan, with the intention of reviving the 'bread basket' idea again. Therefore, the Arab Fund's programme is still relevant to some extent. It assumed that, in the long term, 9 million feddans could be irrigated and that 71 million feddans could be cultivated under rain-fed conditions; currently, only 2.4 million feddans and 17–20 million feddans, respectively, are utilized. Internal and external factors were such that the 'bread basket' strategy was never implemented as planned (Hansohm and Wohlmuth 1985*b*). The main problem with Sudan's horizontal expansion agricultural development strategy is that the two basic assumptions are no longer valid—first, the availability of surplus land under-utilized and under-populated (relative to the natural resource base) and second, the assumption that the expansion of mechanized farming is socially, economically, and ecologically feasible (Wohlmuth and Hansohm 1984). In its *Strategy for Development of Rain-fed Agriculture*, the Sudanese Government acknowledges the problem of increasing land scarcity and increasing competition for land in various regions. In a statement by the Agricultural Planning Administration of the Ministry of Agriculture and National Resources (MANR 1985), the new position is presented: vertical expansion should increase in importance relative to horizontal expansion, by introducing and supplying improved seeds, by supplying adequate inputs and essential agricultural services, by efforts to improve crop handling, marketing channels, and storage facilities, and by developing an appropriate rural infrastructure.

In the fourth period of recovery and rehabilitation initiated in 1980, the

Sudanese Government has, in association with the World Bank, worked out an export action programme which has as its main aims: the provision of spare parts and of machinery needed to reverse the deterioration of the capital equipment; the allocation of more foreign exchange to finance needed inputs; the financing of rehabilitation projects; policy reforms and a revision of incentive systems to stimulate production. The first agricultural rehabilitation programme of 1980 was financed by the World Bank to support the Sudanese Government in its export action programme. It provided urgently needed foreign exchange for high priority imports for the irrigated agricultural subsector, but it also encouraged and supported macro-economic and sectoral policy reforms such as the introduction of cost recovery and accounting systems in irrigated agriculture, already implemented by the Sudanese Government. Such measures are intended to ensure a more rational use of inputs and a distribution of earnings according to efficiency and performance. Reforms of cotton pricing, of the system of payments for cotton delivery and of the parastatals dealing with the export crops have also been implemented. The second agricultural rehabilitation programme of 1983 aimed to support further policy and institutional reforms and to provide critically needed inputs. Another programme, the Gezira rehabilitation project of 1983 refers directly to the rehabilitation of the most important modern sector project in Sudan, aiming to raise production and yields considerably within five years.

The Government's three-year public investment programmes reflect increasingly the drive towards recovery and rehabilitation. The second three-year public investment programme (1980/1–1982/3) was the first to adopt the new strategy as its objectives, but funds were allocated exclusively for irrigated agriculture. The third public investment programme mentioned, among other goals, the need to design a strategy for rain-fed agriculture. This was also a reflection of the fact that in 1981/2 rain-fed crops represented 66 per cent of the total crops produced (FAO 1986, p. 146). The fifth three-year public investment programme (1984/5–1986/7) is aimed at the rehabilitation of projects, the improvement of incentive systems and production relations, and changes in exchange rate, taxation, and pricing policies to ensure better returns to producers and an optimum supply of inputs.

Until now, however, the recovery strategy has not changed the colonial bias towards irrigated agriculture. It has, in fact, been reinforced by the nature of the current rehabilitation programme which is motivated by the expectation that concentrating funds on irrigated agriculture can generate foreign exchange more quickly than channelling funds towards rain-fed agriculture which has huge potential, but would require long-term investment programmes.

We can conclude, therefore, that since Independence in 1956 the Sudan has followed an agricultural policy first biased towards irrigated agriculture, since 1973 more and more biased towards mechanized farming, and since 1980 biased towards the rehabilitation of irrigated agriculture alone. It is obvious that agricultural policy has been unbalanced in its emphasis and the fundamental interconnections between the agricultural subsectors have been disregarded.

Agricultural policies and subsectoral resources allocation

It is necessary to examine the implications of the Government's agricultural policy objectives (Government of Sudan 1986; FAO 1986, p. 155) for the use of resources in various agricultural subsectors. The policy objectives are as follows: expansion of agricultural exports; achievement of self-sufficiency in agricultural imports; diversification of agricultural production and exports; increased and more equitably distributed agricultural incomes; integration of the traditional sector into the economy, and improvement in the balance of regional development.

According to official sources in 1985/6, of the total 200 million feddans of cultivable land, only 20.9 million feddans were actually cultivated: 2.4 million feddans in the irrigated subsector, 8.6 in the mechanized farming sector and 9.9 in the traditional farming sector (ILO 1987, pp. 49–74). The irrigated subsector with 200 000 tenancies (each cultivating about 12 feddans on average) employs an additional 400 000 persons as permanent wage-labour and 500 000 as seasonal labour. The rapid expansion of the mechanized farming sector (which has doubled its area under cultivation within 10 years, and also its potential for sorghum production) implies that 10 000 large-scale farms employ up to 1 million seasonal labourers and in years of good yield provide an important marketable surplus for export and for urban consumption. The traditional sector has a labour force of 4 million on 2 million smallholder farms and provides 2 million migrant labourers on a seasonal basis to urban areas and modern agricultural sectors. Although this sector shows a disappointing record in terms of production, yield, and area expansion, it is important as a source of labour supply and as a source of foreign exchange earned from the export of cash crops. As recent data on comparative crop yields show, mechanized farming is not doing that much better than traditional farming (MANR 1986).

In order to balance the development within and between subsectors, the Sudanese Government has to assess the relative contribution of the subsectors to the achievement of the overall objectives set out above for the economy as a whole and for agriculture as a sector. This is necessary because the subsectors compete for resources (land, capital, credit, public investment, labour, energy, foreign exchange, administrative capacity, transport and communications infrastructure, and skilled labour). On the other hand, the subsectors contribute quite differently to the generation of foreign exchange, public revenues, and savings, to rural development and basic needs provision, to the conservation of natural resources, to the creation of linkages with other sectors (especially industry), to the reduction of external dependencies, and to employment creation, income generation, and income redistribution. An overall assessment of the costs and benefits of promoting different subsectors cannot be presented here (ILO 1987, pp. 55–8), but some evidence may highlight the tradeoffs and policy issues involved in subsectoral resource allocation. Sudan's debt and balance of payments crisis has meant that a series of devaluations and increases in the costs of imported inputs have improved the comparative advantage of traded outputs from rain-fed agriculture.

Data on the relative capital intensity of agricultural subsectors (World Bank 1983, pp. 24ff.) show that despite sharp cost increases for imported and capital-intensive inputs and installations, the capital-intensity of the major irrigation schemes has increased further in the 1980s, in part as a consequence of rehabilitation programmes. This is a perverse response to the underlying factor proportions of the Sudanese economy but is obviously rooted in major price distortions and rigidities. The development budgets reveal that the overwhelming share of public investment is still going to the irrigated subsector. The tendency for a rising capital-intensity is sometimes explained by referring firstly to the labour shortage argument and secondly to the argument that mechanization will allow for better cultivation practices and higher yields. However, no firm evidence is available on these two aspects and, from the point of view of capital allocation, the subsectoral distribution seems to be far from optimal.

The foreign exchange intensity of operations in irrigated agriculture is higher than in the mechanized rain-fed sector and incomparably higher than in the traditional sector. The high potential of traditional agriculture as a net foreign exchange earner is demonstrated in a study by D'Silva (1983, pp. 19ff.) which shows, for 1980 and 1982, the net value of exports (gross value of exports minus value of imported inputs) and the shares of the respective subsectors. The share of the traditional sector (excluding livestock and forestry) increased from 34.8 per cent in 1980 to 51.0 per cent in 1982 while the irrigated subsector experienced a spectacular decline from 42.8 per cent to 19.1 per cent over the same period. The mechanized farming sector ranks second as a net foreign exchange earner. With ageing irrigation systems and increasing costs of inputs due to rapidly increasing costs of spraying and crop protection, it is unlikely that this tendency can be reversed in the near future.

Both capital intensity and the foreign exchange intensity in agriculture have been reinforced by the lending policies of the Agricultural Bank of Sudan (ABS). This Bank is supporting the expansion of mechanized farming by giving subsidies in various forms, by granting low nominal rates of interest on credits in an inflationary environment, and by incurring losses because of credits not being repaid. Thereby incentives are given to use capital and foreign exchange more intensively. However, only 0.5 per cent of the smallholder farms (2 million) benefit from ABS loans; in recent years funding of ABS operations by some donors has provided support to provide programmes for traditional farmers, but usually only 9 per cent of the overall ABS funds are allocated to the small farmers.

Government policy will have to do much more to channel credit to the smallholders in more flexible forms. Pilot projects, indicating some support by the ABS for co-operatives, are underway in the western Sudan, but these moves are limited by factors such as the number of borrowers, the volume of loans, the limited number of crops covered by projects—only sesame and groundnuts, and the area of cultivation.

The competition for land is becoming a major policy issue for the Sudan which already faces population pressure on land, intensified by population growth, ecological degradation, and drought. Figures aimed at demonstrating a large reserve of uncultivated land should be interpreted very carefully. The horizontal agricultural expansion strategy has imposed costs on the traditional producer which are not covered or compensated for by the Government or by private large-scale investors. It was argued in the 1986 land tenure task force study (Republic of the Sudan 1986) that the traditional producer is forced to resume cultivation of land which has had little time to recover after previous cultivation, so that it cannot provide him with sufficient food for subsistence. Uncontrolled mechanization could not be confined to clay soils and has moved into the sandy *qoz* soils which are cultivated by traditional producers. Also increasing is the competition for land between mechanized farming, newly-irrigated farming, and livestock-raising.

Energy consumption varies significantly between subsectors (D'Silva 1983, pp. 15ff.). 18 per cent of all gasoil consumption is used in agriculture, 63 per cent for agricultural operations, 32 per cent for irrigation, and 5 per cent for canal maintenance. The figure of 18 per cent does not include energy use for transport related to agriculture. Irrigated cotton is the dominant consumer of energy (using 4.1–8.1 gallons per feddan for all agricultural operations, and 13 gallons per feddan for irrigation). This exceeds consumption of energy per feddan by all other agricultural modes of production. On a per feddan basis the energy use in irrigated sugar-cane production exceeds that in irrigated cotton production, but it is still grown on a smaller scale. Irrigated agriculture consumes 83 per cent of all the gasoil consumed in agriculture, while the traditional sector generates most of the fuel-wood and charcoal energy, providing 80 per cent of Sudan's total energy base. Development policies for the traditional sector are, therefore, additionally effective in terms of local energy resources development and reducing dependency on imported energy sources.

Where there is competition between subsectors for scarce public investments funds, the neglect of traditional agriculture relative to irrigated agriculture is extremely pronounced. An analysis of the development budget for the year 1986/7 (MFEP 1986*c*) shows an allocation of only 9 per cent aimed directly for the traditional sector, 48 per cent allocated to irrigated agriculture, 2 per cent to mechanized farming (a sector benefiting from public sources more indirectly via low nominal land rents, subsidized credits and incentive exchange rates), 16.1 per cent to animal resources development, 11.8 per cent for agricultural services, 10 per cent for natural resources development and combatting desertification; other allocations have a share of 3.1 per cent. This implies that a greater emphasis is placed on traditional agriculture now than in the 1970s, especially if we add the shares for animal resources development, natural resources conservation, and provision of agricultural services. However, it is not easy to say exactly what will be spent by the Government for the improvement of existing farming and livestock raising systems in the traditional sector. When the public subsidies and the support mechanisms for mechanized and irrigated farming are considered as well, the actual proportion of the Budget for traditional agriculture may still be assessed as being extremely low.

Administrative capacity, and transport and communications infrastructure are extremely unevenly distributed between the subsectors and the facilities available to rain-fed producers are extremely poor. This situation leads to problems of controlling the spread of mechanized farming activities and it also constrains severely the market integration of traditional producers. Despite some regional moves and considerable investment into the transport and communications system, the bias towards irrigated agriculture remains unchanged.

Allocation of skilled labour across the subsectors is also an issue of concern. The Sudan Gezira Board (SGB) has a relatively strong force of qualified manpower; the MFC on the other hand suffers from a lack of skilled labour to supervise the rapidly expanding areas under cultivation; the Agricultural Extension Service dealing with traditional rain-fed agriculture is extremely understaffed and inappropriately staffed in relation to its objectives and responsibilities; the same is true for the institutions of Natural Resources Conservation and the Veterinary Services.

Unskilled labour may also become a factor affecting the subsectoral balance. Labour shortages have often been reported in some regions and for some modern sector projects in the Sudan, but in the years to come such shortages may also occur in traditional agriculture, because drought and far-reaching socio-economic changes may have led to more permanent forms of migration, fuelling the growth of the informal sector. This may also affect the off-farm activities in the rural areas and the income generation there. Neither in mechanized farming nor in irrigated farming is there any evidence of crops remaining unharvested because of labour shortages, but mechanized farming may attract labour for weeding and harvesting activities more easily than irrigated farming because of marginally higher wages.

The neglect of traditional agriculture by inappropriate subsectoral resource allocations is obvious; with this method of allocating resources, neither the objectives of agricultural policy nor the overall developmental objectives can be realized and an erosion of the long-term development potential of the country takes place. Irrigated agriculture is becoming an increasing burden for Sudan's economy, while mechanized farming in its present form is threatening the environment and is imposing long-term costs on the Government in order to combat environmental degradation. In good crop years, mechanized farming provides a source of income for up to 1 million seasonal labourers, but this is not a stable source of income as shown by the sharp decline in production and employment in 1984/5 (ILO 1987, p. 66). Savings accrued by disposing of a marketable surplus produced in this sector are not channelled back to agriculture, but are invested in urban areas and in services-related activities, so it may not be easy to sustain this type of farming for long without public subsidies and inducements for private action. The *Strategy for Development of Rain-fed Agriculture* is an expression of the need for pronounced subsectoral resource shifts, towards traditional agriculture and it must be concluded that the above-mentioned agricultural policy objectives and development objectives can be realized in a cost-efficient way only by giving priority to policies which promote traditional agriculture.

Policies for transforming traditional agriculture: priority for integrated development

The concept of balanced development for agricultural subsectors requires that new models for the support of traditional agriculture are designed and implemented. The traditional sector in Sudan is recovering now after years of drought (ILO 1987, p. 59), but decades of neglect and the draining off of resources from this sector still limit its actual growth performance. There is a basic dilemma: the need to maintain the competitiveness of Sudan's export crops on the world market results in low wages for migrants from the traditional sector and low returns to cash-crop producers in the traditional sector, a prerequisite for the future of Sudan's export economy. This type of involvement in the world market leads to a precarious balance between destruction and preservation for Sudan's traditional agricultural sector. Until the 1970s, the balance was maintained more or less, but the policies of the 1970s and early 1980s together with the recent drought years have led to large-scale destruction of this sector. To develop this subsector it is not just a question of providing adequate resources (e.g. capital, skills, technology, and know-how). It is also necessary to deal with the main structural constraint of Sudan's current development model, that cheap labour and, to an increasing extent, fertile land from the traditional agricultural sector must be mobilized in order to maintain Sudan's export economy which is facing unfavourable world market conditions (Wohlmuth 1980). As a result, market growth and productivity growth remain stagnant in the traditional sector. In the long-term there is no way out of this dilemma other than to change the role of the export sector by introducing a development strategy based more on internal demand (generated by increasing incomes in rural areas) and on linkages between agriculture and industry, thus reducing external dependencies. On the elements of such a strategy see Wohlmuth and Hansohm (1984, pp. 78–91) and Wohlmuth (1986).

There is a long history of developing cash-crop production in Sudan's traditional sector by the establishment of cotton schemes for smallholders. There has been no serious attempt for decades to improve existing farming systems. Four approaches may be considered for transforming traditional agriculture (ILO 1976, pp. 31ff.):

1. The single limiting factor approach assumes that it is possible to identify a single limiting factor for improving traditional agriculture, for example, land tenure, credit availability, extension services or pricing policies. Some programmes based on this assumption have been tried in the Sudan, but with little overall impact.
2. The minimum package programme assumes that it is possible to improve traditional agriculture by making a few essential improvements, for example, improved cultivation practices, provision of inputs and credit facilities, improved infrastructure, and better training and extension services. Experiences here have proved that the approach is too limited to change traditional agriculture significantly.
3. The nucleus plantation approach aims at involving subsistence farmers in cash cropping, where some industrial processing of crops is required. The Talanga tea project in southern Sudan is an example where smallholders are settled around the tea processing unit, but experiments of this kind remain isolated and limited in prospects.
4. The integrated regional development programmes are based on the assumption that package programmes for a whole region are required in order to promote broad-based agricultural development. These programmes operate at a level which covers about three to five districts and they include economic and social development components. In the Sudan this approach has become the most important path for a reform of traditional agriculture. The integrated regional development schemes which are most important for reference are in western Sudan: the Jebel Marra rural development project and the Western Savannah development project, both in Darfur and the Nuba Mountains rural development project in Kordofan.

The Western Savannah development project comprises six administrative districts in Southern Darfur Province and covers a total area of 135 000 km^2 with a human population of about 1.18 million, including about 200 000 nomads. The complex and integrative approach (covering settlements, veterinary services, range and pasture improvement, forestry and rangeland rehabilitation, research and extension, input distribution, training, strengthening of local institutions, building of wateryards, roads, and tracks) makes this programme a testing ground for generating experiences, so that it could be replicated elsewhere. According to a recent evaluation, some progress has been made.

Another project in Darfur is the Jebel Marra rural development project which has contributed significantly to the improvement of local living standards. Although the projected increases in agricultural production did not materialize fully, mainly because of insufficient research, extension, and training facilities, the results of the evaluation have, on the whole, been positive (ILO 1986, p. 61). Technology appeared to be crucial for success, and better co-operation between research and extension, emphasis on on-farm trials, enhanced contact between the extension service and the farmers through contact farmers, and improved varieties of sorghum and millet adapted to local conditions were some of the recommendations for a technological upgrading and extension package. All the experiences show that these problems can be overcome in time, but that concerted efforts are necessary and a better integration of research, extension, training, and infrastructural development is required.

The Nuba Mountains rural development project in Southern Kordofan is concerned with agricultural development in the smallholder sector and the control of ecological degradation by reafforestation. The project which promotes the use of animal traction is still small, covering only 0.7 per cent of the farm households in Southern Kordofan. However, some important questions have emerged. Farmers participating in the project own more livestock and have better access to alternative sources of income, through their activities as traders and merchants, which may suggest a project bias towards wealthier farmers in the area. New measures, such as group credit programmes aimed at including poorer farmers have, therefore, been introduced.

The above examples show that mixed results and experiences must be acknowledged and that the transformation of traditional agriculture in the Sudan is a complex and long-term task requiring integrative efforts at various levels and among many people. Such integrated regional development programmes have to be incorporated more fully into multi-regional and/or national strategies for the traditional sector. The Kordofan rehabilitation development strategy (MFEP 1986*a*) and the Darfur rehabilitation programme (MFEP 1986*b*) are examples of a comprehensive regional strategy. Integrated rural development programmes cannot be a substitute for national systems of services for the agricultural subsectors and a long-term strategy for the build-up of such services is required. There is no serious initiative in Sudan in this regard.

The complex and integrative strategy for transforming traditional agriculture requires the design of new policies in six policy areas:

(1) Land tenure and land-use policy;

(2) Institutions and institutional reform;

(3) Production incentives, pricing, marketing, and finance;

(4) Technology, research and extension;

(5) Infrastructural development;

(6) Linkages with other productive sectors.

In its *Strategy for Development of Rain-fed Agriculture*, the Sudanese Government is taking up some of these issues.

Policy on land use and land tenure must acknowledge that land in the Sudan is no longer a plentiful resource, while policy on institutions must take into consideration the large number of institutions (policy-making institutions, service institutions, control institutions, public production and area development institutions, private sector institutions, including co-operatives and farmers' unions, and financial institutions) of relevance for transforming traditional agriculture effectively, and that appropriate co-ordination and institution-building is indispensable for the success of any strategy.

As regards production incentives, it is necessary to understand first that the whole marketing system is biased against the small farmer, that new models of input provision and distribution are required, that appropriate low-cost storage facilities are necessary and that all explicit and implicit forms of taxing the agricultural producer have to be reconsidered.

Policy on technology, extension, and agricultural research must consider, primarily, that traditional farmers' production and income generation is interrelated with cash-crop markets (especially sesame, groundnuts, and gum arabic), with labour markets (via migration), and with livestock raising. In this area, policy must also address the problems of declining productivity and land exhaustion, and the issue of appropriate production technologies. The current links between research and extension are inadequate, so new models must be developed to support the small farmer effectively.

Infrastructure policy requires first the design of a strategy of rural infrastructure development. Some experience with improving rural infrastructure (rural feeder roads, afforestation projects, filter wells and village infrastructure) has been gained on public works programmes in White Nile Province. The Rural Water Corporation, which is applying locally-adapted technologies for the provision of drinking water, is another good example of successful work. The integration of such activities within a rural infrastructure framework, comprising roads, water, shelter, health, and education, may be envisaged.

To create and maximize linkages between traditional agriculture and industry, policy must relate to the maximization of backward linkages—for example, producing animal traction equipment on a larger scale, forward linkages—by processing local products in small urban centres and villages, and final demand linkages—the off-farm employment and income opportunities associated with agricultural growth. In rural areas, there is potential for workshops and there are local capabilities and skills (ILO 1987, pp. 75–84). Given some support from the Sudanese Government, private capital to invest could be mobilized. An important aspect of linkages is the creation of final demand which allows the utilization of the increased market potential in rural areas for the development of rural and regional industries (Wohlmuth 1986, pp. 23–7). This aspect is dealt with in the concluding sections.

Changes in the six policy areas discussed are not only important for the transformation of traditional agriculture, they also have relevance for the mechanized farming sector and the irrigated agricultural sector (ILO 1987, pp. 49–74). There are so many inter-connections between policy areas that only improvements in all six policy areas will lead to overall successful agricultural policy changes.

Conclusions: towards agricultural demand-led industrialization in Sudan

It has been demonstrated that the development of the agricultural sector is closely connected with the dynamics of Sudan's export economy. As long as low wages and low prices for crops exported from the traditional sector are a prerequisite for Sudan's export prospects, the development of traditional agriculture will remain constrained severely. The integration of Sudan's traditional sector into a modern and dynamic economy requires that industrialization is related to overall agricultural growth and productivity increases—what Adelman (1984) calls 'agricultural demand-led industrialisation' (ADLI). The ADLI strategy requests a shift of public investments towards agriculture to increase productivity and to maximize linkages, thereby creating a domestic mass market for intermediate and final industrial products in rural areas. Such a policy shift has to be complemented by income redistribution measures so that the production of more 'wage goods', such as food and textiles, meets sufficient purchasing power. The ADLI strategy can be considered simultaneously as a programme to accelerate growth, investments, and employment, a programme to improve rural income distribution, basic needs provision, and food security, and as a programme to gain foreign exchange savings and to improve the overall balance of payments situation. In the Sudanese context, this strategy requires the support of traditional producers because of their higher labour intensity and their greater use of locally-produced consumer goods and domestic implements. Employment prospects and market conditions for industrial products would improve, if agricultural productivity increases were stimulated by a strategy incorporating policy changes in the six areas discussed in the preceding section. At the macro-level ADLI requires that production incentives are kept neutral, that they are neither biased in favour of exports or imports, nor in favour of manufactured goods or agricultural/livestock products. As Sudan's economic incentives are still heavily biased against agriculture (Acharya 1971), policy reforms at the macro-level have to be an important part of an ADLI package. Structural adjustment policies for the Sudan based on ADLI-type strategies have been outlined by Kursany (1983) and Wohlmuth and Hansohm (1984). Complementary are measures to promote rural small industries and crafts (Hansohm and Wohlmuth 1985*a*). The advantages of supporting these industries are significant in terms of local resources utilization, technological skill acquisition, employment and income effects, and the attainment of regionally balanced development. Nevertheless, up to now the Sudanese Government has not made the decisions which are necessary to integrate the traditional sector into the national economy in this way. Projections for 1991/2 give the traditional sector overwhelming importance, even for the generation of exports. If the *Strategy for Development of Rain-fed Agriculture* can really be implemented, the traditional rain-fed sector might be expected to export not less than $500.2m. (78 per cent) of the total $637.2m. for all rain-fed sector exports (total rain-fed crops and livestock products). Support for traditional sector agriculture implies a drastic shift of Sudan's export structure and a favourable development of the net foreign exchange earnings, a precondition for any renewed development effort. The positive effects on the balance of payments and on managing indebtedness would reinforce the chances of an ADLI package.

References

Acharya, S. N. (1979). *Incentives for resource allocation: a case study of Sudan*. Washington, IBRD.

Adelman, I. (1984). Beyond export-led growth. *World Development*, **12**, 937–49.

Arab Fund for Economic and Social Development (1976). *Basic programme for agricultural development in the Democratic Republic of the Sudan, 1976–1985*. Arab Fund, Kuwait.

Awad, H. (1983). *Why is the bread basket empty?*, Development Studies and Research Centre Seminar, No. 40. University of Khartoum.

D'Silva, B. C. (1983). Sudan's agricultural production 'potential'. Prospects and policy choices. Unpublished Paper, Annual African Studies Association Meeting, Boston, Massachusetts, 7–10 December 1983.

Eltom, A. (1986). Towards a long-term agricultural development strategy and related policies. Paper, First National Economic Conference, Agricultural Sector Conference, Khartoum, 18–20 February 1986. Agricultural Information Centre, Agricultural Extension Department, Ministry of Agriculture, Khartoum.

FAO (Food and Agriculture Organization) Technical Cooperation Programme (1986). *Agricultural policy development. Sudan, technical report, agricultural policy/planning development in the northern region of the Sudan*. FAO, Rome.

Government of Sudan. The Government Statement delivered by the Prime Minister before the Constituent Assembly on Monday, 7 July 1986. Unpublished.

Hansohm, D. and Wohlmuth, K. (1985*a*). *Promotion of rural handicrafts as a means of structural adjustment in Sudan, with special reference to Darfur Region*, Sudan Economy Research Group Discussion Paper No. 7, University of Bremen.

Hansohm, D. and Wohlmuth, K. (1985*b*). *East–south and south–south economic co-operation of the Democratic Republic of the Sudan*, Sudan Economy Research Group Discussion Paper No. 3, University of Bremen.

ILO (International Labour Organization) (1976). *Growth, employment and equity. A comprehensive strategy for the Sudan*. ILO, Geneva.

ILO (1987). *Employment and economic reform: towards a strategy for the Sudan*, Report of the ILO/JASPA Mission to the Sudan, August–September 1986. ILO, Geneva.

Kursany, I. (1983). A strategy for the transformation of the precapitalist part of the Sudanese society. *Development and Peace*, **4**, 167–89.

MAFNR (Ministry for Agriculture, Forestry and Natural Resources) (1977). *Food investment strategy 1977–1985*. MAFNR, Khartoum.

MANR (Ministry of Agriculture and Natural Resources) (MANR) (1985). *Statement of strategy for rain-fed agriculture*, Agricultural Planning Administration, MANR, Khartoum.

MANR (1986). *Agricultural situation and outlook, Annual Report 1985–1986*, MANR, Khartoum.

Ministry of Finance and Economics (1961). *The ten year plan of economic and social development, 1961/2–1970/1*. Ministry of Finance and Economics, Khartoum.

MFEP (Ministry of Finance and Economic Planning) and UNDP-Khartoum (1986*a*). *Kordofan rehabilitation development strategy*, Vols I and II. MFEP/UNDP, Khartoum.

MFEP (1986*b*). *Reassessment rehabilitation programme Kordofan and Darfur. Final report*. Euroconsult, Arnhem, The Netherlands.

MFEP (1986*c*). *Development budget for the fiscal year 1986–1987*. MFEP, Khartoum.

MNP (Ministry of National Planning) (1977). *The six year plan of economic and social development, 1977/78–1982/83*, Vols 1 and 2. Ministry of National Planning, Khartoum.

Oesterdiekhoff, P. (1980). Agrarpolitische orientierungen: phasen, tendenzen und alternativen. In *Der Sudan. Probleme und Perspektiven der Entwicklung* (ed. R. Tetzlaff and K. Wohlmuth), pp. 143–256. Alfred Metzner Verlag GmbH, Frankfurt am Main.

Oesterdiekhoff, P. and Wohlmuth, K. (1983*a*). The 'bread basket' is empty: the options of Sudanese development policy. *Canadian Journal of African Studies*, **17**, 35–67.

Oesterdiekhoff, P. and Wohlmuth, K. (ed.) (1983*b*). *The development perspectives of the Democratic Republic of Sudan. The limits of the bread-basket strategy*. Weltforum Verlag, Munchen.

Ottaway, M. (1987). Post-Numeiri Sudan: one year on. *Third World Quarterly*, **9**, 891–905.

Salam, M. M. A. (1986). Agricultural policy formation and administration. In *The agricultural sector of Sudan. Policy and systems studies* (ed. A. B. Zahlan and W. Y. Magar), pp. 409–23. Ithaca Press, London.

Sudan, Republic of (1986). *Strategy for development of rainfed agriculture*. MFEP, Khartoum.

Sudanow (1987*a*). Sudan's debt crisis, *Sudanow*, May, 8–11.

Sudanow (1987*b*). Salvation in sight? *Sudanow*, May, 17–8.

Wohlmuth, K. (1980). Die Produkt- und Regionalstruktur des sudanesischen Aussenhandels. Potentielle Märkte für traditionelle sudanesische Exportprodukte. In *Der Sudan. Probleme und Perspektiven der Entwicklung* (ed. R. Tetzlaff and K. Wohlmuth), pp. 1–42. Alfred Metzner Verlag GmbH, Frankfurt am Main.

Wohlmuth, K. (1986). *Sudan's industrialisation after independence. A case of Africa's crisis of industrialisation*. Unpublished Paper, Eighth International Economic Association World Congress, New Delhi, India, 1–5 December 1986, 37 pp.

Wohlmuth, K. and Hansohm, D. (1984). *Economic policy changes in the Democratic Republic of the Sudan*. Research Report, University of Bremen, 98 pp.

Wohlmuth, K. and Hansohm, D. (1986). *Sudan: a case for structural adjustment policies*, Sudan Economy Research Group Discussion Paper, No. 8. University of Bremen.

World Bank (1983). Sudan. Pricing policies and structural balances, Vol. III: Agriculture in Sudan. World Bank, Washington D.C.

World Bank (1985). *Sudan. Prospects for rehabilitation of the Sudan economy, Volume I: Main report*. World Bank, Washington D.C.

Zahlan, A. B. and Magar, W. Y. (ed.) (1986) *The agricultural sector of Sudan. Policy and systems studies*. Ithaca Press, London.

Glossary

a'mara	usufruct rights to grazing and cultivated land, to water, and to camp sites
abré	a beverage; a sour whitish broth of fermented, dehulled sorghum flour
abu ishreen	watering channels in the Gezira
abu sitta	watering channels in the Gezira
adur	*Sorghum aethiopicum*, a common weed of sorghum
agar	long-season groundnuts
aneese	*Sorghum purpureosericeum*, a common weed of sorghum
angaya	a plot of 0.714 feddans of irrigated land in the Gezira
aradeib	tamarind (*Tamarindus indica*)
araki	an illegal alcoholic drink, prepared from fermented date and sorghum beverages
asl	collective ownership rights in a territory and its natural resources
assaleya	a beverage made from millet and sorghum
assida	a porridge-like preparation, made from fermented or fresh batters of sorghum or millet flour
atamur	soils of the rises, where *qozes* interdigitate locally with basin clays; sandy areas good for millet
atang	spherical container of woven grass or Borassus palm
badob	local name for vertisol; dark cracking clay soil
baqr	cattle
baladi	from *balad*, a village, often used to describe the common local variety of a crop
bersim	lucerne (*Medicago sativa*), *Trifolium alexandrinum* is rarely grown in the Sudan
bildat	in Blue Nile Province, a small plot of land, not an individual farm but one of its constituent small scattered plots used for shifting cultivation
buda	the parasitic weed *Striga hermonthica* that attacks *dura*
dahrat	plains of cracking clay on both sides of the Blue Nile River
dakkai	wine produced from whole date fruits
damin	broker
dar	the territory of a tribe; traditional grazing area
darat	dry season
diwab	basic landrights-owning unit; small lineage segment composed of men who trace their decent in the male line from a common ancestor 5–8 generations back
dubbai	intermediate buyer

duber	good quality *dura*, consumed by people with medium and high incomes
dukhn	bulrush millet (*Pennisetum americanum*)
dura	sorghum (*Sorghum vulgare*)
faseikh	fish products preserved in brine
feddan	an area of land approximately equal to an acre; 1 feddan = 1.038 ac = 0.420 ha
feterita	the cheapest *dura*, consumed by the poorer urban classes
foul masri	faba bean (*Vicia faba*), a common food dish
fuda	cracking clay soil with a surface of granular mulch
gadwal	water channel
gardud	non-cracking clays of pediments around the base of hills
gariss	fermented camel's milk
gassab	the straw of *dura*, a very important animal fodder
gerif	sloping, silty land in the bed of the Nile, cultivated as the waters subside after the annual flood
ghaffir	guard
gibna baida	white cheese kept in brine prior to use
gineina	garden
gurrassa	pancake-like preparation made from wholemeal or decorticated wheat flour
gizu	a cover of juicy plants on which stock can survive without water
gol	family cattle hearths
gugus	mud and dung woven baskets, supported on a low platform of wooden poles, with a thatched roof of elephant grass and other grasses, used for storing grain
haboob	a strong wind, usually associated with thick dust
hafir	a large pit excavated out of impermeable clay, which fills during the rainy season and acts as a reservoir
harig	system of periodic burning followed by cultivation
hashab	*Acacia senegal*, a shrubby tree, the source of the best quality of commercial gum
hilu mur	a drink consisting of a brownish spicy extract of fermented, malted sorghum batter
hod	a piece of irrigated land in the Gezira, a constituent part of an *angaya*
howasha	a holding of land; in the Gezira a 10-feddan plot worked by a tenant
hukkam	the rulers in Islamic kingdoms
jallaba	merchants/traders from northern Sudan, originally slave traders
jebel	hill or mountain; rocky hill outcrop
kajaik	sun-dried fish

kankar	white to greyish calcium carbonate nodules in soils
kantar	a yield of crop equal to 141.523 kg; 1 kantar $feddan^{-1}$ = 337 kg ha^{-1}
karkadé	*Hibiscus sabdariffa*
katiri	a roughly circular area of around 200–400 m radius cleared for cultivation by a clan of six to eight families and cultivated for 2–3 seasons then left fallow for 5 years or more
kawal	a food product from fermented *Cassia obtusifolia* leaves
kaya	a Dinka reciprocal work party
keffir	fermented milk products
kerrib	badlands; gullied land along the Blue Nile and Atbara Rivers
kharif	rainy season
kholi	specialist gardener
khor	a seasonal watercourse, usually smaller than a *wadi*, but both words used almost synonymously in the Sudan
kissra	an unleavened, fermented batter made from sorghum flour and water, cooked into thin sheets
koumiss	fermented milk products
kutkut	a Dinka work party where labour is hired in exchange for food and beer
kwer	a long-handled tool for land preparation
lebbad	fine alluvial land in the Gash Delta
luak	a Dinka cattle byre
lubia	hyacinth bean (*Lablab purpureus*), grown mostly for fodder but sometimes for food
lugma	porridge
madaffara	a cheese of Syrian origin
maia't	natural depressions which accumulate water
maloda	a short-handled tool for land preparation
meloha	fish products preserved in brine
masakin	'poor people'; a class of subject commoners
mashandow	sandy land in the Gash Delta
molokhiya	leaves and tender shoots of young jute (*Corchorus olitorius*) plants
merissa	a weak alcoholic drink
mufruka	a sauce/stew
muawin idari	nomadic chiefs
nabit	a wine produced from date syrup
nafir	work parties
naga'a	sandy clay loam; soils of the basins, where *qozes* interdigitate locally with basin clays
nasha	a thin porridge of sorghum or wheat

number	a 90-feddan field
nushug	wet season
ombaz	oilseed cake made from sesame and cotton seed
qatáti	Shukriya hut
qoz	sand dune; applied commonly to the fixed sand dunes occurring in the Darfur, Kordofan, and Northern Regions
qurer	salty land along the bank of the Nile, just above high flood level
rakuba	small stack of straw
rigla	purslane (*Portulaca oleracea*)
robe	fermented milk
safra	good quality *dura*, consumed by people with medium and high incomes
saqqiya	the Persian water-wheel; an animal-operated device for lifting water
seid	nutgrass (*Cyperus rotundus*)
seluka	digging stick; also applied to land cultivated by *seluka*
semn	clarified butter
shaduf	a hand-operated water-lifting device using a counter-weighted bucket
sharia	Islamic law
sheil	credit, a system of money lending repaid at harvest with a specific quantity of crop yield equivalent to the loan
sherbote	wine produced from date syrup
shermute	sun dried strips of meat
siha	*Blepharis* sp., a highly nutritious grass
sunt	*Acacia arabica*
suq	market
tabikh	stew/sauce
tagnet	a low ridge between *howashas*
tahjir	a policy of elite hostage taking
tahl	*Acacia sayal*
takyr	cracked clay
terekeka	short-season groundnuts
terkeen	fish products preserved in brine
terus	earth bunds built for impounding rain water and increasing percolation into the soil
tiang	the korrigum (*Damaliscus korrigum*)
tibin	groundnut fodder
toich	seasonally flooded grazing land along the watercourses draining into the Sudd
tukl	the main homestead
wadi	seasonal watercourse, usually wider than a *khor*, but both words are used almost synonymously in the Sudan

wakeel	nominated agent
weika	okra (*Hibiscus esculentus*)
wut	cattle camps
zahafa	a steel bar drawn by a tractor, used for levelling land
zariba system	system by which northern *jallaba* gunmen exploited southern Sudan through the export of slaves and ivory
zereiba	a compound enclosed by a thorn fence for protecting stock from wild animals.

Abbreviations

AAAID	Arab Authority for Agricultural Investment and Development
ABS	Agricultural Bank of Sudan
ACSAD	Arab Centre for the Studies of Arid Zones and Dry Lands
ADLI	Agricultural demand – led industrialization
AOAD	Arab Organization for African Development
ARC	Agricultural Research Corporation
ARC	Agricultural Research Council
CMV	Cassava mosaic virus
COD	Cash on delivery
cv	Cultivar
DECARP	Desert Encroachment Control and Rehabilitation Programme
FAO	Food and Agriculture Organisation of the United Nations
GDP	Gross Domestic Product
IBRD	International Bank for Reconstruction and Development (World Bank)
IBS	Industrial Bank of Sudan
ICARDA	International Centre for Agricultural Research in Dry Areas
ICRISAT	International Crops Research Institute for the Semi-Arid Tropics
IDA	International Development Association
ILO	International Labour Organisation
IMF	International Monetary Fund
ITD	Intertropical discontinuity
IUCN	Internation Union for the Conservation of Nature
MAFNR	Ministry of Agriculture, Food and Natural Resources
MANR	Ministry of Agriculture and Natural Resources
MEM	Ministry of Energy and Mining
MFC	Mechanised Farming Corporation
MFEP	Ministry of Finance and Economic Planning

MNP	Ministry of National Planning
NRC	National Council for Research
NTC	National Tobacco Company
NUP	National Unionist Party
OAU	Organisation of African Unity
ODA	Overseas Development Administration
ODNRI	Overseas Development and Natural Resources Institute
OPEC	Organisation of Petroleum Exporting Countries
PE	Potential evapotranspiration
SCSEU	Sudanese Craftsmen and Small Enterprises Union
SDEC	Sudan's Desert Encroachment Control
SDIT	Southern Development Investigation Team
SGB	Sudan Gezira Board
SPLA	Sudan Peoples Liberation Army
SPS	Sudan Plantations Syndicate
SRDC	Sudan Rural Development Co Ltd
SSB	Sudanese Savings Bank
TOE	Tons of oil equivalent
UNDP	United Nations Development Program
UNESCO	United Nations Educational, Scientific, and Cultural Organisation
UNHCR	United Nations High Commission for Refugees
UNIDO	United Nations Industrial Development Organisation
USAID	United States Agency for International Development
WHO	World Health Organization
WSARP	Western Sudan Agricultural Research Project

Index